Transcultural Research – Heidelberg Studies on Asia and Europe in a Global Context

Series editors
Madeleine Herren
Thomas Maissen
Joseph Maran
Axel Michaels
Barbara Mittler

More information about this series at http://www.springer.com/series/8753

Andrea Bréard

Nine Chapters
on Mathematical Modernity

Essays on the Global Historical
Entanglements of the Science of Numbers
in China

 Springer

Andrea Bréard
Faculté des Sciences d'Orsay
Université Paris-Sud
Orsay, France

ISSN 2191-656X   ISSN 2191-6578   (electronic)
Transcultural Research – Heidelberg Studies on Asia and Europe in a Global Context
ISBN 978-3-319-93694-9   ISBN 978-3-319-93695-6   (eBook)
https://doi.org/10.1007/978-3-319-93695-6

This Springer imprint is published by the registered company Springer Nature Switzerland AG
The registered company address is: Gewerbestrasse 11, 6330 Cham, Switzerland

Dedicated to Shirley, Clee and Madeley,
the spring flowers on green meadows

Preface

The title of the book and its number of chapters is a pun on the canonical *Nine Chapters on Mathematical Procedures* (*Jiu zhang suan shu* 九章算術), compiled during the first century CE. The classic and its commentaries played an important scientific and political role in the mathematical endeavours of nineteenth- and twentieth-century mathematicians working in China, not in isolation but connected to the outside world. The *Nine Chapters* served as a model for writing mathematics algorithmically; they were the epitome of China's tradition, comparable to the Euclidean canon, and even became a source of inspiration for alternative proof techniques in the mid-twentieth century. Each of the nine chapters in this book illustrates how Chinese scholars mediated between new mathematical objects and discursive modes, and China's autochthonous scientific roots. Actors developed diverse strategies to situate themselves within or against the foreign scientific knowledge systems that they encountered in the emerging global setting following the Opium Wars in the mid-nineteenth century. They shaped the "science of numbers" (*shuxue* 數學), as mathematics had been called in Chinese since the turn of the twentieth century, as a discipline and they gradually loosened, but sometimes also strengthened, its ties to the authority of the past.

The nine chapters of this book grew out of a project idea that was sketched out during a train ride from Zürich to Heidelberg in 2015, after my intervention at Prof. Harald Fischer-Tiné's inspiring research colloquium on Extra-European History and Global History at the ETH. An exciting visit to Zürich turned into this book, which has posed a major challenge not only to the author, but probably also to its readers. The main problem I encountered is that, to date, not a single Chinese mathematical writing is available in its entirety in a foreign language. Most of the research for this book, however, is based on original texts and Chinese secondary sources. To make the material digestible for a non-Chinese public with varying levels of mathematical literacy required a new format of publication. Using the advantages of online publishing, the book is meant to be read by chapter, with or without the technical details and the translations in the appendices. This does not mean that the technical level is equal in all chapters, Chaps. 2, 3 and 5 are certainly more difficult than others, but I hope that the many illustrations will help to make the technical aspects of my argument more accessible.

In bringing this project to fruition, I wish to thank first and foremost my colleague Andrea Hacker, managing editor of open-access publications of the *Cluster of Excellence Asia and Europe in a Global Context* in Heidelberg, who has welcomed my book with great enthusiasm and has provided much encouragement and logistical support along the way. Without her, it may have never come to an end. I would also like to extend my heartfelt thanks to the numerous colleagues and strangers in the audiences who listened and asked questions when I presented my research over the last years. I am grateful for the generous intellectual and institutional support of Fabio Acerbi, Catherine Jami, Joachim Kurtz, Michael Lackner and Bernard Vitrac who led the projects in which I participated and whose ideas provided tremendous inspiration for my work. Rui Magone was also crucial in many ways, giving backstage support whenever I needed it most. The insightful comments of anonymous reviewers also strengthened this work and helped to sharpen my arguments; I am most grateful for their vigilance and enthusiasm. John Day kept a close eye on the entire manuscript, and I feel lucky to count him among my best friends. In its final stage, the manuscript greatly benefited from professional editing by Angela Roberts, who proofread the entire manuscript and offered valuable suggestions to improve my English. Any inaccuracies remaining in the final version of this book, however, are entirely my own or are perhaps due to the birds in the trees, whose singing and twirling was often a most welcome distraction. Finally, I would like to express my love and appreciation for my children, Max-Emanuel and Sarah-Lou, who have patiently followed me across many cultures and languages during the formative years of their youth while I conducted the research for what they see as a "cool" but unreadable book.

Heidelberg, Germany   Andrea Bréard
September 2017

Contents

List of Figures

List of Tables

Chapter 1
Visions of Antiquity

Contents

The project of telling a history of modern mathematics that claims explicitly to extend across geographical parameters might seem awkward at first glance. Does modern mathematics not define itself as universal in nature? Is modern mathematics not, as the outcome of pure and abstract thought written down in formulaic language, independent of national, linguistic, social or political context? Hasn't its pre-modern history always been analysed culturally, approached comparatively and on a global scale? That such is not the case—at least not in view of the recently developing field of global history—becomes clear when we look closely at how categories in historical narratives have been constructed.

Often, the apparently self-evident fact that mathematics were invented in Europe and globally distributed, was taken as evidence that there is no political and historical geography to mathematical knowledge at all. This led to diffusionist stories of how mathematics did spread and how it was received and accepted outside of Europe: emerging in the Greek world, axiomatic-deductive mathematics was transmitted to the Islamic world and back to Europe, where it took its definite normative shape at latest in the early twentieth century with Hilbert.[1] In the particular case of China, such diffusionist approaches have framed much of the

---

[1] Starting with Mehrtens (1990), the advent of mathematical modernity has been problematized by a number of historians of Western science. For a recent overview and discussion of definitions of modernity and modernism in mathematics, see Gray (2008) chapter 1.

asymmetrical narratives about the so-called first encounter of Europe and China; that is, the arrival of the Jesuit missionaries at the emperor's court at the end of the sixteenth century. But considering global processes and historical entanglements with a de-centred approach makes it possible to understand how actors drew global maps of diffusion for various political reasons; how for example the categories of "Western mathematics" became equated with what nowadays is commonly seen to characterize "modern mathematics" and how the invention of "indigenous knowledge" became a knowledge category labelled as "non-Western knowledge."

One prominent example of a historian countering the Eurocentric grand narratives is Joseph Needham (1900–1995), a Cambridge biochemist and Marxist, who worked much of his life on demonstrating in a series of studies on *Science and Civilisation in China* that "the older streams of science in different civilizations like rivers flowed into the ocean of modern science."[2] In his volume on *Physics*, we find a genealogical map (see Fig. 1.1) of the development of the magnetic compass depicting the various elements that contributed to the materialized end product of the modern compass. A world-wide web of cosmologically laden divination techniques, games of chance with cards, dice or dominoes and board-games from Amerindia to East Asia tendentiously suggests paths of transmission and evolution, whereby unlike Europe, only China invents, yet never develops modern science. The Needham puzzle and other Why not?s—why modern science did not develop in China or why the Scientific Revolution did not happen there—were the questions which until recently remained preeminent in writing the history of modern science in China, and they were even paralleled by continuing scholarly efforts in other fields attempting to explain why China did not develop capitalism or democracy before Europe.[3]

Needham's explorations concerning the mathematical sciences stop precisely when the Jesuits arrive and Western science was introduced into China albeit tinged with commercial and religious interests.[4] It is then that a period of four centuries begins during which scholars and practitioners of mathematics in China respond to the West, rediscover their own past, synthesize, naturalize or appropriate, reject and finally accept Western science. All of these verbs are terms that characterize ways within which scientific transactions between the drastically different and culturally incommensurable China and Europe were framed and modelled historically.

But China and the West are not analytical categories nor imagined civilizations[5] reserved to recent historical writing. They were part of the officially sanctioned historiography of the Ming dynasty, compiled after its ending in 1644, which laid the ground for the so-called theory of a "Chinese origins of Western science" (*Xixue Zhongyuan* 西學中源). This theory served as a political means to minimize the

[2]Needham (1970) p. 397.
[3]Pomeranz (2000).
[4]Needham (1959).
[5]As claimed in Hart (2013).

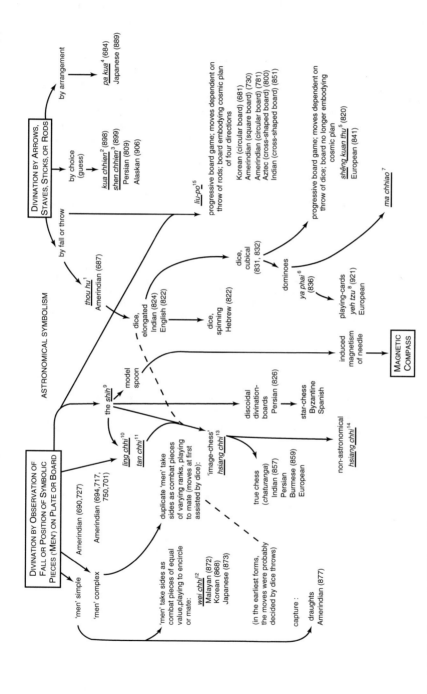

Fig. 1.1 Needham et al. (1962) p. 331

significance of the scientific contributions of the Jesuits,[6] but it was also a kind of apology for the comparatively little developed sciences in China and thus allowed the partial acceptance of parts of what the Jesuits taught and translated from Latin sources:

Scholars from the Western Seas who come to China, designate themselves as "Europeans." Their astronomy resembles the Islamic Astronomical System (*Huihui li* 回回曆),[7] but it is more precise. The study of precedent dynasties shows that most people from far away countries, who were familiar with calendrical methods, came from Western regions. But no mention was made of people coming from the East, the South or the North. The Nine Upholders Calendar (*Jiuzhi li* 九執曆) from Tang dynasty,[8] the Myriad Year Calendar (*Wannian li* 萬年曆) from Yuan dynasty[9] and the Islamic Astronomical System translated during the Hongwu reign[10] are all from Western regions. The reason is that Yao[11] has sent [the two pairs of brothers from the] Xi and He [families with first names] Zhong and Shu all over the world: Xi Zhong, Xi Shu and He Shu respectively to Yüyi,[12] Nanjiao[13] and Shuofang.[14] Only He Zhong has been ordered to 'settle in the West' without any indication of a territorial limit. Wasn't it at that time that the glory of our learning was transmitted far to the West? At the end of the Zhou dynasty,[15] the students and disciples of our mathematicians and astronomers were dispersed everywhere.[16]

西洋人之來中土者，皆自稱甌羅巴人，其曆法與回回同，而加精密．嘗考前代，遠國之人言曆法者多在西域，而東南北無聞．唐之九執曆，元之萬年曆，及洪武間所譯回回曆，皆西域也．蓋堯命羲和、仲叔分宅四方，羲仲、羲叔、和叔則以嵎夷、南交、朔方為限，獨和仲但曰「宅西」，而不限以地，豈非當時聲教之西被者遠哉．至於周末，疇人子弟分散．西域、天方諸國，接壤西陲，非若東南有大海之阻，又無極北嚴寒之畏，則抱書器而西征，勢固便也．甌羅巴在回回西，其風俗相類，而好奇喜新競勝之習過之．故其曆法與回回同源，而世世增修，遂非回回所及，亦其好勝之俗為之也．羲、和既失其守，古籍之可見者，僅有周髀．而西人渾蓋通憲之器，寒熱五帶之說，地圓之理，正方之法，皆不能出周髀範圍，亦可知其源流之所自矣．夫旁搜博採以續千百年之墜緒，亦禮失求野之意也，故備論之．

The "Chinese origins of Western science" theory still had its repercussions in the early twentieth century when, for example, social statistical theories were imported from Germany via Japan into China. The collection of numerical data for the throne

---

[6]See Martzloff (1993).

[7]Chinese translation of a Persian astronomical handbook with tables, a so-called *zīj*, prepared in 1383 and available at the Islamic Astronomical Bureau in Beijing. See Dalen (2002) p. 336–339.

[8]618–807. The Nine Upholders Calendar was a calendrical system "adapted into Chinese from Indian sources at the beginning of the eighth century." See Martzloff (2016) p. 109.

[9]1279–1368. Based on Jamāl al-Dīn's work, a Muslim astronomer from Bukhara. See Sivin (2009) p. 52 and Dalen (2002) p. 330.

[10]1368–1398.

[11]A legendary emperor, ca. 2300 BCE.

[12]An East Shandong peninsula.

[13]The south-eastern Asian region of Indochina.

[14]In the extreme North, now Inner Mongolia.

[15]Eleventh century—256 BCE.

[16]Translated from Zhang et al. (1975, vol. 2, p. 544–545) scroll 31 卷三十一, Memoires 7 志第七, Calendrical Systems 1 曆一, chap. "On Calendar Reform" (*Lifa gaige* 曆法沿革).

had been used as a tool of statecraft since early times, as traditionalist factions underlined who defended a programme of institutional reforms within traditional bureaucratic structures at the turn of the twentieth century.

The case of applied, but also of pure mathematics in China, is thus a rich field for studying the global historical processes of conceptual flows in science. Yet the recent literature on globalization and modern science has either left out the field of mathematical knowledge and its transformations in nineteenth- and early twentieth-century China,[17] or has portrayed modernization as a process of progressive Westernization.[18] This book is a first attempt to fill this historiographic lacuna, and I hope it will stimulate more lively discussions about the cultural relativity of modern mathematical knowledge.

1.1 About This Book

Because the modernization of mathematics in China was neither a unified process nor a conscious and organized movement in China, it can best be understood through a patchwork of individual stories and anecdotal history. I have thus chosen not to organize the *Nine Chapters* of this book strictly chronologically—except for a timeline of the major events from 1600 to 1950 in Appendix A—but, rather, along topical lines illustrated through a series of individual portraits of mathematical go-betweens. In each chapter, I give a voice to the protagonists of the history of mathematics in China through many quotes from original sources, translated for the first time here into a Western language.[19] As alluded to in the title of this book, the canonical Han dynasty compilation, *Nine Chapters on Mathematical Procedures* (*Jiu zhang suan shu* 九章算術, ca. first century CE),[20] will play a prominent role in most of the nine essays brought together here. Whether symbolically as a landmark of indigenous tradition, as a paradigm for mathematical style or as a benchmark against which new, foreign knowledge was measured, the classic and its many layers of commentaries remained the texts of reference in mathematical learning until the turn of the twentieth century and had an agency comparable to that of Euclid's *Elements* in the West.[21]

---

[17] See, for example Tsu and Elman (2014) and Renn (2012).

[18] See Tian (2005).

[19] In particular, the first and integral translation of Li Shanlan's *Methods for Testing Primality* (*Kao shugen fa* 考數根法, 1872) is provided in Appendix B.

[20] See the editions in Qian (1963) vol. 1, p. 83–258, Guo (1990) and Li (1993). Annotated English translations of the text, including its commentaries, can be found in Shen et al. (1999) and Guo (2013); Chemla and Guo (2004) provide a French translation with commentaries and a Chinese edition of the text.

[21] The long-lasting impact of the *Nine Chapters* was not limited to China, but spread to other Chinese-language based cultures. About the *Nine Chapters* in nineteenth-century Korea, for example, see Ying (2011).

   The first and the last of my *Nine Chapters* analyses the nationalist and ideological aspects of some nineteenth- and twentieth-century self-assertive discourses that connect Chinese tradition and modernity with backwardness and progress in science. The other seven chapters, in a somewhat experimental way, propose other possible themes with which to approach the global intellectual history related to numbers. Chapter 4 will relate to the seemingly most universal object of mathematical endeavours: the symbolic language of mathematics and, in relation to it, the individual mediators who established connections between discursive and formal modes of scientific discourse.[22] Situated in China at the turn of the twentieth century, I describe the power-laden processes of the creation of a global mathematical symbolism that allowed elimination of seemingly insurmountable borders between scientific and linguistic cultures. Chapter 7 suggests another theme that is both non-technical in nature and global-historically relevant: the institutional history of collecting numbers in late Qing imperial bureaucracy. The Statistical Department of the China Maritime Customs Service, a Chinese government institution run mainly by foreigners from the end of the Opium Wars to the 1949 Communist Revolution, might have made a very good case for an international institution where power relations and the rationalization of the production of trade statistics were negotiated in global constellations. But due to the closure of the No. 2 Historical Archives in Nanjing, where most of the Customs papers are stored, I shifted my focus to numbers in China's first central Statistical Bureau, created in 1907. Chapter 2 uses the case of the ellipse as a geometrical object introduced to China in the context of Aristotelian mechanics and Keplerian astronomy and shows the reliance on the classics and the conceptual changes when, for example, Dong Youcheng 董 祐誠 (1791–1823) in 1821 proposed a solution to the problem of calculating its circumference by applying a problem from the *Nine Chapters on Mathematical Procedures*. Chapter 3 analyses Li Shanlan's 李善蘭 (1811–1882) original attempt to disclose the Euclidean canon by adding an extra procedure for testing the primality of a number. Li is a particularly interesting figure since he worked and taught mathematics in a compartmentalized fashion: either in traditional algorithmic Chinese style or by adopting a syncretistic symbolism, integrating Chinese linguistic components into algebraic formalism.

   Moving from the linguistic to the discursive level, Chap. 5 takes a look at the changing modes of argumentation. Whether it is inductive proof, visual tools or arguments by analogy in number theoretical contexts, new standards seem to emerge under the social pressure of claiming the validity of the obtained results. Abstract science and rational belief in the objectivity of its results had gradually brought into question the whole future of Chinese cosmological and divinatory traditions. Chapter 6 describes attempts to prove their rational foundations: one, by Jiao Xun 焦循 (1763–1820), grounding the sequence of the divinatory hexagrams of the

---

[22]Studying intermediaries, translations and networks has been proposed in Moyn and Sartori (2013) p. 9–16 as one possible framework for global intellectual history. See also Schaffer et al. (2009).

Classic of Changes, also known under its Chinese title *Yijing* 易經, in combinatorial theories; the other, by Yuan Shushan 袁樹珊 (1881–1952?), using polar coordinates to model mathematically on a unit circle the cosmological transformations between the Five Elements, Heavenly Branches, Earthen Stems, etc. Two chapters relate to statistics: its role played on the path to a modern state as a means of quantifying China's social and economic reality in an entangled world (Chap. 7) and the conceptual and political discourses about statistics as an applied science (Chap. 8). The modernization of statistics in China is a good case study in complex modes of translation, instigated as it is by political and social change and involving both the adoption of foreign institutional models and the emergence of new areas of knowledge in the curriculum of higher education institutions. Of particular interest is the legitimating strategy of the reformers in appealing to the authority of a native past. Official and unofficial discourse during the reforms at the end of the Qing dynasty abounds in justifications of the new by referencing the old, and disciplinary distinctions between social, mathematical and administrative statistics were often associated with national characteristics.

1.2 Saving the Nation Through Mathematics

"Saving China through science" was a famous slogan of a group of idealistic young Chinese studying at Cornell University in the United States.[23] There, in 1915, they founded the Science Society of China and launched the publication of their journal *Science* (*Kexue* 科學), which in its early years was imbued with scientific nationalism.[24] "What will continue our heritage, bring back glory to the forest of learning of the Divine Land [a poetic name for China] and provide salvation to all living creatures, that is only science, and science only!" claims Ren Hongjun 任鴻雋 (1886–1961), one of the founding fathers of the Society, in his editorial note published in the first issue of *Science*.[25]

The idea of securing China a respectful place in the world by building upon science was not entirely new. Already some 20 years earlier, intellectuals had developed strategies that would make China a "rich and powerful" (*fuqiang* 富強) nation.[26] Teaching or pursuing research in mathematics and the writing of the history of mathematics were not exempt from participation in such efforts. Mathematics even figured among the first, and is considered the most important discipline to be included in encyclopaedias which contained knowledge that is

---

[23] See the foundational articles Lan (1915, 1916).

[24] See Wang (2002).

[25] 繼茲以往. 代興于神州學術之林. 而為芸芸眾生所托命者, 其唯科學乎, 其唯科學乎! Quoted from the editorial reproduced in Fan and Zhang (2002) p. 18.

[26] A compound expression for building "a rich country and a strong army" (*fuguo qiangbing* 富國強兵). On the emerging concept of nation in late nineteenth-century China, see Matten (2012).

explicitly praised in the title and prefaces for contributing to the strengthening of the nation's economic and political power.[27] Not all of the titles point to a Western origin however; on the contrary, most of them refer to traditional themes of Chinese mathematics. Such is the case, for example, in the *Collection of Books of Western Learning [to make the state] Rich and Powerful* (*Xixue fuqiang congshu* 西學富強叢書) from 1896. "This work in 48 volumes tried to assemble all information relevant for making the nation 'rich and powerful,' including volumes on such fields as mathematics,"[28] electricity, chemistry, astronomy, history, law, mining, engineering and military affairs. That it "seems to consist only of translations of Western textbooks"[29] certainly does not apply to the mathematics section.[30] The same holds for the *Complete Compendium of Sequels to the Collection of Books from the Studio for [making the state] Rich and Powerful* (*Fuqiang zhai congshu xu quanji* 富強齋叢書續全集) from 1901. It contains first of all a treatise on written computations, followed by a text that demonstrates the equality between Western and Chinese procedures (*Zhongxi tongshu* 中西通術), two texts giving explanations and detailed workings to problems borrowed from the *Nine Chapters*,[31] an exegesis to the Chinese translation of William Wallace's (1768–1843) *Algebra*[32] as well as a text on traditional Yuan dynasty algebra,[33] and the Chinese translation of Euclid's *Elements* and a book on conic sections.[34]

These compilations that link mathematical learning to the notions of a modern nation date from the turn of twentieth-century China, when projects of various kinds to introduce mathematical disciplines such as algebra or calculus were not uncontested. The history of science, too, was affected to varying degrees by domestic power struggles and ideological debates that unfolded along the central themes of China versus the West, tradition versus modernity and socialism versus

---

[27]See, for example, Yuan (1901) 凡例, p. 1A.

[28]Quoted from Doleželová-Velingerová and Wagner (2014) p. 10.

[29]Janku (2014) p. 333n13.

[30]Zhang (1896) contained twelve titles on mathematics (*Suanxue* 算學) of which only two are translations: 勾股六術一卷 (by Xiang Mingda 項名達), 九數外錄一卷 (by Gu Guanguang 顧觀光), 算式集要四卷 (original by Charles Haynes Haswell 哈司韋, transl. by John Fryer 傅蘭雅 and Jiang Heng 江衡), 衍元要義一卷 (by Xie Jiahe 謝家禾), 弧田問率一卷 (by Xie Jiahe), 直積囬求一卷 (by Xie Jiahe), 割圜連比例術圖解三卷淸 (by Dong Youcheng 董祐誠), 橢圜求周術一卷 (by Dong Youcheng), 斜弧三邊求角補術一卷 (by Dong Youcheng), 堆垛求積術一卷 (by Dong Youcheng), 三統術衍補一卷 (by Dong Youcheng), 器象顯四卷 圖一卷 (*The Engineer and Machinists Drawing Book* by V. Lebland and J. Armengaud, transl. by John Fryer and Xu Jianyin 徐建寅).

[31]For example, two mice digging under a wall from opposite directions, or two plants growing at different speed, as explained by Cui Chaoqing 崔朝慶 in the two titles *Pu guan bing sheng cao* 蒲莞並生草 and *Liang shu chuan yuan cao* 兩鼠穿垣草 in Yuan (1901).

[32]Wu Cheng's 吳誠 *Detailed Explanations of Algebra* (*Daishu shu xiangjie* 代數術詳解) discuss the Chinese translation of Wallace (1853) in Fu and Hua (1873).

[33]Jiao Xun's *Explanation of the One Celestial Element* (*Tianyuan yi shi* 天元一釋), 2 vols. already compiled a century earlier in 1800.

[34]Li and Edkins (1898), see p. 30.

capitalism. For mathematics, the problem of China's relation to the outside world and the role of the West in China's search for national greatness through scientific achievements were particularly important. China had its own mathematical past, and it wanted to win the race of nations, by proving, for example, its primacy in determining the most precise approximation for π.

There were only few who did not consider mathematics part of the useful knowledge that could help to save the crumbling dynasty through reforms aimed at enriching and strengthening the nation. For example, in a 1867 memorial Woren 倭仁 (1804–1871), a Mongol high official, protested against Western influence on racial and cultural grounds. He argued that it was preferable for a nation to be based on ritual ceremonies and ethical values rather than on tactics and clever contrivances. In other words, instead of acquiring technical skills, one should focus on the cultivation of heart and mind:

> Mathematics, one of the [Confucian] six arts, should indeed be learned by scholars as indicated in the Imperial decree, and it should not be considered an unworthy subject. But according to the viewpoint of your slave, astronomy and mathematics are of very little use. If these subjects are going to be taught by Westerners as regular studies, the damage will be great. [...] Your slave has learned that the way to establish a nation is to lay emphasis on propriety and righteousness, not on power and plotting. The fundamental effort lies in the minds of the people, not in techniques. Now, if we seek trifling arts and respect barbarians as teachers regardless of the possibility that the cunning barbarians may not teach us their essential techniques — even if the teachers sincerely teach and the students faithfully study them, all that can be accomplished is the training of mathematicians. From ancient down to modern times, your slave has never heard of anyone who could use mathematics to raise the nation from a state of decline or to strengthen it in time of weakness. The empire is so great that one should not worry lest there be any lack of abilities therein. If astronomy and mathematics have to be taught, an extensive search should find someone who has mastered the technique. Why is it limited to barbarians, and why is it necessary to learn from the barbarians?[35]

Woren's objection to Western learning and "barbarian" science teachers was vehemently rebutted by the ministers. "Because firearms have their roots in astronomy and geometry and are developed by trigonometry and mathematics, so that their guns can be cleverly discharged and marvellously hit the mark [...], they all agreed that the clever methods for manufacturing must begin with mathematics."[36] As a consequence, in the first modern government schools in late Qing China, mathematics were taught by European and Chinese instructors alike.

I will return, in several chapters of this book, to one such instructor, who can be seen as a go-between operating within Western and Chinese conceptual frameworks in parallel, Li Shanlan. He is a particularly interesting figure for my global history of mathematics, because he not only translated and developed conceptually further English mathematical writings, but also because he extended and taught the content of the canonical *Nine Chapters on Mathematical Procedures* and other classical Chinese works, not for their historical value but because they

---

[35] Teng and Fairbank (1979) p. 76.
[36] *Idem* p. 78.

provided a methodological alternative to foreign mathematics. Often considered the first professional mathematician in China who made his living out of teaching mathematics to students in an institutional setting, he was not the last to defend the idea of teaching Chinese mathematics. The idea emerged in a 1907 proposal to establish a School for National Essence (*Guocui xuetang* 國粹學堂). The various curricula were all based on resources from Chinese traditional knowledge, but were classified along the disciplinary lines of Western sciences. In the suggestion for calendrical and mathematical sciences (*lishuxue* 曆數學), Western mathematics was only marginally present in juxtaposition to the Chinese tradition in a single course on "Differences and similarities between Chinese and Western mathematics" (*Zhongxi suanxue yitong* 中西算學異同). All other courses pertained to the Chinese past, introducing the different schools of mathematics and the *Nine Chapters*, the so-called "celestial element" method for one unknown (*tianyuan yi fa* 天元一法), the thirteenth-century Yuan dynasty algebra for four unknowns, calendrical science[37] and a final course on the general meaning of mathematics.[38]

1.3 Mathematics as History

Although the idea of a School for National Essence remained half-baked, it was during the first two decades of the twentieth century that "the assumption that Chinese mathematics was an indispensable part of 'National Essence' and 'National Studies' was quite widespread."[39] Research on China's culture and its scientific heritage within National Studies (*Guoxue* 國學) was characterized by efforts to construct a past that supported claims about a unique Chinese cultural identity. Its essentialist approach was criticized by those who considered China's past as an obstacle to the creation of a modern nation.[40]

Within this intellectual movement, Chinese mathematics were no longer studied *per se*, but began to be approached from a specific historical point of view. One of the pioneers of the study of the history of mathematics in China, Li Yan 李儼 (1892–1963), did research in this new field in imitation of foreign scholars who had

---

[37]Lit. *Lifa* 曆法. The expression does not only refer to the calendar, but also to the computational system underlying astronomy more generally. See Martzloff (2016) p. 16–19 on the technical meaning of the term *li*.

[38]See Anonymous (1907) and Amelung (2014) p. 52.

[39]Amelung (2014) p. 52.

[40]Due to the critiques of the National Studies movement, the notion of National Studies was conceptualized in various ways and evolved over time. It reemerged in 1993, a point that I will return to in the final Chap. 9. On the diverse facets of *guoxue* over the last century, see the collection of articles in the journal *Perspectives Chinoises* No. 2011/1. Online edition at: https://perspectiveschinoises.revues.org/5723. See also the programmatic article (Zheng 1929) defending the idea that scientific knowledge and methodology is necessary to pursue National Studies.

begun to study the topic earlier, i.e. Alexander Wylie, Yoshio Mikami 三上義夫 and David Eugene Smith.[41] Li claimed that it was existential for Chinese to study their own scientific past, since otherwise National Studies, i.e. a national identity, would be lost:

> Slowly growing up, I read European books and realized that there were writings about the mathematics of my country. I would deeply lament that if National Studies fell from grace, foreigners would pick up on the subject [of history of Chinese mathematics] (*shen tan guoxue duo wang fan wei wairen suo shi* 深嘆國學墮亡反為外人所拾).[42]

Similar arguments were forwarded by Ye Qisun 葉企孫 (1898–1977),[43] who considered one chapter of the *Nine Chapters* as China's National Essence (*Shenzhou guocui* 神州國粹). The chapter which "cannot but not be known"[44] was precisely the one on the "Estimation of Workload" (*Shang gong* 商工) containing many problems related to public works, including the building of dikes or fortifications in certain three-dimensional geometric shapes. Ye analyses it in order:

> to show the readers that the mathematical writings from my country [China] are excellent in themselves and unlike those recently translated books, those jumbled writings which are hard to understand![45]

A glimpse at a four-volume annotated bibliography of articles considered important reading in the context of National Studies in the mid-1930s confirms the rejection of all Chinese mathematical writings that showed discursive modes of foreign influence. Research on Western mathematics had nothing to contribute to the resurrection of China's authentic scientific past, whereas the indigenous, ancient tradition, marginalized as a consequence of the influx of foreign knowledge, needed to be studied in depth. Several of the recommended readings were written by Li Yan whose reputation as a historian was already well established even before 1936 when the list concerning science appeared (see Table 1.1).[46]

Of one item in the reading list, discussing a chapter from Zhu Shijie's 朱世傑 1303 *Jade Mirror of Four Elements* (*Si yuan yujian* 四元玉鑑), it is said that:

> therein is established a theory of algebraic equations. Unveiling its genuine geometric meaning, [this article] takes the nineteen problems of the book [chapter] in order to prove that in our country the invention of algebra was indeed prior to the Europeans in the West.[47]

---

[41] See Wylie (1897), Mikami (1912) and the correspondence between D. E. Smith and Li Yan in Xu (2015). The collaboration between the latter two ended abruptly in 1917 because Smith was unsatisfied with the many historical imprecisions in Li's writings and the little he could learn from Li Yan.

[42] Li (1917) p. 238.

[43] "One of China's foremost physicists," who "in his youth published several articles on the history of mathematics." Amelung (2014) p. 53.

[44] See Ye (1916) p. 59.

[45] *Idem.*

[46] See, for example, the introduction to the Eighth Section: A Primer of the History of Mathematics in China (第八門 數學小史內篇之部) in Zhao (1923) p. 721.

[47] Liu (1936) p. 325–326.

Table 1.1 Entries on Mathematics (Suanxue 算學) in the *Index to Essays on National Studies* 國學論文索引 (Liu 1936)

Title	Author(s)	Published in	Volume
中國的數理 Mathematical Principles of China	Li Yan 李儼	Wenhua jianshe 文化建設	1-1
中國古代算學 Mathematics in Ancient China	Yan Zhongduo 嚴忠鐸	Jiaoda banyuekan 交大半月刊	2-1
九九傳說及九九表 On the Legend of "Nine Nine" and the Nine-Nine [multiplication] Table	Sun Wenqing 孫文青	Xueyi zazhi 學藝雜誌	13-7
九章算術篇目考 A Study of the Chapter Titles of the *Nine Chapters*	Sun Wenqing 孫文青	Jinling xuebao 金陵學報	2-2
中國隋唐前圓周率之研究 Resarch on π in China before the Sui and Tang Dynasties	Cui Hong 催宏	Beiqiang yuekan 北強月刊	1-5
敦煌石室算經一卷之序 A Mathematical Classic from the Dunhuang Grottoes in one Scroll with a Preface	Li Yan 李儼	Guoli Beiping tushuguan kan 國立北平圖書館刊	9-1
介紹『鎖套吞容』十九問 Introducing the Nineteen Problems in the Chapter 'Figures' within Other Figures'[a]	Zhang Peiyuan 張培元	Quanxue 勸學	2
『測圓海鏡』批校 A Critical Edition of the *Sea Mirror of Circle Measurements*[b]	Li Yan 李儼	Guoli Beiping tushuguan kan 國立北平圖書館刊	8-2
清代數學教育制度 The Mathematical Education System During the Qing	Li Yan 李儼	Xueyi 學藝	13-4,5,6
清代文集算學類論文 About the Category of Mathematics in Qing Dynasty Collected Works	Wang Zhongmin 王重民 & Li Yan 李儼	Xuefeng 學風	5-2
算盤發明小考 A Short Study of the Invention of the Abacus	Wu Mao 吳卯	Beiping chenbao yipu 北平晨報藝圃	24年8月24日
廿年來國人對於珠算的研究述要 Précis of Chinese Studies on the Abacus during the Last 20 Years	Cao Richang 曹日昌	Zhonghua jiaoyu shijie 中華教育世界	2-10,11
大小數命位分節之商權 Discussion of the Division of Positions Assigned to Small and Large Numbers	Gao Mengdan 高夢旦	Dongfang zazhi 東方雜誌	31-3
讀大小數命位分節之商權後 A sequel to "Discussion of the Division of Positions Assigned to Small and Large Numbers"	Xiong Conglin 熊從棻	Dongfang zazhi 東方雜誌	31-9

[a]Chin. *Suotao tunrong* 鎖套吞容, a chapter in Zhu Shijie's *Jade Mirror of Four Elements*, edition, modern Chinese and English translation in Zhu (2006) vol. 2, p. 496–541

[b]Chin. *Ceyuan haijing* 測圓海鏡, written by Li Ye 李冶 (1192–1279) in 1248

Another popular topic that would allow temporal comparisons of inventions was the number of correct figures in approximations for π, or, in Chinese terms, the "ratio of the circle circumference" (*yuan zhou lü* 圓周率). A comparison with foreign methods, in particular with Archimedes, was turned into a competition between nations. The search for the winner of the race even took a patriotic form in certain early writings on the history of Chinese mathematics:

In recent years, after having read [books on] Western mathematics, normal people, especially ordinary high school students, have been as surprised and excited as those astronomers and mathematicians (*chouren* 疇人) during Qing times who had seen the "nine procedures" (*jiu shu* 九術) of Master Du.[48] They all exclaimed in unison: "Oh how intelligent are these foreigners! How did the Westerners manage to invent a discipline of this degree of refinement which is so wonderful?" Wait a moment. Do not overstress the high aspirations of others and cancel out our own achievements! Our China, the China with its 5000 years of civilization! Is it really possible that there was nobody who knew about mathematics? Is it really possible that there was nobody who was able to think in such a refined way? Yet there was! There was not only one person, but, in fact, there were several persons who have discovered important mathematical theorems and methods and these are our country's astronomers and mathematicians!

Sir Mao Yisheng 茅以昇[49] said the following fine words: "Where my homeland's mathematics can win glory against Europe and the West is above all the ratio of the circle circumference [i.e. π]." I think that when Mao wrote this, he definitely had this Zu Chongzhi 祖沖之 in mind, because the discovery of the ratio of the circle circumference in my country, even if it can be dated very early, the real achievement is only due to this Master Zu. Zu Chongzhi's discovery of the circle ratio not only has the greatest glory in the history of Chinese mathematics, but it also occupies its place within the world history of mathematics. When nowadays many people do research on the principles of the circle, they certainly use that value of π ($= 3.1416$). Deep inside, they still think that it is a creation of foreigners. Who of them would guess that, on the contrary, this value is indeed a product of their native country and that it does not carry the slightest foreign flavour? Having written this, my mouth cannot refrain from saying: "Chinese people are so intelligent!"[50]

Although none of the original texts written by Zu Chongzhi, a fifth-century mathematician and astronomer, were transmitted, Zu was transformed into a national hero and his portrait now hangs in probably every science museum in China. In 1951, he figured in a series of articles on "China Being First in the World." Its explicit goal was "to develop the spirit of patriotism" (*Mudi shi weile fayang aiguo zhuyi jingshen* 目的是为了发扬爱国主义精神).[51] After the establishment of the People's Republic of China in 1949, patriotic education became the paradigm for history teaching in schools, and Chinese scientific inventions or discoveries serve

---

[48] Formulas for the summation of infinite trigonometric series brought to China by the French Jesuit Pierre Jartoux (1669–1720). See also chap. 2, p. 23.

[49] Mao Yisheng (1896–1989) is regarded as the founder of modern bridge engineering in China. He obtained his master's degree from Cornell University and was the first ever to obtain a PhD in engineering at the Carnegie Institute of Technology in 1919.

[50] See Yan (1936) p. 37, translated partly to German in Amelung (2014) p. 54.

[51] See *Renmin ribao* 人民日報 (People's Daily) (1951.08.16).

the nourishment of patriotic feeling. As Qian Baocong 錢寶琮 (1892–1974), a well-known historian of Chinese mathematics in 1951 points out:

> When studying mathematics, we should venerate the outstanding achievements of our homeland!
> 我們學習數學, 應當尊重祖國數學家的偉大成就.[52]

In the same year, Hua Loo-keng 華羅庚 (1910–1985), a self-trained mathematician, who had taken up an invitation to Princeton in 1946 and returned to China in October 1949, wrote a fervent anti-imperialist article in the *People's Daily* newspaper demonstrating that "Mathematics is the discipline my country's people excel in." Wanting to be part of a new epoch after the 1949 Revolution, he had decided to return permanently to China, where he subsequently held research director positions in major academic institutions. Examples from the Chinese mathematical tradition were dated prior to comparable results by Western mathematicians: the "Pythagoras theorem," the value of π, modular arithmetic, the arithmetic triangle and systems of linear equations. Although they were never a research theme nor a source of inspiration in Hua's own scientific work, these examples were showcased in this 1951 article (Fig. 1.2) to demonstrate that China did have a highly developed scientific past, that would put it on a par with foreign achievements and could counter the imperialists' biased binary vision of good and bad races:

> The country was stripped bare politically and economically by the imperialists, leaving the population as mere slaves to a colonial and half-feudalistic country. At the same time the self-respect and self-confidence of the people were undermined by the cultural influence of missionaries through schools, hospitals and other so-called charitable organizations. We paid for the political invasion with blood, we felt the economic invasion as loss of wellbeing. The cultural invasion enveloped us like a luxurious overcoat making us forget our own ancestors in favour of the incoming thieves. The problem with such an invasion is that we may not even realize that we have been invaded and we may accept the notion that we really were underdeveloped and not as capable as other peoples. Without doubt, such poisonous ideas would eventually reach our souls, destroying our competitive spirit and placing us in an irrecoverable position.
> In actual fact, our country is a great nation with a tremendous record of achievement in human history. We were let down by these blinkered people who believed that science was not our forte and that we should not follow others into such an unrewarding empty enterprise [...] Now let me speak briefly on some parts of mathematics that I know. I shall give some examples and let you readers judge fairly and objectively for yourselves whether we are what the imperialists call a "backward race." You may then decide whether you agree with those who have been so poisoned with notions of inferiority that they believed science to be not "what we excelled in." [53]

The above historiographic snapshots reveal different ways of instrumentalizing research on China's mathematical tradition, all of them in the context of the growing political and economic entanglements of China with the outside world after the

---

[52]Qian (1951) p. 1043.

[53]Translated from Hua (1951) in Wang and Shiu (1999) p. 162 and quoted in Amelung (2003) p. 254–255.

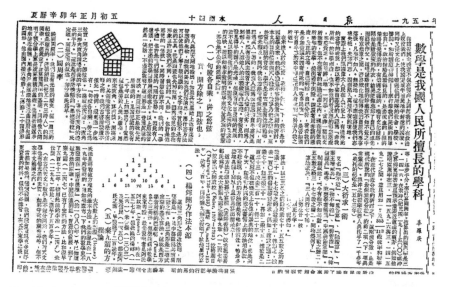

Fig. 1.2 Hua Loo-keng 華羅庚, "Mathematics is the discipline my country's people excel in" Hua (1951)

Opium Wars. Whether for enriching the country, for preserving cultural identity or for patriotic education, the transmitted mathematical texts were selectively interpreted to serve a specific ideological purpose and rarely were read in view of developing further new mathematical theories *per se*. For the latter half of the twentieth century, Wu Wen-Tsun 吳文俊 (Pinyin: Wu Wenjun, 1919–2017) is a rare exception. A major topologist in twentieth-century China, he had actively used ancient Chinese mathematical sources to develop a theory of mechanical proof. By negotiating the conflict between tradition and progress, he found a path that not only ensured his political and personal survival during and after the Cultural Revolution, it also brought him renown as a mathematician of international status who claimed to stand outside the dominant Western tradition of mathematics.[54]

The appeal of a radical scientific "modernity"[55] outside of China nevertheless erased efforts like Wu's to revive China's national tradition in contemporary mathematical research and follow conceptually and methodologically an alternative path to the West. As for mathematics education, its history has taken a different course, one which will be discussed in a future book: due to the general interest in a problem-oriented approach to learning and the relatively high ranking of Chinese students in International Mathematical Olympiads, the Chinese algorithmic tradition regained new momentum in the late twentieth century.

---

[54]On Wu Wen-Tsun's politically motivated change in research orientation, see Hudecek (2012) and the excellent biography (Hudecek 2014).

[55]I shall come back to the problem of "modernity" in the last chapter of this book.

References

Amelung, Iwo (2003). Die "vier grossen Erfindungen": Selbstzweifel und Selbstbestätigung in der chinesischen Wissenschafts- und Technikgeschichtsschreibung. In Iwo Amelung, Matthias Koch, Joachim Kurtz, Eun-Jeung Lee, and Sven Saaler (Eds.), *Selbstbehauptungsdiskurse in Asien: Japan — China — Korea*, 243–274. München: Iudicium.

Amelung, Iwo (2014). Historiography of Science and Technology in China. The First Phase. In Jing Tsu and Benjamin Elman (Eds.), *Science and Technology in Modern China, 1880s–1940s*, Number 27 in China Studies, 39–65. Leiden: Brill.

Anonymous (1907). Nishe guocui xuetang qi (fu biao) 擬設國粹學堂啓 (附表) (A Proposal for Opening a School for National Essence). *Guocui xuebao* 國粹學報 *26*, 1A–4B.

Chemla, Karine and Shuchun Guo (2004). *Les neuf chapitres sur les procédures mathématiques. Le classique mathématique de la Chine ancienne et ses commentaires*. Paris: Dunod.

Dalen, van, Benno (2002). Islamic and Chinese Astronomy under the Mongols: a Little-known Case of Transmission. In Yvonne Dold-Samplonius, Joseph W. Dauben, Menso Folkerts, and Benno van Dalen (Eds.), *From China to Paris: 2000 Years Transmision of Mathematical Ideas*, Volume 46 of *Boethius*, 327–356. Stuttgart: Steiner.

Doleželová-Velingerová, Milena and Rudolf G. Wagner (Eds.) (2014). *Chinese Encyclopaedias of New Global Knowledge (1870–1930). Changing Ways of Thought*. Transcultural Research — Heidelberg Studies on Asia and Europe in a Global Context. Berlin, Heidelberg: Springer.

Fan, Hongye 樊洪业 and Jiuchun Zhang, 张久春 (Eds.) (2002). *Kexue jiuguo zhi meng: Ren Hongjun wencun* 科學救國之夢: 任鴻儁文存 (The Dream of Saving China through Science. Writings of Ren Hongjun). Shanghai: Shanghai keji jiaoyu chubanshe 上海科技教育出版社.

Fu, Lanya 傅蘭雅 (Fryer, John) and Hua Hengfang 華蘅芳 (1873). *Daishu shu* 代數術 (Algebra). Shanghai: Jiangnan jiqi zhizao zongju 江南機器製造總局. Original: (Wallace 1853), or probably the earlier edition (Wallace 1842) which was identical.

Gray, Jeremy J. (2008). *Plato's Ghost: The Modernist Transformation of Mathematics*. Princeton, NJ: Princeton University Press.

Guo, Shuchun 郭書春 (Ed.) (1990). *Jiu zhang suan shu* 九章算術   (Nine Chapters on Mathematical Procedures). Shenyang: Liaoning jiaoyu chubanshe 遼寧教育出版社.

Guo, Shuchun 郭书春 (Ed.) (2013). *Jiu zhang suan shu: Han Ying duizhao* 九章算术 汉英对照 (Nine Chapters on the Art of Mathematics. Chinese–English), 3 vols. Library of Chinese Classics 大中华文库. Shenyang: Liaoning jiaoyu chubanshe 辽宁教育出版社. English critical edition and translations, with notes by Joseph W. Dauben and Xu Yibao.

Hart, Roger (2013). *Imagined Civilizations*. Baltimore, Md.: Johns Hopkins University Press.

Hua, Luogeng 華羅庚 (1951, February 10). Shuxue shi wo guo renmin suo shanchang de kexue 數學是我國人民所擅長的學科 (Mathematics is the Discipline my Country's People Excel in). *Renmin ribao* 人民日报 (People's Daily) *3*, 3.

Hudecek, Jiri (2012). Ancient Chinese Mathematics in Action: Wu Wen-Tsun's Nationalist Historicism after the Cultural Revolution. *East Asian Science, Technology and Society: an International Journal 6*, 41–64.

Hudecek, Jiri (2014). *Reviving Ancient Chinese Mathematics: Mathematics, History, and Politics in the Work of Wu Wen-Tsun*. London: Taylor & Francis.

Janku, Andrea (2014). "New Methods to Nourish the People": Late Qing Encyclopaedic Writings on Political Economy. In Milena Doleželová-Velingerová and Rudolf G. Wagner (Eds.), *Chinese Encyclopaedias of New Global Knowledge (1870–1930). Changing Ways of Thought*, Transcultural Research — Heidelberg Studies on Asia and Europe in a Global Context, 329–365. Berlin, Heidelberg: Springer.

Lan, Zhaoqian 藍兆乾 (1915). Kexue jiuguo lun 科學救國論 (About Saving China through Science). *Liu Mei xuesheng jibao* 留美學生季報 (The Chinese Students' Quarterly) *2*(2), 63–73.

Lan, Zhaoqian 藍兆乾 (1916). Kexue jiuguo lun er 科學救國論二 (About Saving China through Science, part 2). *Liu Mei xuesheng jibao* 留美學生季報 (The Chinese Students' Quarterly) *3*(2), 1–7.

Li, Jimin 李繼閔 (Ed.) (1993). *Jiu zhang suan shu jiao zheng* 九章算術校証 (Critical Edition of the *Nine Chapters on Mathematical Procedures*). Xian: Shanxi keji chubanshe 陝西科技出版社.

Li, Shanlan 李善蘭 and John Edkins, 艾約瑟 (1898). Yuanzhui quxian shuo 圓錐曲線説 (Theory of Conic Sections). In Liu Duo 劉鐸 (Ed.), *Gujin suanxue congshu* 古今算學叢書 (Compendium of Mathematics, Old and New) (Weiboxie 據微波榭本等石印 ed.), Volume 3 (象數第三). Shanghai: Shanghai suanxue shuju 上海算學書局. Originally published as an Appendix to the translation of William Whewell's *Elementary Treatise on Mechanics*, first published in 1859.

Li, Yan 李儼 (1917). Zhongguo shuxueshi yulu 中國數學史餘錄 (Further Notes about [my book] *History of Chinese Mathematics*). *Kexue* 科學 *3*(2), 238–241.

Liu, Xiuye 劉修業 (Ed.) (1936). *Guoxue lunwen suoyin sibian* 國學論文索引四編 (Index to Essays on National Studies, Fourth Instalment), Volume 2 of *Zhonghua tushuguan xiehui congshu* 中華圖書館協會叢書. Beiping: Zhonghua tushuguan xiehui 中華圖書館協會.

Martzloff, Jean-Claude (1993). Eléments de réflexion sur les réactions chinoises à la géométrie euclidienne à la fin du XVIIe siècle. *Historia Mathematica 20*, 160–179.

Martzloff, Jean-Claude (2016). *Astronomy and Calendars: The Other Chinese Mathematics 104 BC–AD 1644*. Berlin, Heidelberg: Springer.

Matten, Marc A. (2012). "China is the China of the Chinese": The Concept of Nation and its Impact on Political Thinking in Modern China. *Oriens Extremus 51*, 63–106.

Mehrtens, Herbert (1990). *Moderne Sprache Mathematik. Eine Geschichte des Streits um die Grundlagen der Disziplin und des Subjekts formaler Systeme*. Frankfurt am Main: Suhrkamp.

Mikami, Yoshio (1912). *Development of Mathematics in China and Japan*, Volume 30 of *Abhandlungen zur Geschichte der Mathematik*. Leipzig: Teubner.

Moyn, Samuel and Andrew Sartori (Eds.) (2013). *Global Intellectual History*. Columbia Studies in International and Global History. New York: Columbia University Press.

Needham, Joseph (1959). *Mathematics and the Sciences of the Heavens and the Earth*, Volume 3 of *Science and Civilisation in China*. Cambridge: Cambridge University Press.

Needham, Joseph (1970). The Roles of Europe and China in the Evolution of Oecumenical Science (1966). In Joseph Needham (Ed.), *Clerks and Craftsmen in China and the West: Lectures and Addresses on the History of Science and Technology*. Cambridge: Cambridge University Press.

Needham, Joseph, Ling Wang, and Kenneth G. Robinson (1962). *Physics*, Volume IV.1 of *Science and Civilisation in China*. Cambridge University Press.

Pomeranz, Kenneth (2000). *The Great Divergence*. The Princeton Economic History of the Western World. Princeton, NJ: Princeton University Press.

Qian, Baocong 錢寶琮 (1951). Zhongguo gudai de shuxue weida chengjiu 中國古代數學的偉大成就 (The Outstanding Achievements of Mathematics in Ancient China). *Kexue tongbao* 科學通報 (Chinese Science Bulletin) *10*, 1041–1043.

Qian, Baocong 錢寶琮 (Ed.) (1963). *Suan jing shi shu* 算經十書 (Ten Books of Mathematical Classics), 2 vols. Beijing: Zhonghua shuju 中華書局.

Renn, Jürgen (Ed.) (2012). *The Globalization of Knowledge in History*, Volume 1 of *Studies*. Berlin: Max Planck Research Library for the History and Development of Knowledge.

Schaffer, Simon, Lissa Roberts, Kapil Raj, and James Delbourgo (2009). *The Brokered World: Go-Betweens and Global Intelligence, 1770–1820*. Watson.

Shen, Kangshen, J. N. Crossley, and A. W.-C. Lun (1999). *The Nine Chapters on the Mathematical Art: Companion and Commentary*. Beijing: Oxford University Press and Science Press.

Sivin, Nathan (2009). *Granting the Seasons. The Chinese Astronomical Reform of 1280, With a Study of its Many Dimensions and an Annotated Translation of Its Records*. Sources and Studies in the History of Mathematical and Physical Sciences. New York, NY: Springer.

Teng, Ssu-yü and John King Fairbank (1979). *China's Response to the West*. Cambridge: Harvard University Press. With a new preface. Includes index.

Tian, Miao 田淼 (2005). *Zhongguo shuxue de xihua lichen* 中国数学的西化历程 (The Westernization of Mathematics in China). Jinan: Shandong jiaoyu chubanshe 山东教育出版社.

Tsu, Jing and Benjamin Elman (Eds.) (2014). *Science and Technology in Modern China, 1880s–1940s*. Number 27 in China Studies. Leiden: Brill.

Wallace, William (1842). Algebra. In *The Encyclopædia Britannica or Dictionary of Arts, Sciences, and General Literature* (7th ed.), Volume II, 420–502. Edinburgh: Adam and Charles Black.

Wallace, William (1853). Algebra. In *The Encyclopædia Britannica or Dictionary of Arts, Sciences, and General Literature* (8th ed.), Volume II, 482–564. Edinburgh: Adam and Charles Black.

Wang, Yuan 王元 (1994). *Hua Luogeng* 華羅庚. Beijing: Kaiming chubanshe 開明出版社.

Wang, Yuan and Peter Shiu, (Trans.) (1999). *Hua Loo-Keng*. Singapore: Springer. Chinese original title (Wang 1994).

Wang, Zuoyue (2002). Saving China through Science: The Science Society of China, Scientific Nationalism, and Civil Society in Republican China. *Osiris 17*, 291–322.

Wylie, Alexander (1897). *Chinese Researches*. Shanghai.

Xu, Yibao (2015). Correspondence of 李儼 Li Yan and David Eugene Smith. In David E. Rowe and Wann-Sheng Horng (Eds.), *A Delicate Balance: Global Perspectives on Innovation and Tradition in the History of Mathematics: A Festschrift in Honor of Joseph W. Dauben*, Trends in the History of Science, 245–273. Springer.

Yan, Dunjie 嚴敦傑 (1936). Zhonggou suanxuejia Zu Chongzhi jiqi yuanzhou lü 中國算學家祖仲之及其圓周率之研究 (The Chinese Mathematician Zu Chongzhi and his Ratio for π). *Xueyi* 學藝 *15*, 37–50.

Ye, Qisun 葉企孫 (1916). Kaozheng shang gong 考正商功 (A Textual Criticism of the Chapter "Estimation of Workload"). *Qinghua xuebao* 清華學報 *2*(2), 59–87.

Ying, Jia-Ming 英家銘 (2011). The *Kujang sulhae* 九章術解. Nam Pyong-Gil's Reinterpretation of the Mathematical Methods of the *Jiuzhang suanshu*. *Historia Mathematica 38*, 1–27.

Yuan, Junde 袁俊德 (Ed.) (1901). *Fuqiang zhai congshu xu quanji* 富强齋叢書續全集 (Complete Compendium of Sequels to the Collection of Books from the Studio for [making the state] Rich and Powerful) (Xiaochuang shanfang 小創山房 ed.).

Zhang, Tingyu 張廷玉 et al. (1975). *Xin jiaoben Mingshi* 新校本明史 (New Edition of the History of the Ming Dynasty, 1368–1644), 12 vols. Dingwen shuju 鼎文書局.

Zhang, Yinhuan 張蔭桓 (Ed.) (1896). *Xixue fuqiang congshu* 西學富强叢書 (Collection of Books of Western Learning [to make the state] Rich and Powerful). Shanghai: Hongwen shuju 鴻文書局.

Zhao, Liao 趙繚 (1923). *Shuxue cidian* 數學辭典 (A Dictionary of Mathematics) (3rd ed.). Shanghai: Qunyi shushe 群益書社.

Zheng, Zhenduo 鄭振鐸 (1929). Qie mantan suowei 'guoxue' 且慢談所謂"國學" (Musings on so-called 'National Studies'). *Xiaoshuo yuebao* 小说月报 (Fiction Monthly) *20*(1), 15–20.

Zhu, Shijie 朱世杰 (2006). *Si yuan yujian: Han Ying duizhao* 四元玉鉴 汉英对照 (Jade Mirror of the Four Unknowns. Chinese–English), 2 vols. Library of Chinese Classics 大中华文库. Shenyang: Liaoning jiaoyu chubanshe 辽宁教育出版社. Translated into Modern Chinese by Guo Shuchun 郭书春, translated into English by Chen Zaixin 陈在新, revised and supplemented by Guo Jinhai 郭金海.

Chapter 2
The Ellipse Seen from Nineteenth-Century China

Contents

When Xia Luanxiang 夏鸞翔 (1823–1864) gives a summary of his work on conic sections he introduces the subject matter of his book as follows:

> Heaven is big and round. When describing the objects of heaven, there is none that does not relate to the circle. Although "circle" is only a unique name, there are a myriad of species. When following the circle for one round, curves are generated. Westerners divide [the different kinds of] lines according to the order of generation into the following categories: lines of the first order, [this class consists of] the straight line only. Lines of the second order, [this class comprehends] four species: the circle, the ellipse, the parabola and the hyperbola. Lines of the third order have eighty different kinds. Lines of the fourth order have more than five thousand kinds. Beyond the lines of the fifth order, it is such that it is impossible to investigate them.[1] Here, I explore the four kinds of lines of the second order and trace their origins, and additionally, I add explanations to all the parabolas of higher order.[2] Although there are a myriad variations, their principle is all from one strain. All the equations of conics are entirely provided for on the solid [i.e. the surface] of a cone. That is the reason why the cone is the mother of the curves of the second order.[3] For the ellipse one uses congregation, for the parabola one uses extension, for the

---

[1] This passage is reminiscent of Isaac Newton's classification of curves that is related in more detail by Elias Loomis (1811–1889) in his *Elements of Analytical Geometry, and of Differential and Integral Calculus* in 1859 (Loomis 1851) and translated into Chinese by Li Shanlan and Alexander Wylie (chin. 偉烈亞力, 1815–1887). See Li and Wylie (1859) vol. 8, p. 6B–7B.

[2] I.e. the cubical parabola, the biquadratic parabola, etc.

[3] Cf. Loomis (1851) p. 103: "the only curves whose equations are of the second degree, are the circle, parabola, ellipse, and hyperbola."

© Springer International Publishing AG, part of Springer Nature 2019
A. Bréard, *Nine Chapters on Mathematical Modernity*, Transcultural
Research – Heidelberg Studies on Asia and Europe in a Global Context,
https://doi.org/10.1007/978-3-319-93695-6_2

hyperbola one uses dispersion (*tuoyuan liyong ju paowuxian liyong yuan shuangquxian liyong san* 橢圓利用聚拋物綫利用遠雙曲綫利用散), but their principles all stem from the plane circle. If indeed, we unite what they have in common, then we "build tools to imitate the cosmos."[4] By "bowing [to the earth] and looking upwards [to the heaven], by observing [the sky] and investigating [the ground],"[5] their applications are without limitations (*gouhui qi tong ze zhiqi shang xiang fuyang guancha wei yong wu qiong yi* 苟會其通則制器尚象俯仰觀察為用無窮矣). I will now explain this one by one as follows.[6]

It is perhaps surprising that a Chinese mid-nineteenth-century mathematician who was familiar with translations of English works on analytic geometry, differential and integral calculus, as well as with the works of his Chinese predecessors, frames philosophically a complex mathematical topic that he deals with using a technically original and novel approach. That Xia wanted to promote his work among Confucian scholars with a preface that proved that he was versed in the classics and familiar with the *Classic of Changes*, is only a guess, but Xia certainly wanted to understand mathematics as a numerical science that studies change. Analogous to the transformations of broken into unbroken lines in the divinatory hexagrams,[7] conics, as he underlines in the above-quoted passage, can be obtained from a single origin, the cone, and transformed one into another: the ellipse can be obtained by joining together the two extremities of the parabola, and if one extends one endpoint of the major axis of the ellipse to infinity, it is transformed into a hyperbola.[8] This was, of course, Xia's philosophical thought experiment, since strictly speaking mathematically, in order to perform a mapping from an ellipse to a hyperbola or parabola a projective transformation is needed. No affine transformation can change a bounded curve into an unbounded one.[9] In Europe, Johannes Kepler (1571–1630) had already expressed in 1604 an idea, similar to Xia, of unifying the different kinds of conic sections "in a short note of some four pages on conics" in the context of optics.[10] Instead of treating them separately, he

---

[4]The expression *Zhiqi shangxiang* 制器尚象 is borrowed from *The Great Appendix* to the *Classic of Changes* 《周易》易傳 繫辭下. For a translation, see Legge (1963) app. III, chap. X.59, p. 367–369.

[5]The common phrase *yang guan fu cha* 仰观俯察 has its origins in *The Great Appendix* to the *Classic of Changes* 《周易》易傳 繫辭上: "仰以觀於天文, 俯以察於地理." Translated in Legge (1963) app. III, chap. IV.21, p. 353:

> (The sage), in accordance with (the Yî), looking up, contemplates the brilliant phenomena of the heavens, and, looking down, examines the definite arrangements of the earth.

[6]Translated from the *Diagrammatic Explanations [of Procedures] for Curves* (*Zhiqu tujie* 致曲圖解, 1861) in Xia (2006b) p. 438.

[7]See Chap. 6 p. 150 for a short explanation of the Chinese hexagrams.

[8]On Xia's ideas concerning a comprehensive treatment of conics, see Liu (1990) p. 13.

[9]Projective transformations map lines to lines but do not necessarily preserve parallelism, whereas affine transformations respect parallel lines. The latter allow, for example, mapping a circle to an ellipse, but not an ellipse to a parabola or hyperbola.

[10]Swinden (1954) p. 45. It is uncertain if Xia Luanxiang had access to Kepler's text (Kepler 1604), which had been part of the Beitang library according to Verhaeren (1949) p. 559 item no. 1901,

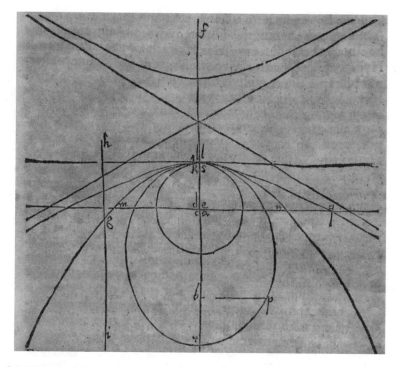

Fig. 2.1 J. Kepler, *Ad Vitellionem Paralipomena* ... (Kepler 1604)

established analogies based upon geometric transformations. By moving the foci of an ellipse closer together, for example, when they coincide at the greatest possible distance from the curve one may obtain a circle, but by moving them further apart at infinite distance, one may obtain a parabola (see Fig. 2.1 from Kepler (1604) p. 94).

What Xia's biographers took notice of were not the mathematical technicalities related to conics that Xia developed throughout his treatises on conics. Instead, they retained from his earlier writing, the *Diagrammatic Explanations of Procedures for Thoroughly Understanding the Square (Dongfang shu tujie* 洞方術圖解, 1857), a cosmological passage (nearly) identical to the one at the end of the above-translated preface which links mathematical transformations to cosmological change.[11] After his death, Xia was more generally remembered as an innovative mathematician

---

and bound together with Kepler's essay *Strena seu de nive sexangula* (1611) on the six-cornered snowflake.

[11] See the *History of the Qing Dynasty (Qingshi* 清史), scroll 506 卷五百六, Biographies 292 列傳 二百九十二, Mathematician-Astronomers 2 疇人二 in *Qingshi* (1961) vol. 7, p. 5501–5502 and the entry on Xia Luanxiang in the anonymous *Qingdai xueren liezhuan* 清代學人列傳 (*Biographies of Learned Men from Qing Dynasty*), online http://ctext.org/wiki.pl?if=en&chapter=127281, §85:

inspired by Western learning, who organized mathematical knowledge according to
a traditional philosophical framework. That the characteristics of the circle are based
upon the principles of the square was one major philosophical concern in mathematics in China, and the unity of method applicable to a vast variety of problems was
another. Xia was remembered for having succeeded in achieving both.[12] But his
success was also measured in comparison with Western mathematics, a knowledge
system taken into account by Xia and politely revered but not considered superior
by the author of the biographical entry in the *Draft History of the Qing Dynasty*
(*Qingshi gao* 清史稿):

> Luanxiang died in the year 1864. Because he solved the problem of finding a proce-
> dure for the chord with differences of rectangular products (*fangji* 方積),[13] he jumped
> onto the bandwagon of the Westerners. But in differential [calculus], the constants that
> are neglected are just like the square and the corner [coefficients] in the [method] of
> rectangular products. The variable that is sought, is just like the difference between
> the sequential sums of two border [coefficients]. When applying these procedures to
> curves, there is none that cannot be understood. Luanxiang has established procedures,
> treating all categories one by one. It is so that one cannot but yield prominence to the
> foreigners. Yet Western methods for extracting roots, from equations of third degree on,
> are all branches and segments, they do not equal the consistency of the Chinese method

> 鸞翔同治三年卒。因方積之較而悟求求弦矢之術，駸駸乎駕西人而
> 上之，然微分所棄之常數，猶方積之方與隅也。所求之變數，猶兩廉
> 遞加之較也。其術施之曲綫，無所不通，鸞翔猶待逐類立術，是則不
> 能不讓西人以獨步。然西法開方，自三次式以上，皆枝枝節節為之，
> 不及中法之一貫。鸞翔又於中法外獨創捷術，非西人所能望其項背
> 雲。[14]

Beginning with Xia Luanxiang's writings in relation to "foreign" knowledge on
conics, this chapter will use the example of the calculation of the circumference and
the surface area of the ellipse in order to analyse how earlier authors in late imperial
China, in particular before the introduction of differential and integral calculus,
approached the rectification and quadrature of curves.

> All the principles of the ellipse and curves stem from the circle. If indeed, we unite what
> they have in common, then we "build tools to imitate the cosmos." By "bowing [to the
> earth] and looking upwards [to the heaven], by observing [the sky] and investigating [the
> ground]," their applications are without limitations.
> 楕圓及諸曲線其理皆出於平圓，苟會其通，則制器尚象，俯仰觀察，為用無窮。

[12] See the entry on Xia Luanxiang in the anonymous *Biographies of learned men from Qing dynasty*
(*Qingdai xueren liezhuan* 清代學人列傳) (*Biographies of learned men from Qing dynasty*), online
http://ctext.org/wiki.pl?if=en&chapter=127281, §83–85.

[13] The term *fangji* used here is ambiguous, since it refers on the one hand to rectangular surfaces,
but on the other to the binomial coefficients used in solving numerically polynomial equations for
which Xia has developed a unique method. It is likely that the authors played on this ambiguity
here in order to stress the relation between geometric objects of curved lines (here a chord of a
circle) and the square (here more generally rectangular figures).

[14] Text quoted and translated from the *History of the Qing Dynasty* (*Qingshi* 清史), scroll 506 卷
五百六, Biographies 292 列傳二百九十二, Mathematician-Astronomers 2 疇人二 in *Qingshi*
(1961) vol. 7, p. 5501–5502.

In a first section, I will discuss how Xia represented, used and transformed Western knowledge on conics and what role epistemic values played in the organization of procedures related to them. I will argue that a shift of analysis in Xia's work was triggered by a change in scales of observation. Thereby, scale is not only understood in a geographic sense and in relation to the fact that Xia was exposed to knowledge that has its origins beyond the Chinese territory. Change of scale here also regards the treatment of conics from a panoptic point of view as compared to observing the ellipse as an isolated object. In a second part, I will go back in time and give an overview of some earlier Chinese writings related to the ellipse before the introduction of calculus to China in the eighteenth and early nineteenth century.[15]

2.1 Xia Luanxiang and Conics

Following the encounter in 1859 with a Chinese translation of Elias Loomis's (1811–1889) *Elements of Analytical Geometry, and of Differential and Integral Calculus*,[16] Xia had access to a mathematical treatise that systematically discusses the equations related to each of the conic sections and, through a series of examples, applies the integral calculus to the problem of rectification and quadrature of curves, surfaces and solids of revolution. In his first publication on conics, Xia organized knowledge according to a similar order of the different kinds of conics: the circle, the ellipse, the parabola, the hyperbola and their three-dimensional counterparts resulting from revolution around their respective axes. Yet within each of these categories he juxtaposed foreign and "newly determined" procedures. Thereby, his foreign sources are twofold: he either refers to the Chinese translation of the *Elements of Analytical Geometry, and of Differential and Integral Calculus*, or to a certain Master Du. Du Demei 杜德美 was the Chinese name of the Jesuit missionary Pierre Jartoux (1668–1720) who taught astronomy and geometry to the Emperor Kangxi (r. 1661–1722). In his *Fast Methods for the Circle Division and the Precise [Circle] Ratio (Geyuan milü jiefa* 割圜密率捷法), Ming Antu 明安圖 (ca. 1692–ca. 1763), Xia Luanxiang's teacher, added six more to the "Three procedures of Master Du" (*Du Shi san shu* 杜氏三術).[17] All procedures are power series expansions of trigonometrical functions allowing calculations

---

[15] A list of Chinese writings related to conics can be found in Tables 2.2 and 2.3 p. 27–28.

[16] The *Dai wei ji shi ji* 代微積拾級 (Li and Wylie 1859), translated by Li Shanlan in collaboration with the British missionary Alexander Wylie (1815–1887) was based on Loomis (1851).

[17] Ming (1774). As Chen (2015) p. 493n7 points out that "no currently known documentation, Chinese or Mongolian, indicates Ming's birth and death years." On Jartoux and Ming's work on infinite series see Martzloff (1997) p. 353–359.

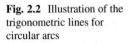

Fig. 2.2 Illustration of the
trigonometric lines for
circular arcs

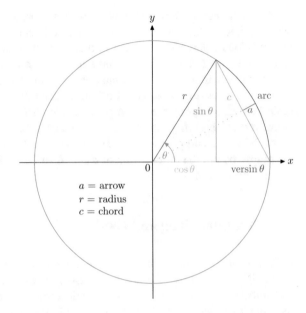

a = arrow
r = radius
c = chord

of the circumference and any arc, chord, "arrow,"[18] sine or versed sine of a
circle.[19]

But what Xia labels explicitly as foreign procedures is neither copied from
Loomis's book, nor does it always apply his analytical methods to reconstruct what
one finds in the indirectly quoted Jesuit sources. Xia's procedures are a rewriting in
his own (recursive) terms for the sake of unity:

> In Du's procedures, finding the arc with the sine is solved directly, whereas for finding
> the arc with the versed sine[20] one first determines its square (*suan* 筭).[21] The procedures
> used are not uniform. I have always had some doubts. In the year 1861, in old age, I
> incidentally used the Westerners' integrals and differentials to deduce another for finding
> the arc with the versed sine in which it is not necessary to determine the square (*suan* 筭)
> of the arc. Instead, one obtains the arc directly. I therefore thought that there also must
> be another procedure to find the arc with the sine. Indeed, by first determining the square

---

[18]The "arrow" (chin. *shi* 矢) here is not what in the history of mathematics was also called the
"sagitta," a Latin word meaning "arrow," and which came to be used interchangeably with "versed
sine." See Newton (1999) p. 306–307. In the Chinese context, the "arrow" is rather the versed sine
of half the arc under consideration. See Fig. 2.2.

[19]For a complete list of the nine formulas, see Jami (1990) p. 58–60. For the procedure to find the
arc with the versed sine at the bottom of page 59 there is a mistake. The series yields the square of
the arc, not the arc itself. See Ming (1774) scroll 1 卷上, p. 5A.

[20]Nowadays an outdated trigonometric function, it is defined as versin $z = 1 - \cos z$.

[21]In this text, Xia uses an unusual term for expressing the square of a number: *suan* 筭, a variant of
the character *suan* 筭, literally meaning "to calculate"; in its nominal form it designates a "counting
rod". In the present text, for example, the expression *zhengxian suan* 正弦筭 stands for the square
of the sine, the expression *banjing suan* 半徑筭 is the square of the radius.

Table 2.1 Table of Content of Xia Luanxiang's 夏鸞翔 *Diagrammatic Explanations [of Procedures] for Curves* (*Zhiqu tujie* 致曲圖解), 1861

Contents
1. All curves begin from one point and end at one point 論諸曲綫始於一點終於一點第一
2. On foci of all expressions for curves 論諸曲綫式之心第二
3. On the directrix 論諸曲綫式皆有準綫第三
4. On the *gui*-lines 論諸曲綫式皆有規綫第四
5. On both major and minor axis 論諸曲綫式之橫直二徑第五
6. On conjugate diameters 論諸曲綫式之兌徑第六
7. On the distance between two focal points 論諸曲綫式之兩心差第七
8. On normal and tangent lines 論諸曲綫式之法綫切綫第八
9. On the diameter of curvature 論諸曲綫式之斜規綫第九
10. On equations referred to [rectangular] co-ordinates 論諸曲綫之縱橫綫式第十
11. All expressions for curves are mutually proportional 論諸曲綫式互為比例第十一
12. On the eight [trigonometric] lines 論八綫第十二

(*suan* 筭) of the arc, and by expanding, one can obtain another procedure. The number of multiplications and divisions are all pairwise identical. Therefore, when juxtaposing the four procedures, one sees the principle that connects them. What is extremely combined and alternated is also extremely orderly and perfect![22]

What Xia obviously worries about is the inconsistency and incompleteness of Jartoux's nine procedures, which do not uniformly yield the arc, but also, in the case where the versed sine is given, the square of the arc. He thus sets out to complement these lacunae and develops procedures yielding the square of the arc or the arc "directly," depending on what is already provided for by Ming Antu. In the case of the procedure for the arc with the versed sine given, he finds the missing procedure by applying a technique which he found in Loomis's applications of the differential calculus to the rectification of plane curves.[23]

As underlined at the beginning of this chapter, a unified presentation was a major preoccupation for Xia. In his second work on conics, the *Diagrammatic Explanations [of Procedures] for Curves* (*Zhiqu tujie* 致曲圖解), he chooses a different organization of knowledge for discussing conics, bringing them closer together by investigating their common properties. The major magnitudes involved in characterizing conics are discursively explained (without algebraic equations) and illustrated for all of the four conics, the circle, the ellipse, the parabola and the hyperbola, in a series of 12 chapters (See Table 2.1).[24]

---

[22]Translated from Xia (1908) 2B, textually identical to the manuscript edition reprinted in Xia (2006a) p. 426.

[23]For the mathematical details, see Appendix C.

[24]As for Chap. 4, there is no correspondence in the Western tradition to the concept of Xia's *gui*-lines. For a short discussion of these, see Liu (1990) p. 15–16.

Finally, in his third and last book related to conics, Xia returned to the format of his first work, giving, one by one, more than 130 procedures for algebraic curves of first to third order and for transcendental curves.[25] This large scope, indicated by the title *A Myriad Images—a Single Origin* (*Wan xiang yi yuan* 萬象一原),[26] is justified philosophically from the start. In his preface, Xia begins with a quote from a Han dynasty mathematical book, revealing a geometric interpretation of his book title before tracing earlier Chinese and foreign contributions to the field of curves:

> "The circle comes from the square" (*yuan chu yu fang* 圓出於方).[27] Yet circular forms are not uniform. The names of curves are therefore of a myriad variety. What the Ancients are said to have had procedures for, were only the plane circle. Concerning elliptic curves, something was handed down from earlier times. My teacher, Sir Xiang Meilü 項梅侶[28] has thoroughly thought about this and established a procedure to find the circumference of the ellipse. Following him, Master Dai Eshi 戴鄂士[29] and Master Xu Junqing 徐君青[30] have each established another procedure, thus making the circumference of the ellipse into a form with a method. Nevertheless, one could only find the circumference of the ellipse, but one could not find the arc of a section of the ellipse. Also, one could not find the arc, the surface area and the volume for all the curves. What a regrettable affair! From the two schools of Newton 奈端 and Leibniz 來本 on, curves were mastered through coordinates. This created what is named the differential and the integral. From then on, what in ancient times had no method now all has a method. Although the forms are myriad, the method is unique. They are indeed the heroes of mathematics! What a happy affair for humankind![31]

As for the conics,[32] Xia labelled the indicated procedures according to their origin. He distinguishes between procedures based on Xiang Mingda, Dai Xu, Xu Youren, Jartoux, the Chinese translation of Loomis's work and on Li Shanlan's *Expounding the Secrets of Square and Circle* (*Fangyuan chanyou* 方圓闡幽). He also provided "new procedures" and "self-invented" ones, partly referring back to his first work where certain procedures "can already be found." All of the procedures are given rhetorically except for two series expansions in the first scroll, where the development of functions into a series, as an application of the differential calculus, is discussed through several examples. One of them is the development of the function $\sqrt[n]{a+x}$ and the particular case of $\sqrt{a+x}$. Only the expansion of the latter

---

[25]For some examples of Xia's procedures and a discussion of errors, see Song and Bai (2008).

[26]The title is probably a pun on a book written by his teacher, Xiang Mingda 項名達 (1789–1850), entitled *Images and Numbers—a Single Origin* (*Xiang shu yi yuan* 象數一原), prefaces from 1843 and 1849. Reprint in Guo et al. (1993) vol. 5, p. 473–602.

[27]Quoted from *The Gnomon of Zhou* (*Zhoubi suanjing* 周髀算經), reprinted in Anonymous (1931), *juan* 上, p. 1B. English translation and commentary in Cullen (1996) p. 174.

[28]Style name of Xiang Mingda 項名達 (1789–1850).

[29]Style name of Dai Xu 戴煦 (1805–1860).

[30]Style name of Xu Youren 徐有壬 (1800–1860).

[31]Translated from Xia (1898a) Preface p. 1A.

[32]*Idem* scrolls 4–7.

Table 2.2 Overview of writings and translations on conic sections and the ellipse in particular

Year	Authors/translators	Original source
1614	Matteo Ricci & Li Zhizao 李之藻, *Yuanrong jiaoyi* 圜容較義 (The Meaning of Compared [figures] Inscribed in a Circle) (Ricci and Li 1614)	Euclid's *Elements* and other sources
1674	Ferdinand Verbiest (chin. 南懷仁, 1623–1688), *Xinzhi lingtai yixiang zhi* 新製靈臺儀象志 (Records on the Newly Built Astronomical Instruments from the Beijing Observatory)	Gardener construction of the ellipse (see Fig. C.1)
1723	*Shuli jingyun* 數理精蘊 (Essence of Numbers and their Principles), scroll 卷 12, chapter on 'Curvilinear figures' (*Quxian xing* 曲線形) in Guo et al. (1993) vol. 3, p. 678–679	*De Conoidibus et Sphaeroidibus*, propostions 5–7 in Archimedes and Eutocius (1544) or Clavius (1611) *Geometria Practica*, Book 4, Prop. V
1742	*Yuzhi lixiang kaocheng houbian* 御製曆象考成後編 (Later Volumes of the Thorough Investigation of Calendrical Astronomy Imperially Composed) (Qianlong 乾隆 1773)	multiple;[a] for ellipse construction possibly (Dechales 1674)
1774	Ming Antu 明安圖, *Geyuan milü jiefa* 割圓密率捷法 (Fast Methods for the Circle Division and the Precise [circle] Ratio) (Ming 1774)	–
1819	Dong Youcheng 董祐誠, *Geyuan lianbili shu tujie* 割圓連比例術圖解 (Diagrammatic Explanations of Continued Proportions in Circle Division) (Dong 1819)	–
1821	Dong Youcheng 董祐誠, *Tuoyuan qiu zhou shu* 橢圓求周術 (Procedure to Find the Circumference of an Ellipse) (Dong 1821)	Solution based on *Nine Chapters*
before 1840	Xu Youren 徐有壬, *Tuoyuan zhengshu* 橢圓正術 (Correct Procedures for the Ellipse), reprint in Xu (1985) p. 33–42	–
1848	Xiang Mingda 項名達, *Tuoyuan qiu zhou shu* 橢圓求周術 (Procedure to Find the Circumference of an Ellipse) (Xiang 1888)	–
1857	Dai Xu 戴煦, *Tuoyuan qiu zhou tujie* 橢圓求周圖解 (Diagrammatic Explanations [to the Procedure] to Find the Circumference of an Ellipse) (Dai 1888)	Reaction to Xiang Mingda's *Tuoyuan qiu zhou shu* 橢圓求周術 (Procedure to Find the Circumference of an Ellipse) (Xiang 1888)

[a] See Shi (2008a)

Table 2.3 Overview of writings and translations on conic sections and the ellipse in particular (cont.)

Year	Authors/translators	Original source
1857	Xia Luanxiang 夏鸞翔, *Dongfang shu tujie* 洞方術圖解 (Diagrammatic Explanations of Procedures for Thoroughly Understanding the Square) in Xia (1873)	–
1859	Li Shanlan 李善蘭 & Joseph Edkins, *Yuanzhui quxian shuo* 圓錐曲線説 (Theory of Conic Sections), appended to *Zhongxue* 重學 (Mechanics) (Li and Edkins 1898)	Unidentified
1859	Li Shanlan 李善蘭 & Alexander Wylie, *Dai weiji shiji* 代微積拾級 (Elements of Analytical Geometry, and of Differential and Integral Calculus) (Li and Wylie 1859)	(Loomis 1851)
1861	Xia Luanxiang 夏鸞翔, *Zhiqu shu* 致曲術 (Xia 1898b), (Xia 1908), (Xia 2006a)	–
1861	Xia Luanxiang 夏鸞翔, *Zhiqu tujie* 致曲圖解 (Diagrammatic Explanations [of Procedures] for Curves) (Xia 1898c), (Xia 2006b)	–
1862	Xia Luanxiang 夏鸞翔, *Wan xiang yi yuan* 萬象一原 (A Myriad Images—a Single Origin) (Xia 1898a) *juan* 4 on second order curves	–
1867	Li Shanlan 李善蘭, *Tuoyuan zhengshu jie* 橢圓正術解 (Explanations of the Correct Procedures for the Ellipse) (Li 1867c)	Reaction to Xu Youren's *Tuoyuan zhengshu* 橢圓正術 (Correct Procedures for the Ellipse) (Xu 1985)
1867	Li Shanlan 李善蘭, *Tuoyuan xinshu* 橢圓新術 (New Procedure for the Ellipse) (Li 1867b)	–
1867	Li Shanlan 李善蘭, *Tuoyuan sheyi* 橢圓拾遺 (Picking Up on Omissions Concerning the Ellipse) (Li 1867a)	–
1888	John Fryer 傅蘭雅, *Quxian xuzhi* 曲線須知 (Essentials of Conics) (Fu 1888)	–
1898	J. H. Judson 求德生 and Liu Weishi 劉維師, *Yuanzhui quxian* 圓錐曲線 (On Conic Sections) (Qiu and Liu 1898)	Loomis (1877)

is given in the syncretistic formulaic language developed by late nineteenth-century translators of Western mathematical books (see Fig. 2.3), whereas the former, the general case, is stated rhetorically. Figure 2.4 shows a later transcription of Xia's verbal procedure for the expansion of $\sqrt[n]{a + x}$ into formulaic language.[33]

For calculating an arc of the ellipse with the sine or versine, Xia reproduces in his work *A Myriad Images—A Single Origin* the same four procedures already given in the *Procedures for Curves* (*Zhiqu shu* 致曲術), insisting again in his commentary

---

[33] For mathematical details, see the Appendix p. 257.

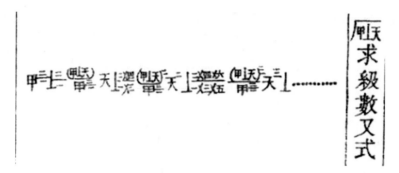

Fig. 2.3 Expansion of $\sqrt{a+x}$ in Xia (1898a)

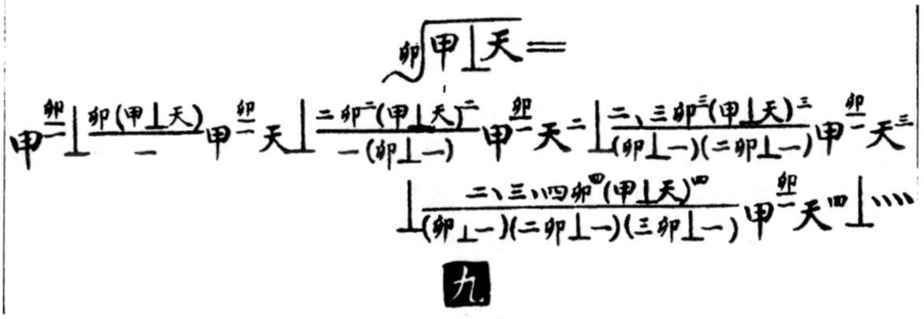

Fig. 2.4 Expansion of $\sqrt[n]{a+x}$ in Lu (1902)

upon his original contribution to their harmonization through application of the differential calculus:

> I therefore acclaim the ingeniousness of recent Western methods. When determining the arc with the sine one can also use the square of the sine to begin the calculation. The two procedures using the square compared to the two procedures using the root, [the former ones] both add an extra operation for extracting the root. By maximizing their uniformity, one can know that numbers have their roots in nature, they cannot be surpassed![34]

Although Xia looks at the calculus with admiration, he nevertheless does not copy its solutions blindly. This is also the case for the quadrature, where the integral calculus leads to an easy formula for the area of an ellipse A with a and b denoting its semi-major and semi-minor axis respectively:[35]

$$A = \frac{b}{a} \cdot a^2\pi = ab \cdot \pi$$

[34]Xia (1898a) *juan* 3 p. 4A.

[35]For the derivation of this formula, see Loomis (1851) p. 225.

Rather than stating this equality, Xia, as for all other magnitudes discussed in relation to conics, prefers to indicate recursively the terms of an infinite series:[36]

$$A = 2a \cdot 2b \cdot \underbrace{\left(1 - \frac{1}{2 \cdot 3} - \frac{1}{8 \cdot 5} - \frac{1}{16 \cdot 7} - \cdots \right)}_{= \frac{\pi}{4}}$$

So far, I have concentrated my discussion of conics in China on Xia Luanxiang, since he is an interesting actor in the transcultural processes of knowledge transfers and the first to take into account foreign knowledge in the field of calculus in his own mathematical work.[37] But Xia was not the first to be interested in conics in China before 1859. He was a disciple of Xiang Mingda, whom he often cites in relation to procedures for the circumference of the ellipse. Xiang was part of an earlier generation of mathematician-astronomers in China who were interested in ellipses only for astronomical reasons but not in conics more generally.[38] The change of reference around 1850 reflects a move from a focus on Western astronomy to Western mechanics, or more specifically to ballistics. Trajectories of missiles were studied in the context of military technologies and approximated not only by elliptical paths but also by the hyperbola and the parabola. Li Shanlan, as a translator, had specifically appended a short treatise in Chinese language on conics in general to a translation of William Whewell's (1795–1866) *Elementary Treatise on Mechanics* (*Zhongxue* 重學) in 1859.[39] Yet in his own mathematical writings, Li continued to concentrate on the ellipse.[40] Dissatisfied with the content of the translated work on conics for its use in astronomy, he uses the ellipse as a paradigmatic representative of second order curves:

> The formerly translated *Theory of Conic Sections* (*Yuanzhui quxianshuo* 圓錐曲線説) still omits many meanings. Because the ellipse is constantly used by astronomers, it is all the more pressing to complement it. For the two curves, hyperbolic and parabolic (*shuangqu paowu* 雙曲拋物), one can infer analogically (*ke litui* 可例推).[41]

---

[36] See Xia (1898a) *juan* 4 p. 3B.

[37] For later reactions to Li and Wylie (1859), see for example Horng (1991) p. 342–353.

[38] See for example Xu Youren's 徐有壬 (1800–1860) *Correct Procedures for the Ellipse* (*Tuoyuan zhengshu* 橢圓正術), posthumously published in 1872, in Xu (1985) p. 33–42, here p. 33:

> The new [astronomical] methods all calculate with the ellipse for the expansion and contraction, acceleration and deceleration [of the planets], but calculation of their revolving trajectories are complicated. They all use approximations but no correct procedures.

[39] Translated together with John Edkins (chin. 艾約瑟, 1823–1905), see Li and Edkins (1898) and Whewell (1859). The original source of the three chapters on conics, the *Theory of Conic Sections* (*Yuanzhui quxianshuo* 圓錐曲線説) has not yet been identified. See Wang (1996) p. 353 and Han (2009) p. 104.

[40] See Li (1867c), Li (1867b) and Li (1867a), of which the first is a sequel to Xu Youren's *Correct Procedures for the Ellipse* in Xu (1985) p. 33–42. For a discussion of Li's work on conics, see Horng (1991) p. 353–362, Gao (2009).

[41] Li (1867a) p. 1a.

Being confronted with a new and imported, geometrical object—the ellipse as a conic section—did not mean that late imperial mathematicians blindly followed the conceptual framework provided by the fragmentary knowledge they received from the West. In order to better understand the efforts they made to integrate their own past, I will discuss in the following part of this chapter some earlier attempts to calculate certain magnitudes of the ellipse. As we shall see, the first-century canonical *Nine Chapters on Mathematical Procedures* (*Jiu zhang suan shu* 九章算術) also served here as a resource and reference for them.

2.2   The Global Fate of Conics

There are no traces of curves or curvilinear surfaces analysed geometrically in Chinese sources before the arrival of the Jesuits except for the circle and the sphere, and their various parts.[42] The area of a circle segment is approximated by different algorithms in diverse sources, and one finds a peculiar form in Qin Jiushao's 秦九韶 *Mathematical Book in Nine Chapters* (*Shushu jiu zhang* 數書九章) from 1247. The procedure solving the problem does not rely on earlier formulas for the segment of a circle calculated by the chord and the arrow, nor does it provide a rationale that would allow for a clear definition of the shape of the elongated leaf considered (see Fig. 2.5):

> Finding the surface area of a banana [-shaped] field.
> The problem is the following: One has one specimen of a field [in the shape] of a banana leaf (*jiaoye tian* 蕉葉田), with a middle length of 576 *bu* [= *a*] and its breadth in the middle being 34 *bu* [= *b*]. When one does not know its perimeter, how much is the surface area in terms of *mu*?
>
> **The answer says:** The surface area of the field is 45 *mu* 1 *jiao* and 11 5213/63070 *bu*.[43]
>
> **The procedure says:** One adds length and breadth together, and multiplies this twice by itself. Furthermore one multiplies this by ten,[44] which makes the *dividend*. Half the breadth and half the length are both multiplied by themselves. One subtracts the results from another, which makes the *joined divisor*. The unit makes the *joined corner*. By extracting the square root and dividing by two one obtains the surface area.[45]

---

[42] The *Nine Chapters* give procedures for calculating a ring and segments of the circle and the sphere. See Chemla and Guo (2004) p. 141–143.

[43] 1 *mu* = 240 *bu* [square], 1 *jiao* = 60 square *bu* [square].

[44] The value 10 is probably an approximation for π^2.

[45] Translated from Qin (1842) p. 499. The three expressions in italic represent technical terms for the coefficients of a polynomial equation in the Chinese mathematical tradition. Here, the positive root *x* of the quadratic equation gives the double of the surface area of the banana-leaf.

Fig. 2.5 A banana-shaped
field in Qin (1842)

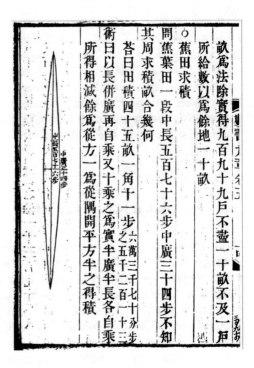

The algorithm described here corresponds to the resolution of the following
quadratic equation:

$$\underbrace{1}_{\text{joined corner}} \cdot x^2 + \underbrace{\left[\left(\frac{a}{2}\right)^2 - \left(\frac{b}{2}\right)^2\right]}_{\text{joined divisor}} \cdot x - \underbrace{10 \cdot (a+b)^3}_{\text{dividend}} = 0$$

It is not clear how Qin has arrived at this solution procedure, nor does it reveal if
the thought of the banana leaf as being delimited by a circular arc or eventually by an
elliptic arc. He certainly did not use the approximate value of a circular segment as
given by the *Nine Chapters*, which would have produced a surface area of $\frac{ab}{2} + \frac{b^2}{4}$.[46]

It might seem surprising that nothing explicit concerning the ellipse can be
found in China before the arrival of Western knowledge.[47] But although the

---

[46]Libbrecht (1973) p. 109 remarks that Qin's formula was probably meant as a better approximation for the case of a circular segment where the arrow is relatively small compared to the chord
sustaining the arc. For the corresponding problems, algorithm and proof in the *Nine Chapters*, see
problems I.35 and I.36 in Chemla and Guo (2004) p. 191–193.

[47]For Japan, where it was commonplace to pose highly original problems about the ellipse in
early nineteenth-century *sangaku* temple geometry riddles, foreign influence is uncertain. See for
example Fukagawa and Rothman (2008) problems 19 and 21.

Jesuits came to China from the early seventeenth century on and brought many scientific books and *savoir faire* with them, this did not necessarily imply that they generously shared their geometric knowledge. Jean-Francois Foucquet's (1665–1741) plan, in particular, was to maintain the Chinese astronomers in a state of dependency upon the missionaries. Conic sections played an important role in this strategy and the transmission of their theory was consciously partial, as he explains:

> [I]n defending the known defects of Tychonic astronomy, one could not but succumb. If on the contrary, the Europeans were able to agree and substitute it with the theory based on recent observations, if they declared for example that the path of planets is elliptic or approaching the ellipse [...], nothing more would be needed to disconcert the Regulo's [Yinshi's] men, and to reduce them again to the condition of students of the Europeans; with the first books of Euclid and a few propositions from Archimedes on solids that have been added to them since, no matter how they knew them, it was not to be feared that they could by themselves penetrate the secrets of conic sections, or of higher geometry; thus the modern theories of the Sun, of the Moon and of other planets would be a sealed language for them[48];

The ellipse first appears in China in the context of mechanics in the *Diagrams and Explanations of Magnificent Machines [from the Far West]* (*[Yuanxi] qiqi tushuo* 遠西奇器圖説, 1627), an encyclopaedic work compiled on the basis of a number of Western language sources and traditional Chinese knowledge by the Jesuit missionary Johannes Schreck (1576–1630) and the Chinese scholar Wang Zheng 王徵 (1571–1644). There, theorem 18 reads (Fig. 2.6):

> For a circle and a "circular form of a little chicken" (*jizi yuanxing* 雞子圓形), the centre of gravity is the same as the geometric centre.[49]

According to the historian Yang Zezhong 楊澤忠, an even earlier mention of an ellipse in China was made in *The Meaning of Compared [figures] Inscribed in a Circle* (*Yuanrong jiaoyi* 圜容較義, 1614) by Matteo Ricci and Li Zhizao 李之藻 (1565–1630), who in their commentary refer to a certain *Book about the Circle* (*Yuanshu* 圓書).[50] Yang assumes that this is the Archimedean *De Circuli Dimensione*, which he says "contains a defintion of the ellipse and a method to determine the surface area of an ellipse." In fact, the *De Circuli Dimensione* contains only three propositions concerning the circle but nothing about the ellipse. The drawing related to the relevant figure in the problem that Yang refers to is not an ellipse, but, as the commentator underlines, "the circle 卯辰,[51] because we

---

[48]Quoted from Jami (2012) p. 289.

[49]Deng and Wang (1830) *juan* 1 p. 24B. As pointed out in Zhang et al. (2008) vol. 1, p. 98, this theorem was copied from Federico Commandino's *Liber de Centro Gravitatis Solidorum*. See Theorema IIII in Commandino (1565) p. 6, where also a detailed proof is given:

In circulo & ellipsi idem est figuræ & gravitatis centrum.

[50]According to Yang (2004) p. 47n3 in particular in problems 17 and 18 of Ricci and Li (1614).

[51]The circle with a diameter from the point 卯 to the point 辰, shown in the upper left figure in Fig. 2.7.

Fig. 2.6 Diagrams and
explanations of magnificent
machines from the Far West,
1627 Deng and Wang (1830)

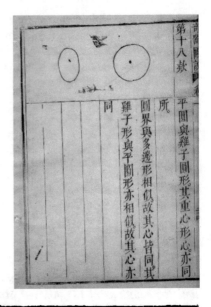

Fig. 2.7 Proof of Propostion
18 in *The Meaning of
Compared [figures] Inscribed
in a Circle* (*Yuanrong jiaoyi*
圓容較義) (Ricci and Li
1614)

wish to make apparent the triangle, is drawn flattened, but it is in fact a regular
circle."[52]

---

[52]Ricci and Li (1614) p. 20B in Fig. 2.7.

Several Archimedian texts were brought to China by the first Jesuits,[53] as well as indirect accounts of Archimedean propositions, in particular in Christophorus Clavius's (1538–1612) writings, which were well known to the Jesuit missionaries at the emperor's court.[54]

Ricci's Archimedean source corresponding to the *Book about the Circle* (*Yuanshu* 圓書) is in fact *De Sphaera et Cylindro*,[55] which again does not concern the ellipse. Where Ricci and Li might have found something is *De Conoidibus et Sphaeroidibus*,[56] where Archimedes proves three fundamental theorems for the surface area of an ellipse:

> Proposition 4: The area of any ellipse is to that of the auxiliary circle as the minor axis to the major.[57]
> Proposition 5: If AA', BE' be the major and minor axis of an ellipse respectively, and if d be the diameter of any circle, then
> (area of ellipse) : (area of circle) $= AA' \cdot BE' : d'^2$.
> Proposition 6: The areas of ellipses are as the rectangles under their axes.[58]

Chinese mathematicians would probably have preferred to see a more familiar, algorithmic approach to the ellipse, such as the one found in the *Metrica* by Heron from Alexandria (probably second half of first century).[59] Referring to Archimedes in the solution, Heron gives the following problem:

> Soit à mesurer une ellipse dont d'une part le grand axe est de 16 unités, d'autre part le petit [axe] de 12 unités.
> Alors, puisqu'il a été démontré par Archimède dans les *Conoïdes* que le [rectangle contenu] par les axes équivaut à un cercle égal à l'ellipse, il faudra, multipliant les 16 par les 12, de

---

[53] According to the catalogue (Verhaeren 1949) item 2612 the Jesuits had, for example, a 1597 edition by Adriaen van Roomen (1561–1615) of a book entitled *In Archimedis circuli dimensionem Expositio & Analysis. Apologia pro Archimede ad Clariss. virum Iosephum Scaligerum* ... (Roomen 1597). Verhaeren (1949) item 4000 is the *editio principes* of the collection of Archimedean texts, Archimedes and Eutocius (1544).

[54] Verhaeren (1949) item 1288, for example, lists Clavius's collected mathematical writings, the *Opera mathematica* (Clavius 1611).

[55] A comparison of reference in the commentary of proposition 18 to "*Yuanshu* 圓書 Book I proposition 31" for example, shows clearly that *De Sphaera et Cylindro* is what Ricci and Li consulted when they made a statement about the fourfold surface area of the sphere with respect to a circle with the same radius. See Archimedes and Eutocius (1544) *De Sphaera et Cylindro* Liber Primus vol. I, p. 31–32 and Ricci and Li (1614) p. 20B.

[56] Contained in Archimedes and Eutocius (1544).

[57] In the proof of Prop. 4 it is assumed that this auxiliary circle is the one having the major axis as radius.

[58] Paraphrases of the enunciations (which are longer and more involved) according to Heath (1897) p. 113–115, these propositions correspond to Propositions 5 to 7 in Archimedes and Eutocius (1544) p. 52–53.

[59] On the problem of dating Heron, see Héron d'Alexandrie (2014) p. 15–26.

ceux-ci, prendre les $^{11}/_{14}$: et c'est 150 $^6/_7$; déclarer que l'aire de l'ellipse est autant que cela.[60]

But since the only extant manuscript of Heron's text was used for the first and last time in fifteenth-century Constantinople, where it remained until it was rediscovered at the very end of the nineteenth century, it is only counterfactual speculation that a Chinese audience would have felt more familiar with his approach. In the context of the early Jesuit missionaries' education in Rome, Clavius was an important reference for Western mathematical knowledge. Clavius refers to the above Archimedean propositions in his *Geometria Practica* and proves that the area of an ellipse is the same as the area of a circle whose diameter is the mean proportional between the minor and major axis of the ellipse (Fig. 2.8):

> Aream propositæ Ellipsis indagare.
> Lubet denique librum hunc quartum duobus problematibus terminare, quæ ab Archimede Syracusano acutissime inventa sunt, ac demonstrata. Unum est de area Ellipsis; alterum de area Parabolæ. Sit ergo Ellipsis ABCD, cuius maior diameter BD, & minor AC, secans maiorem in E, bifariam. Inueniatur HI, media proportionalis inter BD, & AC: & circuli circa diametrum HI, descripti area inquiratur, per ea, quæ cap. 7. huius lib. scripsimus. Dico hanc aream areæ Ellipsis ABCD, esse æqualem.[61]

Clavius's formulation is equivalent to the one that made its way into the *Later Volumes of the Thorough Investigation of Calendrical Astronomy Imperially Composed* (*Yuzhi lixiang kaocheng houbian* 御製曆象考成後編, 1738), a publication that resulted from a calendrical reform compiled by the Jesuits Ignaz Kögler (chin. Dai Jinxian 戴進賢, 1680–1746) and André Pereira (chin. Xu Maode 徐懋德, 1689–1743), as well as the Chinese scholars Mei Juecheng 梅瑴成 (1681–1763) and He Guozong 何國宗 (d. 1766). This publication took into consideration Kepler's first two laws of planetary motion and other new important theories and astronomical constants used by Western astronomers.[62] The first scroll of the book is devoted "to the explanation of the so-called 'Mathematical theory of the sun' (*ri chan shuli* 日躔數理), including discussions on why and how an elliptic orbit was adopted, since Kepler, in the account of the motion of the sun, and in what respects this new model was different from and similar to the old Tychonic model," etc.[63] The surface area of the ellipse is given here explicitly, and is mathematically similar to the above-quoted passage by Clavius, albeit in a constructive manner:

---

[60]See Book I, prop. XXXIV translated in Héron d'Alexandrie (2014) p. 235. As pointed out in *idem* p. 235n310, this is not exactly the Archimedean formulation in *De Conoidibus et Sphaeroidibus*, where Archimedes gives the proportion:

Ellipse : circle = minor axis : diameter.

[61]Clavius (1604) *Geometria Practica* bk. IV, prop. V, p. 224.

[62]On the wider context of the book within the whole history of Jesuit astronomy in China, see Shi (2008a).

[63]Shi (2008b) p. 225. See also Hashimoto (1971).

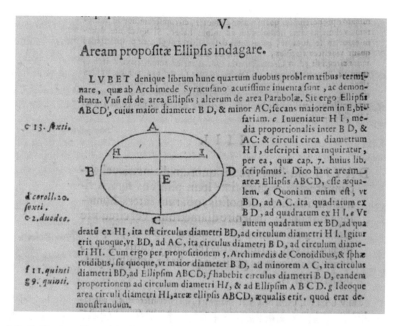

V.

Aream propofitæ Ellipfis indagare.

LVBET denique librum hunc quartum duobus problematibus termi-
nare, quæ ab Archimede Syracufano acutiffime inuenta funt, ac demon-
ftrata. Vnū eft de area Ellipfis; alterum de area Parabolæ. Sit ergo Ellipfis
ABCD, cuius maior diameter B D, & minor AC, fecans maiorem in E, bi-
fariam. c Inueniatur H I, me-
dia proportionalis inter B D, &
AC: & circuli circa diametrum
H I, defcripti area inquiratur,
per ea, quæ cap. 7. huius lib.
fcripfimus. Dico hanc aream
areæ Ellipfis ABCD, effe æqua-
lem. d Quoniam enim eft, vt
B D, ad A C. ita quadratum ex
B D, ad quadratum ex H I. e Vt
autem quadratum ex BD, ad qua
dratū ex HI, ita eft circulus diametri BD, ad circulum diametri H I. Igitur
erit quoque, vt BD, ad AC, ita circulus diametri B D, ad circulum diametri HI. Cum ergo per propofitionem 5. Archimedis de Conoidibus, & fphæ
roidibus, fit quoque, vt maior diameter B D, ad minorem A C, ita circulus
diametri BD, ad Ellipfim ABCD; f habebit circulus diametri B D, eandem
proportionem ad circulum diametri HI, & ad Ellipfim A B C D. g Ideoque
area circuli diametri HI, areæ ellipfis ABCD, æqualis erit. quod erat demonftrandum.

c 13. fexti.

d coroll. 20.
fexti.
e 2. duodec.

f 11. quinti
g 9. quinti.

Fig. 2.8 C. Clavius, *Geometria Practica* Book IV Prop. V Clavius (1604) p. 129

With the mean proportional of the major and minor axis one constructs a circle, its surface
area is equal to the ellipse.[64]
以大小徑之中率作平圓其面積與橢圓等

In the *Later Volumes of the Thorough Investigation of Calendrical Astronomy
Imperially Composed* the surface area of the ellipse is also stated proportionally
with respect to the circle, as in Archimedes' proposition 5.[65] This statement is nearly
identical to what is already included (actually twice) in the *Essence of Numbers and
their Principles* (*Shuli jingyun* 數理精蘊, 1723). This slightly earlier encyclopaedic
work was the mathematical part of a larger imperial compilation project, the purpose
of which was to frame Western and Chinese traditions in a single historical narrative.
As Jami points out, "the core material of the editorial work on the *Essence of
Numbers and their Principles* was the lecture notes that the Jesuits had prepared for
Kangxi in the 1690s" in the fields of arithmetic, algebra and geometry, but "printed

---

[64]Translated from Qianlong 乾隆 (1773) p. 45A. According to Yang (2004) p. 49 this statement is
to be found in *juan* 1 in the second part entitled "Finding the distance between two centers" (*qiu
liang xin cha* 求兩心差), but it is in fact quoted from the section on "Finding the mean proportional
between the major and minor axis of an ellipse" (*qiu tuoyuan da xiao jing zhi zhonglü* 求橢圓大
小徑之中率).

[65]See Qianlong 乾隆 (1773) p. 41B–42A.

mathematical works were also available to the staff."[66] For the ellipse (*tuoyuan xing* 橢圜形), for which, as the commentator points out, "another name is duck-egg shape" (*yadan xing* 鴨蛋形), one finds two occurrences of the general proportional statement and a numerical calculation of its surface area using the rule of three. Considering the surface area of a rectangle circumscribed to an ellipse and a square circumscribed to a circle, the text states in the chapter on "Curvilinear figures" (*quxian xing* 曲線形)[67]:

> In general, the ratio between the surface area of a circle and the surface area of an ellipse is equal to the ratio between the surface area of a square circumscribed to the circle and the surface area of a rectangle circumscribed to the ellipse. See *Elements of Geometry* (*Jihe yuanben* 幾何原本) scroll 8 section 12.[68]

The reference justifying the correctness of the procedure here is not Euclid's *Elements*, but a Chinese adaptation of the *Elemens de geometrie ou par une méthode courte & aisee l'on peut apprendre ce qu'il faut scavoir d'Euclide, d'Archimede, d'Apollonius, & les plus belles inventions des anciens & des nouveaux Geometres* (1673, 2nd ed.), which was originally written by the French Jesuit Ignace Gaston Pardies. The Chinese version was included in the *Essence*, but in Pardies' original nothing corresponds to the twelfth theorem in scroll eight, where the special case of the above proportion with the major axis of the ellipse being equal to the circle's diameter is proved.[69]

2.3   Using the Past to Solve the New

The grand compilation enterprises such as the *Later Volumes* and the *Essence*, undertaken as Sino-European collaborations, are often portrayed as the synthesis of Chinese and Western knowledge traditions. Yet they do not seem to result in a cross-cultural amalgam, nor do they refer back to the authentic Chinese mathematical tradition, at least insofar as the treatment of the ellipse is concerned. On the

---

[66] Jami (2012) p. 320. For possible foreign sources, see also the introduction by Han Qi 韩琦 to the reprint edition of the *Essence* in Guo et al. (1993) vol. 3, p. 1–10.

[67] See Yunzhi 允祉 (1723) scroll 20, p. 17B–18B (Fig. 2.9). The preceding numerical example asks for the surface area of an ellipse with a major axis of 9 *chi* and a minor axis of 6 *chi*. In the solution, the argument by proportion is applied to a circle with a diameter equal to 10^4 and a surface area of 78 539 816, thus assuming a value of $\pi = 4 \cdot 0.785\,398\,16 = 3.141\,592\,64$.

[68] Translated from Yunzhi 允祉 (1723) scroll 20, p. 18A.

[69] See Yunzhi 允祉 (1723) scroll 3, p. 53A–55A:

> In general, when the diameter of a circle is equal to the major axis of the ellipse another name is "duck-egg shape," then the ratio between the surface area of the circle and the ellipse is also equal to the ratio between the rectangles circumscribed to the two figures. The ratio between the surface area of the circle and the ellipse is also equal to the ratio between the diameter of the circle and the minor axis of the ellipse.

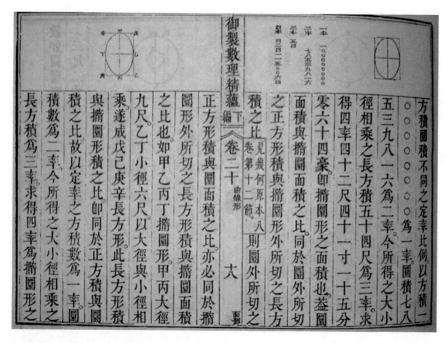

Fig. 2.9 Ratio between the surface areas of ellipse and circle in the *Essence of Numbers and their Principles* (*Shuli jingyun* 數理精蘊) (Yunzhi 允祉1723)

other hand, the writings of Chinese early nineteenth-century mathematicians, the independent thinkers outside the imperial court, are more creative in digesting newly imported geometrical objects such as conics.

Dong Youcheng 董祐誠 (1791–1823), for example, in his *Procedure to Find the Circumference of an Ellipse* (*Tuoyuan qiu zhou shu* 橢圜求周術, 1821) attempts to solve the rectification of the ellipse. This was a non-trivial geometric problem for which he could not find an answer in the hitherto available Chinese sources; thus, he turned to elements from authentic Chinese writings instead. Dong cites explicitly the origin of his methodological approach: a problem found in the *Nine Chapters*. In his short preface from 1821, we can read that Dong's point of departure was to consider the ellipse as a cylindrical section:

Finding the circumference of an ellipse is a procedure that did not exist in Antiquity. Mister Zhu Hong 朱鴻[70] from Xiushui 秀水[71] said that by oblique section of a cylinder one can obtain an ellipse, and with rectangular triangles one can solve the problem. The autumn was chilly, and I had nothing to do. I have thus developed what Mister [Zhu] said. I explain it using a diagram [see Fig. 2.10]. Basically, a circle is similar to the square, an ellipse is

---

[70]Dates unknown, Jinshi in 1801. See Yan (1990) entry 0174, p. 48.

[71]A county in Zhejiang province, now Jiaxing 嘉興.

Fig. 2.10 Dong (1821)

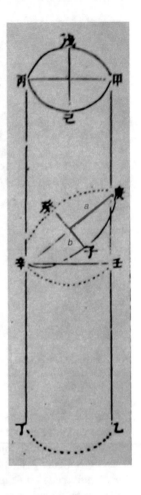

similar to the rectangle. The ellipse has a small diameter, a big diameter, a circumference, an area. One needs to know two [magnitudes] in order to determine the others.[72]

To determine the perimeter p of an ellipse with a major axis a and a minor axis b (see Fig. 2.10), Dong finds an algorithm that corresponds to the following formula:

$$p = \sqrt{\pi^2 b^2 + 4 \cdot \left(a^2 - b^2\right)}$$

He bases his calculations upon an ancient approximation of π, the "more precise ratio" (*mi lü* 密率) $\pi \approx 355/113 = 3.141\,592\,9\ldots$ ascribed to the fifth-century mathematician and astronomer Zu Chongzhi 祖沖之 (429–500).

---

[72]Dong (1821) p. 1A.

Altogether this leads to a series of operations for finding the circumference of the ellipse corresponding to the following formula:

$$p = \sqrt{355^2 \cdot b^2 + 113^2 \cdot 4 \cdot \left(a^2 - b^2\right)} \div 113$$

At the end of his rhetorical list of operations, Dong Youcheng reveals how he found this procedure:

> As for obtaining a circular arc by establishing a right triangle, it is exactly the intention of the procedure for "A kudzu vine winding up a tree" (*ge sheng chan mu* 葛生纏木) in the *gougou* [chapter][73] of the *Nine Chapters*.

Dong refers here to problem 9.5 from the *Nine Chapters on Mathematical Procedures*, which uses the Pythagorean theorem to determine the length of a plant growing steadily up a tree. If we assume that the plant's growth is regular and that it thus takes the shape of a helix, the given procedure models the length of the plant correctly (Fig. 2.11). The problem in the *Nine Chapters* reads as follows:

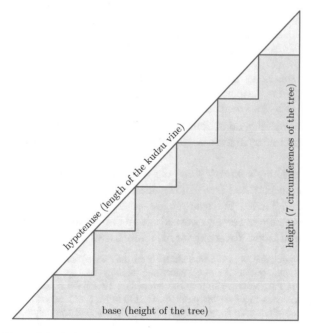

Fig. 2.11 Problem of the kudzu vine in the *Nine Chapters on Mathematical Procedures* (first century AC)

---

[73] Chapter 9 of the *Nine Chapters* concerned with problems related to right triangles.

Suppose that is given a tree 2 *zhang* high and 3 *chi* in circumference. A kudzu vine winds around it 7 times from its root to its top. One asks what is the length of the vine.

Answer: 2 *zhang* 9 *chi*.

Procedure: One multiplies by 7 circuits the circumference, which makes the height (*gu* 股); let the height of the tree be the base (*gou* 句) and find the hypotenuse. This hypotenuse is the length of the vine.

Dong applies this problem to the case where the kudzu vine makes only a single circuit of the tree. He thinks of the circumference of the ellipse with a major axis a as making two half tours of a cylinder with diameter b. By thus 'unfolding' his ellipse onto the cylinder, he resolves the rectification of the ellipse, as in the *Nine Chapters*, with a right triangle whose base corresponds to the "height of the tree," as twice the height of the section of the cylinder:

$$\text{Hypotenuse} = \text{length of kudzu vine} = p$$

$$\text{Base} = 2 \cdot \text{height of cylinder section} = 2 \cdot \sqrt{a^2 - b^2}$$

$$\text{Height} = 1 \cdot \text{circumference of cylinder} = 1 \cdot \pi \cdot b$$

Applying the Pythagorean theorem gives Dong's algorithm for p:

$$p = \sqrt{\pi^2 b^2 + 4 \cdot (a^2 - b^2)}$$

Culturally speaking, Dong is making a strong statement here: by using a procedure from the past he suggests the omnipotence of the *Nine Chapters* in solving complex geometrical problems. Mathematically speaking, he was incorrect. Indeed, when projecting an oblique section of the cylinder on a plane, one does not obtain a straight line as in the case of the helix (see the blue line in Fig. 2.12), but a sinusoid.

The error in Dong's *tour de force* did not remain unnoticed by his contemporaries. Luo Shilin 羅士琳 (1789–1853), one of the last proponents of the Qian-Jia school, which defended the resurrection of the ancient Chinese tradition and disregarded contemporaries who worked within the Western conceptual framework, harshly criticized Dong Youcheng for his faulty use of the past:[74]

> This is not compatible with the procedure [from the *Nine Chapters*]. In the case of the "kudzu vine winding around a tree," when one joins the two opposite surfaces, one will have corners in the points where they come together. This is the reason why one can use rectangular triangles to solve this problem. On the other hand, the ellipse is an oblique section of the cylinder. Thus, one cannot identify the trace of the points where one has cut, these might have a relation to an extended circle, but they cannot have anything in common with rectangular triangles. And the points where they touch are not even corners. Because the points where they touch are not even corners, their forms are not identical. The [resulting] number certainly is constantly smaller than the ellipses' circumference, these methods truly do not communicate!

---

[74]Quoted from Ruan and Luo (1882) scroll 51 卷五十一, entry on Dong Youcheng 董祐誠.

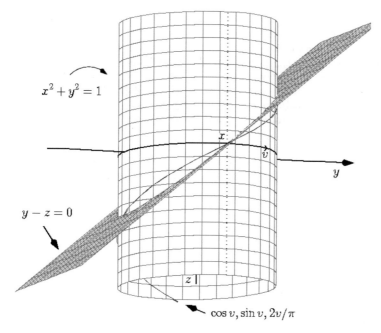

$$x^2 + y^2 = 1$$

$$y - z = 0$$

$$\cos v, \sin v, 2v/\pi$$

Fig. 2.12   Oblique cut of a cylinder compared to the helix (in blue)

於術不通，蓋葛生纏木，若使兩面對纏，其相交處必有角，故可借為勾股形求之。
而橢圓之形，則為斜剖之圓柱，與葛纏者迥異，其受剖處無痕跡可尋，故能有合於
長圓，而不能有合於勾股，以其相交處無角也。夫其相交處無角，則其形不同，其
數必恆小於橢周，信非通法！

Soon after Dong's attempt to solve the rectification of the ellipse with elementary
geometry, a correct procedure was given by Xiang Mingda in his *Procedure to Find
the Circumference of an Ellipse* (*Tuoyuan qiu zhou shu* 橢圓求周術)[75]:

The method: What one obtains when taking the big diameter [of the ellipse, $2a$] as the
diameter to find the circumference of a plane circle, makes the first number. Then, one takes
half of the big diameter of the ellipse to make the first *lü* 率 [a].[76] Half of the small diameter
[b] multiplied by itself and divided by half of the big diameter, this in turn subtracted from
half of the big diameter makes the third *lü* [$a - \frac{b^2}{a}$]. Thereupon put down the first number,
multiply it with the third *lü*, divide it by the first *lü* and divide it by 2 multiplied by itself to
make the second number [$2\pi a \cdot \frac{a^2 - b^2}{a^2} \div 2^2$]. Then, put down the second number, multiply it

[75]Equivalent, but historically unrelated, procedures were found by Colin MacLaurin (1698–1746)
in his *Treatise on Fluxions* (1742) and by the Japanese mathematician Sakabe Kōhan 坂部広胖
(1759–1824) in his *Guide to Tenzan [Mathematical Methods]* (*[Sampō] Tenzan shinan roku* [算
法] 点竄指南錄) in 1815. See MacLaurin (1742) § 806, p. 658–659 and Mikami (1912).

[76] *Lü* is a technical mathematical term conceptually introduced by Liu Hui in his 263 commentary
to the *Nine Chapters* where it designates a number that is in a ratio with another one. See Chemla
and Guo (2004) p. 956–958.

with the third *lü*, divide it by the first *lü*, multiply it by one, multiply it by three, divide it by 4 multiplied by itself to make the third number $[2\pi a \cdot \frac{1^2 \cdot 3}{2^2 \cdot 4^2} \cdot e^4]$.[77] Then, put down the third number, multiply it with the third *lü*, divide it by the first *lü*, multiply it by three, multiply it by five, divide it by 6 multiplied by itself to make the fourth number $[2\pi a \cdot \frac{1^2 \cdot 3^2 \cdot 5}{2^2 \cdot 4^2 \cdot 6^2} \cdot e^6]$. Then, put down the fourth number, multiply it with the third *lü*, divide it by the first *lü*, multiply it by five, multiply it by seven, divide it by 8 multiplied by itself to make the fifth number $[2\pi a \cdot \frac{1^2 \cdot 3^2 \cdot 5^2 \cdot 7}{2^2 \cdot 4^2 \cdot 6^2 \cdot 8^2} \cdot e^8]$. Then, put down the fifth number, multiply it with the third *lü*, divide it by the first *lü*, multiply it by seven, multiply it by nine, divide it by 10 multiplied by itself to make the sixth number $[2\pi a \cdot \frac{1^2 \cdot 3^2 \cdot 5^2 \cdot 7^2 \cdot 9}{2^2 \cdot 4^2 \cdot 6^2 \cdot 8^2 \cdot 10^2} \cdot e^8]$.[78]

Xiang's procedure defines recursively the coefficients of the first five terms of a potentially infinite (hypergeometric) series with a and b denoting the major and minor axis respectively:

First number: $+ 2\pi a$

First *lü*: a

Third *lü*: $a - \dfrac{b^2}{a}$

Second number: (first number · third *lü* ÷ first *lü*) $\div 2^2 =$

$$2\pi a \cdot \frac{a^2 - b^2}{a^2} \div 2^2 = +2\pi a \cdot \frac{1}{2^2} \cdot e^2$$

Third number: (second number · third *lü*) ÷ first *lü* $\cdot 1 \cdot 3 \div 4^2 =$

$$2\pi a \cdot \frac{1^2 \cdot 3}{2^2 \cdot 4^2} \cdot e^4$$

Fourth number: (third number · third *lü*) ÷ first *lü* $\cdot 3 \cdot 5 \div 6^2 =$

$$2\pi a \cdot \frac{1^2 \cdot 3^2 \cdot 5}{2^2 \cdot 4^2 \cdot 6^2} \cdot e^6$$

Fifth number: (fourth number · third *lü*) ÷ first *lü* $\cdot 5 \cdot 7 \div 8^2 =$

$$2\pi a \cdot \frac{1^2 \cdot 3^2 \cdot 5^2 \cdot 7}{2^2 \cdot 4^2 \cdot 6^2 \cdot 8^2} \cdot e^8$$

before concluding:

According to this [pattern], successively multiply and successively divide. The obtained number becomes gradually smaller, one stops when one reaches below the positions of the [desired number of decimal] units. The first number is positive; from the second

---

[77]For convenience only, the expression $\sqrt{\frac{a^2 - b^2}{a^2}}$ is replaced in my transcription by e, the eccentricity of an ellipse, a concept measuring its deviation from circularity not present in Xiang Mingda's text.

[78]Translated from Xiang (1888) vol. 6, p. 27B–28A.

number on all [numbers] are negative. Positive and negative mutually subtracted, this is the circumference of the ellipse [L].[79]

$$L = 2\pi a \left(1 - \frac{1}{2^2}e^2 - \frac{1^2 \cdot 3}{2^2 \cdot 4^2}e^4 - \frac{1^2 \cdot 3^2 \cdot 5}{2^2 \cdot 4^2 \cdot 6^2}e^6 - \frac{1^2 \cdot 3^2 \cdot 5^2 \cdot 7}{2^2 \cdot 4^2 \cdot 6^2 \cdot 8^2}e^8 - \cdots \right)$$

Dai Xu's reconstruction of how Xiang might have found this infinite series refers back to results from the *Later Volumes* and theorems from Euclid's *Elements* alike. It operates merely with statements from elementary geometry[80] and leads to a series equivalent to the one found by Colin MacLaurin (1698–1746) in his work on fluxions in 1742.[81]

With Xia Luanxiang's work, strongly reminiscent of Xiang Mingda's and Dai Xu's way of formulating procedures for infinite series, we can close the loop of this chapter's narrative. Using the ellipse as an example, we have seen how late imperial Chinese authors approached a geometric object that was not part of their panoply of concepts but essential to the newly introduced fragments of astronomical and mechanical theories. Ranging from Xia's global view on conics to such original solutions as Dong Youcheng's, the embedding of the ellipse into a discursive and conceptual context was not culturally innocent.

References

Anonymous (Ed.) (1931). *Song ke suanjing liu zhong* 宋刻算經六種 (Six Books of the [Ten Books of] Mathematical Classics Printed in the Song Dynasty). Beijing: Wenwu chubanshe 文物出版社. Reproduced from the original Song dynasty edition printed in 1213.

Archimedes and Eutocius (1544). *Archimēdous tou Syrakousiou ta mechri nyn sōzomena hapanta = Archimedis Syracusani ... opera, quae quidem extant, omnia : nuncque primum & Graece et Latine in lucem edita. Adiecta quoque sunt Eutocii Ascalanitae in eosdem Archimedis libros commentaria : item Graece et Latine, .../ edidit Thomas Geschauff, cognomento Venatorius.* Basilea: [Hervagius].

Chemla, Karine and Shuchun Guo (2004). *Les neuf chapitres sur les procédures mathématiques. Le classique mathématique de la Chine ancienne et ses commentaires.* Paris: Dunod.

Chen, Jiang-Ping Jeff (2015). Trigonometric Tables: explicating their Construction Principles in China. *Archive for History of Exact Sciences* 69(5).

Clavius, Christophorus (1604). *Geometria practica.* Romae.

---

[79]Translated from Xiang (1888) vol. 6, p. 28A. A different version of Xiang's procedure is given by Dai Xu in his *Diagrammatic Explanations [to the Procedure] to Find the Circumference of an Ellipse (Tuoyuan qiu zhou tujie* 橢圓求周圖解, 1857), which was published as an appendix to the posthumous publication of Xiang Mingda's *Images and Numbers—a Single Origin (Xiang shu yi yuan* 象數一原). See Dai (1888) vol. 7, here p. 53A–53B.

[80]For a formulaic transcription of Dai Xu's work, see Li (1955).

[81]See MacLaurin (1742) § 806, p. 658–659. A more detailed comparison between Xiang Mingda's, Dai Xu's and MacLaurin's approach is in preparation for a future publication.

Clavius, Christophorus (1611). *Opera mathematica V Tomis distributa*. Moguntie [Mainz]: sumptibus Antonij Hierat excudebat Reinhard Eltz.

Commandino, Federico (1565). *Liber de centro gravitatis solidorum*. Bononiæ.

Cullen, Christopher (1996). *Astronomy and Mathematics in Ancient China: The Zhou bi suan jing*. New York: Cambridge University Press.

Dai, Xu 戴煦 (1888). *Tuoyuan qiu zhou tujie* 橢圓求周圖解 (Diagrammatic Explanations [to the Procedure] to Find the Circumference of an Ellipse). In Dai Xu 戴煦 (Ed.), *Xiang shu yi yuan* 象數一原 (Images and Numbers—A Single Origin), 7 vols., 7:1A–53B. Shanghai: Jin'gui Hua shi 金匱華氏. Reprint in (Guo et al. 1993) 5:574–600.

Dechales, Claude François Milliet (1674). *De sectionibus conicis*. In *Cursus seu Mundus mathematicus*, Volume 3 (Tomus Tertius), Tractatus XXXI, 793–863. Lugduni [Lyon]: Anisson.

Deng, Yuhan 鄧玉函 (Schreck, Johannes) and Zheng Wang, 王徵 (1830). *Qiqi tushuo* 奇器圖説 (Diagrams and Explanations of Magnificent Machines) (Lailutang 來鹿堂藏板 ed.). Preface by Wang Zheng from 1627.

Dong, Youcheng 董祐誠 (1819). Geyuan lianbili shu tujie 割圜連比例術圖解 (Diagrammatic Explanations of Continued Proportions in Circle Division). In *Dong Fangli yishu* 董方立遺書 (The Posthumous Works of Dong Fangli), vol. 1 (1823 ed.). Reprint in (Guo et al. 1993) 5:435–460.

Dong, Youcheng 董祐誠 (1821). Tuoyuan qiu zhou shu 橢圜求周術 (Procedure to find the circumference of an ellipse). In *Dong Fangli yishu* 董方立遺書 (The Posthumous Works of Dong Fangli), vol. 1 (1823 ed.). Reprint in (Guo et al. 1993) 5:460–461.

Fu, Lanya 傅蘭雅 (Fryer, John) (1888). *Quxian xuzhi* 曲線須知 (Essentials of Conics), 1 vol, Volume 7 第七冊 of *Gezhi xuzhi* 格致須知. Shanghai: Jiangnan jiqi zhizao zongju 江南機器製造總局.

Fukagawa, Hidetoshi and Tony Rothman (2008). *Sacred Mathematics: Japanese Temple Geometry*. Princeton, NJ: Princeton University Press.

Gao, Hongcheng 高红成 (2009). Li Shanlan dui weijifen de lijie yu yunyong 李善兰对微积分的理解与运用 (Understanding and Application of Calculus by Li Shanlan). *Zhongguo keji shi zazhi* 中国科技史杂志 (The Chinese Journal for the History of Science and Technology) *30*(2), 222–230.

Guo, Shuchun 郭書春 et al. (Eds.) (1993). *Zhongguo kexue jishu dianji tonghui: Shuxue juan* 中國科學技術典籍通彙: 數學卷 (Comprehensive Collection of Ancient Classics on Science and Technology in China: Mathematical Books), 5 vols. Zhengzhou: Henan jiaoyu chubanshe 河南教育出版社.

Han, Qi 韩琦 (2009). Li Shanlan, Ai Yuese yi Hu Weili «Zhongxue» zhi diben 李善兰 艾约瑟译胡威立《重重》之底本 (The Original Edition Underlying Li Shanlan's and John Edkins' Translation of William Whewell's *Elementary Treatise on Mechanics*). *Wakumon* 或問 *101*(17), 101–111.

Hashimoto, Keizō 橋本敬造 (1971). Daenhō no tenkai: Rekishō kōsei kōhen no aiyō ni tsuite 橢圓法の展開 - 『暦象考成後編』の内容について - (The Development of Elliptical Methods. On the Contents of the *Later Volumes of the Established System of Calendrical Astronomy*). *Tōhō Gakuho* 東方學報 *(Journal of Oriental Studies)* 42, 245–272.

Heath, Thomas L. (1897). *The Works of Archimedes: Edited in Modern Notation with Introductory Chapters*. Cambridge: At the University Press.

Héron d'Alexandrie (2014). *Metrica*. Number 4 in Mathematica graeca antiqua. Pisa, Roma: Fabrizio Serra. Introduction, texte critique, traduction française et notes de commentaire par Fabio Acerbi et Bernard Vitrac.

Horng, Wann-Sheng (1991). *Li Shanlan: The Impact of Western Mathematics in China during the Late 19th Century*. Ph. D. thesis, Graduate Center, City University of New York.

Jami, Catherine (1990). *Les méthodes rapides pour la trigonométrie et le rapport précis du cercle (1774)*. Mémoires des Hautes Études Chinoises. Paris: Collège de France.

Jami, Catherine (2012). *The Emperor's New Mathematics*. Oxford: Oxford University Press.

Kepler, Johannes (1604). *Ad Vitellionem Paralipomena, Quibus Astronomiae Pars Optica Traditvr Potißimum De Artificiosa Observatione Et Aestimatione Diametrorvm deliquiorumq[ue] Solis & Lunae ; Cvm Exemplis Insignivm Eclipsivm … Tractatum luculentum de modo visionis, & humorum oculi vsu, contra Opticos & Anatomicos*. Francofvrti.

Legge, James (Trans.) (1899). *The Yî King* (2nd ed.), Volume 16 of *Sacred Books of the East*. Oxford: the Clarendon Press.

Legge, James (Trans.) (1963). *The I Ching* (2nd ed.). The Sacred Books of China. New York, NY: Dover. Unabridged and unaltered republication of (Legge 1899).

Li, Shanlan 李善蘭 (1867a). *Tuoyuan sheyi* 橢圜拾遺 (Picking up on Omissions Concerning the Ellipse), 3 vols. In *Zeguxizhai suanxue* 則古昔齋算學 (Mathematics from the Studio Devoted to the Imitation of the Ancient Chinese Tradition) (Jinling 金陵刻本 ed.), Volume 9.

Li, Shanlan 李善蘭 (1867b). *Tuoyuan xinshu* 橢圜新術 (New Procedure for the Ellipse), 1 vol. In *Zeguxizhai suanxue* 則古昔齋算學 (Mathematics from the Studio Devoted to the Imitation of the Ancient Chinese Tradition) (Jinling 金陵刻本 ed.), Volume 8.

Li, Shanlan 李善蘭 (1867c). *Tuoyuan zhengshu jie* 橢圜正術解 (Explanations of the Right procedure for the Ellipse), 2 vols. In *Zeguxizhai suanxue* 則古昔齋算學 (Mathematics from the Studio Devoted to the Imitation of the Ancient Chinese Tradition) (Jinling 金陵刻本 ed.), Volume 7.

Li, Shanlan 李善蘭 and John Edkins 艾約瑟 (1898). Yuanzhui quxian shuo 圜錐曲線説 (Theory of Conic Sections). In Liu Duo 劉鐸 (Ed.), *Gujin suanxue congshu* 古今算學叢書 (Compendium of Mathematics, Old and New) (Weiboxie 擴微波榭本等石印 ed.), Volume 3 (象數第三). Shanghai: Shanghai suanxue shuju 上海算學書局. Originally published as an Appendix to the translation of William Whewell's *Elementary Treatise on Mechanics*, first published in 1859.

Li, Shanlan 李善蘭 and Alexander Wylie 偉烈亞力 (1859). *Dai weiji shiji* 代微積拾級 (Elements of Analytical Geometry and of the Differential and Integral Calculus) 18 scrolls. Shanghai: Mohai shuguan 墨海書館. Original by Elias Loomis 羅密士 (Loomis 1851).

Li, Yan 李儼 (1955). *Zhongsuanjia de yuanzhui quxian shuo* 中算家的圜錐曲線説 (Explanations of Conic sections by Chinese Mathematicians). In Li Yan 李儼 (Ed.), *Zhongsuanshi luncong* 中算史論叢 (Collected Writings on the History of Mathematics in China), Volume 3 第三集, 519–537. Beijing: Kexuechubanshe 科學出版社. Originally published in *Kexue* 科學 29-4 (1947), 115–120. Reprint in (Li and Qian 1998), 485–508.

Li, Yan 李儼 and Baocong Qian, 錢寶琮 (1998). *Li Yan Qian Baocong kexue shi quanji* 李儼錢寶琮科學史全集 (Complete Works of Li Yan and Qian Baocong), 9 vols. Shenyang: Liaoning chubanshe 遼寧出版社.

Li, Zhizao 李之藻 et al. (1965). *Tianxue chuhan* 天學初函 (First Collectanea of Heavenly Studies), 6 vols (Reprint ed.), scroll 23, First Series (Chubian 初編 卷二十三) of Wu Xiangxiang 吳相湘 (Ed.), *Zhongguo shixue congshu* 中國史學叢書. Taipei: Taiwan xuesheng shuju 臺灣學生書局.

Libbrecht, Ulrich (1973). *Chinese Mathematics in the Thirteenth Century. The Shu-shu chiu-chang of Ch'in Chiu-shao*. Cambridge (Mass.): MIT Press.

Liu, Dun 劉鈍 (1990). Xia Luanxiang dui yuanzhui quxian de zonghe yanjiu 夏鸞翔对圓錐曲线的综合研究 (The Comprehensive Research on Conic Sections by Xia Luanxiang). In Du Shiran 杜石然 (Ed.), *Di san jie guoji Zhongguo kexueshi taolunhui lunwenji* 第三届国际中国科学史讨论会论文集 *(Proceedings of the Third International Conference on the History of Chinese Science)*, 12–18. Kexue chubanshe 科学出版社.

Loomis, Elias (1851). *Elements of Analytical Geometry and of the Differential and Integral Calculus*. New York: Harper & Brothers.

Loomis, Elias (1877). *Elements of Geometry, Conic Sections, And Plane Trigonometry* (Rev. ed.). New York: Harper & Brothers.

Lu, Jing 盧靖 (1902). *Wan xiang yi yuan yanshi* 萬象一原演式 *(Formulas for A Myriad Images—A Single Origin)* 1+9 scrolls (Mianyang Lu shi 沔陽盧氏刊本 ed.). [Hubei?].

MacLaurin, Colin (1742). *A Treatise of Fluxions in Two Books*, Volume 2. Edinburgh: T. W. and T. Ruddimans.

Martzloff, Jean-Claude (1987). *Histoire des mathématiques chinoises; préf. de J. Gernet et J. Dhombres*. Paris [etc.]: Masson.

Martzloff, Jean-Claude (1997). *A History of Chinese Mathematics*. Berlin, Heidelberg [etc.]: Springer. Originally published in French (Martzloff 1987).

Mikami, Yoshio (1912). The rectification of the ellipse by Japanese mathematicians. *Bibliotheca Mathematica 3*, 225–237.

Ming, Antu 明安圖 (1774). *Geyuan milü jiefa* 割圜密率捷法 (Fast Methods for the Circle Division and the Precise [circle] Ratio) (Luo Shilin 羅士琳, Guanwosheng shi huigao 觀我生室彙稿, 1839 ed.). Reprint in (Guo et al. 1993) 4:865–943.

Newton, Isaac (1999). *The Principia. Mathematical Principles of Natural Philosophy*. A New Translation by I. Bernard Cohen and Anne Whitman. Berkeley, Calif. et al.: University of California Press.

Qianlong 乾隆 (Ed.) (1773). *Yuzhi lixiang kaocheng houbian* 御製曆象考成後編 (Later Volumes of the Thorough Investigation of Calendrical Astronomy Imperially Composed), Volume 6 (子部六) of *Chizao tang Sikuquanshu huiyao* 摛藻堂四庫全書薈要. Originally compiled in 1738.

Qin, Jiushao 秦九韶 (1842). *Shu shu jiu zhang* 數書九章 (Mathematical Book in Nine Chapters) (Yijiatang congshu 宜稼堂叢書 ed.). Reprint in (Guo et al. 1993) 1:439–724.

Qingshi (1961). *Qingshi* 清史 (Annals of the Qing Dynasty), 8 vols. Taipei: Guofang yanjiuyuan 國防研究院.

Qiu, Desheng 求德生 (J. H. Judson) and Weishi Liu 劉維師 (Trans.) (1898). *Yuanzhui quxian* 圓錐曲線 (On Conic Sections). Shanghai: Meihua shuguan 美華書館. Partially translated from (Loomis 1877).

Ricci, Matteo and Zhizao Li, 李之藻 (1614). *Yuanrong jiaoyi* 圜容較義 (The Meaning of Compared [figures] Inscribed in a Circle). [N.p.]. New edition Fanyu: Haishan xianguan, 1847 (Haishan xianguan congshu), reprint in (Li et al. 1965) 6:3427–3484.

Roomen, van (1561–1615), Adriaan (1597). *In Archimedis circuli dimensionem Expositio & Analysis. Apologia pro Archimede, ad Clariss. virum Josephum Scaligerum. Exercitationes cyclicae contra Josephum Scaligerum, Orontium Finaeum, & Raymarum Ursum, in decem Dialogos distinctae. Authore Adriano Romano, ...* Wurceburgi.

Ruan, Yuan 阮元 and Luo Shilin 羅士琳 (Eds.) (光緒壬午春 1882). *Chouren zhuan sishiliu juan fu Xu Chouren zhuan liu juan* 疇人傳四十六卷續傳六卷 (Biographies of Astronomers and Mathematicians in 46 Scrolls; Continuation of Biographies of Astronomers and Mathematicians in 6 Scrolls) (Zhang 張氏重校刊 ed.). [N.p.]: Haiyan zhangjing changxing zhai 海鹽張敬常惺齋. Includes *Chouren jie* 疇人解 (Explanation of Astronomers and Mathematicians) by Tan Tai 談泰 (Ed.). Preface to *Chouren zhuan* 疇人傳 dated November 1799 (嘉慶四年十月), preface to *Xu Chouren zhuan* 續疇人傳 dated May 1840 (道光二十年夏四月).

Shi, Yunli (2008a). Reforming Astronomy and Compiling Imperial Science in the Post-Kangxi Era: The Social Dimension of the *Yuzhi lixiang kaocheng houbian* 御製曆象考成後編. *EASTM 28*, 36–81.

Shi, Yunli (2008b). The *Yuzhi lixiang kaocheng houbian* in Korea. In Luís Saraiva and Catherine Jami (Eds.), *History of Mathematical Sciences: Portugal and East Asia III. The Jesuits, the Padroado and East Asian Science (1552–1773)*, 205–229. New Jersey [etc.]: World Scientific.

Song, Hua 宋華 and Bai Xin 白欣 (2008). Xia Luanxiang de weijifen shuiping pingxi 夏鸞翔的微積分水平評析 (Analysis of Xia Luanxiang's Knowledge in Differential and Integral Calculus). *Nei Menggu shifan daxue xuebao (Ziran kexue Hanwen ban)* 內蒙古師範大學學報 (自然科學漢文版) (Journal of Inner Mongolia Normal University, Natural Science Edition) *37*(4), 566–572.

Swinden, B. A. (1954). 2391. Johann Kepler: Paralipomena ad Vitellionem. *The Mathematical Gazette 38*(323), 44–46.

Verhaeren, Hubert (1949). *Catalogue de la Bibliothèque du Pé-T'ang*. Beijing: Imprimerie des Lazaristes. Reprint Paris: Les Belles Lettres, 1969.

Wang, Yusheng 王渝生 (1983). Li Shanlan: Zhongguo jindai kexue de xianquzhe 李善蘭：中國近代科學的先驅者 (Li Shanlan: Forerunner of Modern Science in China). *Ziran bianzheng fa tongxun* 自然辯證法通訊 (Dialectics of Nature) *II*(5), 59–72. Translated to English in (Wang 1996).

Wang, Yusheng (1996). Li Shanlan: Forerunner of Modern Science in China. In Dainian Fan (Ed.), *Chinese Studies in the History and Philosophy of Science and Technology*, Number 179 in Boston Studies in the Philosophy of Science, 345–368. Dordrecht: Kluwer. Originally published in Chinese as (Wang 1983).

Whewell, William 胡威立 (1859). *Zhongxue* 重學 (Mechanics. Translated by Li Shanlan 李善蘭 and John Edkins 艾約瑟). Shanghai: Mohai shuguan 墨海書館 (Inkstone Press).

Xia, Luanxiang 夏鸞翔 (1898a). Wan xiang yi yuan 萬象一原 (A Myriad Images—A Single Origin). In Liu Duo 劉鐸 (Ed.), *Gujin suanxue congshu* 古今算學叢書 (Compendium of Mathematics, Old and New), Volume 3 (象數第三). Shanghai suanxue shuju 上海算學書局. Preface dated 1862.

Xia, Luanxiang 夏鸞翔 (1898b). Zhiqu shu 致曲術 (Procedures for Curves). In Liu Duo 劉鐸 (Ed.), *Gujin suanxue congshu* 古今算學叢書 (Compendium of Mathematics, Old and New), Volume 3 (象數第三). Shanghai: Shanghai suanxue shuju 上海算學書局.

Xia, Luanxiang 夏鸞翔 (1898c). Zhiqu tujie 致曲圖解 (Diagrammatic Explanations [of Procedures] for Curves). In Liu Duo 劉鐸 (Ed.), *Gujin suanxue congshu* 古今算學叢書 (Compendium of Mathematics, Old and New), Volume 3 (象數第三). Shanghai suanxue shuju 上海算學書局.

Xia, Luanxiang 夏鸞翔 (2006a). Zhiqu shu 致曲術 (Procedures for Curves). In Xuxiu siku quanshu bianwei hui 續修四庫全書編委會 (Ed.), *Xuxiu siku quanshu* 續修四庫全書 (Supplement to the Complete Books in the Four Treasuries), Volume 1047, 425–437. Shanghai: Shanghai guji chubanshe 上海古籍出版社. Reprint of the manuscript ed. in the Shanghai Library 影印上海圖書館藏稿本.

Xia, Luanxiang 夏鸞翔 (2006b). Zhiqu tujie 致曲圖解 (Diagrammatic Explanations [of Procedures] for Curves). In Xuxiu siku quanshu bianwei hui 續修四庫全書編委會 (Ed.), *Xuxiu siku quanshu* 續修四庫全書 (Supplement to the Complete Books in the Four Treasuries), Volume 1047, 437–467. Shanghai: Shanghai guji chubanshe 上海古籍出版社. Reprint of the manuscript ed. in the Shanghai Library 影印上海圖書館藏稿本.

Xia, Luanxiang 夏鸞翔 (Guangxu 1875–1908). *Zhiqu shu* 致曲術 (Procedures for Curves) (蟄雲雷齋 Zhiyun leizhai ed.).

Xia, Luanxiang 夏鸞翔 (同治十二年 1873). Dongfang shu tujie er juan 洞方術圖解二卷 (Diagrammatic Explanations of Procedures for Thoroughly Understanding the Square in Two Scrolls). In *Zou Zhengjun yishu (fu Xia shi suanxue si zhong, fu Xu shi suanxue san zhong)* 鄒徵君遺書 (附夏氏算學四種 附徐氏算學三種) (Posthumous Writings of Zou Zhengjun, with Four Mathematical Books by Mister Xia and Three Mathematical Books by Mister Xu Appended) (Zou Daquan shijieyuan 鄒達泉拾芥園 刊本 ed.). [Guangzhou?].

Xiang, Mingda 項名達 (1888). Tuoyuan qiu zhou shu 橢圓求周術 (Procedure to Find the Circumference of an Ellipse). In Dai Xu 戴煦 (Ed.), *Xiang shu yi yuan* 象數一原 (Images and Numbers—A Single Origin), 7 vols., 6:27B–39A. Shanghai: Jin'gui Hua shi 金匱華氏. Reprint in (Guo et al. 1993) 5:567–573.

Xu, Youren 徐有壬 (1985). *Wumin yizhai suanxue* 務民義齋算學 (Mathematics from the Wuminyi Studio). Congshu jicheng chubian 丛书集成初编. Zhonghua shuju 中华书局.

Yan, Wenyu 嚴文郁 (Ed.) (1990). *Qing ruzhuan lüe* 清儒傳略 (Short Biographies of Qing Scholars). Taipei: Taiwan shangwu yinshuguan 臺灣商務印書館.

Yang, Zezhong 楊澤忠 (2004). Ming mo Qing chu tuoyuan zhishi zhi donglai 明末清初橢圓知識之東來 (Knowledge on the Ellipse Coming to the East during the Late Ming and Early Qing). *Shuxue jiaoxue* 數學教學 3, 11, 47–49.

Yunzhi 允祉 (Ed.) (1723). *Yuzhi shuli jingyun* 御製數理精蘊 (Essence of Numbers and their Principles). [Beijing?]: [n.p.]. Reprint in (Guo et al. 1993) 3.

Zhang, Baichun 张柏春, Tian Miao 田淼, Matthias Schemmel 马深梦, Jürgen Renn 雷恩, and Peter Damerow 戴培德 (2008). *Chuanbo yu huitong: "Qiqi tushuo" yanjiu yu jiaozhu* 传播与会通—《奇器图说》研究与校注 (Transmission and Integration—Qiqi tushuo (Illustrations and Descriptions of Extraordinary Devices): New Research and Annotated Edition), 2 vols. Nanjing: Jiangsu kexue jishu chubanshe 江苏科学技术出版社.

Chapter 3
Filling Euclid's Gaps

Contents

In 1872, Li Shanlan 李善蘭 (1811–1882) (Fig. 3.1) wrote in the preface to his *Methods for Testing Primality* (*Kao shugen fa* 考數根法):

> In general, a number that when measured by another number cannot be exhausted, and can only exhaustively be measured by unity, is called a prime number (*shugen* 數根). See the *Elements of Geometry*. But, if we take any number and if we want to decide whether it is a prime number or not then there is no method in Antiquity. I have thought about this carefully and for a long time, and I have found four methods for testing this in order to complement what in the [*Elements of*] *Geometry* was not yet complete.[1]

The definition of a prime number here paraphrases the one found in the translation of the beginning of Book VII of Euclid's *Elements*, which Li Shanlan had completed together with the Protestant missionary Alexander Wylie some years earlier in 1857.[2] Li was obviously unsatisfied with what he found in the number-theoretical books VII to IX of the *Elements* and complemented its content by methods for testing the primality of any given integer number. His findings prior to his 1872 publication made their way out of China and provoked some discussions with respect to Fermat's Little Theorem and ancient Chinese knowledge about prime numbers. Wylie, who closely collaborated with Li not only for the translation of the *Elements* but also for other Western scientific treatises, brought into global

---

[1] Translated from Li (1872a) p. 13A.

[2] See Li and Wylie (1865).

Fig. 3.1 A portrait of Li
Shanlan from *The Chinese
Scientific and Industrial
Magazine* (*Gezhi Huibian* 格
致彙編) (July, 1877)

circulation some elements of—also providing the title of his publication—"A
CHINESE THEOREM":

> The theorem of which the following is a translation was jotted down in my notebook a
> few days ago, by 李善蘭 Le Shen-lan, a native mathematician whose name has been more
> than once before the European public. I have no hesitation in saying that it is a purely
> independent discovery on his part, and as such, think it may be worth publicity in your
> pages. Some of your scientific readers will probably be able to say if an analogous rule is to
> be found in European books.
> *To ascertain if any number is a prime number.*
> "Multiply the given number by the logarithm of 2. Find the natural number of the resulting
> logarithm, and subtract 2 from the same. Divide the remainder by the given number. If there
> be no remainder, it is a prime number. If there be a remainder, it is not a prime.["][3]

The criterion for primality stated here is mathematically equivalent to testing for
a given integer number N, if $(2^N - 2)$ is divisible by N, or, in terms of modular
arithmetic, to determine if $2^N - 2 \equiv 0 \pmod{N}$ holds. Yet Li's statement is false,
since only one of the implications holds; namely, for any prime number N, $2^N - 2$
is divisible by N, or, expressed in modular arithmetic, for any prime number N[4]:

$$2^N - 2 \equiv 0 \pmod{N}.$$

Along its routes of transmission, Li Shanlan's statement turned into the myth that
"the Chinese seem to have known as early as 500 B.C. that $2^p - 2$ is divisible by
the prime p."[5] This chapter will end with a discussion of the related nineteenth-
and twentieth-century historiographies of the above-quoted "Chinese Theorem" in

---

[3]Dated 10 May 1869, quoted from Wylie (1869).

[4]The statement is even valid more generally, that is: if N is a prime number, then for every integer
a, the number $a^N - a$ is divisible by N.

[5]Quoted from Dickson (2002) vol. 1, p. 59. For more details on this myth, see Han and Siu (2008)
and Sect. 3.3 of this chapter.

the context of modern, universal mathematics and Fermat's Little Theorem. At the core of this chapter, I will analyse the entangled history of Li Shanlan's published algorithms for testing primality and their characteristics in relation to Li's and his readers' sources, both ancient and contemporary, Western and Chinese. This will necessitate an excursion into the number-theoretical books in Euclid's *Elements* and their Chinese translation. In addition, a first full translation of Li Shanlan's primality test algorithms with numeric examples is given in Appendix B, where symbolic transcriptions of Li's entirely rhetoric text are added to help understand the mathematical operations involved.

3.1 Beyond the First Six Books of Euclid's *Elements*

In the latter half of nineteenth-century China, no Western or Chinese source on number theory was available in Chinese language, nor were there mathematical journals in Chinese language which could have systematically introduced research themes in any of the mathematical disciplines. Li Shanlan thus most likely first encountered the concept of prime number when he worked together with the Protestant missionary Alexander Wylie on the translation of the last seven books of Euclid's *Elements*. This was a continuation of the work by the Jesuit missionary Matteo Ricci (chin. Li Madou 利瑪竇, 1552–1610) and the Christian convert scholar Xu Guangqi 徐光啓 (1562–1633), who had printed in 1607 a translation of the first six Books[6] (Fig. 3.2) based on the Latin version by the Jesuit Christopher Clavius (1538–1612). Yet they left untranslated the number-theoretical books VII to IX as well as Book X on the theory of irrational lines and regions, Books XI to XIII on stereometry, the so-called method of exhaustion, and the Platonic solids, as well as the last two books that are not ascribed to Euclid.

As for the original on which Li's and Wylie's translation is based on the translation of the last seven books of Euclid's *Elements*, it was established by Xu Yibao, the historian of mathematics, that it was Henry Billingsley's English version, *The Elements of Geometrie of the Most Auncient Philosopher Euclide of Megara* from 1570 (Fig. 3.3).[7] Nevertheless, as we shall see in this chapter, several examples from Book VII reveal that Billingsley's version, if used at all, was not a strict model for the translators.

Billingsley defines prime numbers as follows:

Definition (VII.12) A prime (or first) number is that, which only unity does measure.[8]

---

[6] See Engelfriet (1998).

[7] See Xu (2005) p. 16–24.

[8] Text according to Billingsley (1570) f186r. Adaptation to modern English spelling is mine, the same holds for all following citations from this book.

Fig. 3.2 Ricci and Xu (1607)
Book I, Prop. 1

whereas the Chinese translation gives:

Definition (VII.11) A prime number can only be measured by unity, but other numbers cannot measure it.

第十一界 數根者惟一能度。而他數不能度。 [9]

Li and Wylie's numbering of the definitions at the beginning of Book VII differ from Billingsley's because they did not follow him in adding a definition of oddly even numbers, in addition to the evenly even, the evenly odd and oddly odd numbers.[10] This logical addition with respect to Clavius's Latin version, as Billingsley himself remarks, was a supplementary definition that

---

[9]Chinese text quoted from Li and Wylie (1865) scroll 7, p. 2A, emphasis in red is mine.

[10]On the classification of numbers in Greek mathematics, see Heath (1981) vol. 1, p. 70–74, here p. 73–74:

> Euclid then, as well as Aristotle, includes 2 among prime numbers. Theon of Smyrna says that even numbers are not measured by the unit alone, except 2, which therefore is odd-like without being prime. The Neo-Pythagoreans, Nicomachus and Iamblichus, not only exclude 2 from prime numbers, but define composite numbers, numbers prime to one another, and numbers composite to one another as excluding all even numbers; they make all these categories subdivisions of odd. Their object is to divide odd into three classes parallel to the three subdivisions of even, namely even-even $= 2^n$, even-odd $= 2 (2m + 1)$ and the quasi-intermediate odd-even $= 2^{n+1} (2m + 1)$; accordingly they divide odd numbers into (a) the prime and incomposite, which, are Euclid's primes excluding 2, (b) the secondary and composite, the factors of which must all be not only odd but prime numbers, (c) those which are "secondary and composite in themselves but prime and incomposite to another number," e.g. 9 and 25, which are both secondary and composite but have no common measure except 1. The inconvenience of the restriction in (b) is obvious, and there is the further objection that (b) and (c) overlap, in fact (b) includes the whole of (c).

Fig. 3.3 Title pages of Clavius (1591) and Billingsley (1570)

is not found in the Greek neither was it doubtless ever in this manner written by *Euclid*; which thing the slenderness and the imperfection thereof and the absurdities following also of the same declare most manifestly.[11]

Concerning the definition of numbers, Billingsley thus seems to express a certain dissatisfaction with the incompleteness of the *Elements*. Li was dissatisfied, too, albeit for a different reason. He did not see a lacuna in the list of definitions for different kinds of numbers, but in the lack of a method to test for a given number, whether it is prime or not. I will come back to this point when discussing Proposition IX.20 in more detail, where, in the widely accepted interpretation, Euclid expressed that there are an infinite number of primes. I would also like to underline a philosophical problem with the Greek conceptualization of numbers that was not taken into account in the Chinese version, where no distinction seems to have been made between unity and "other numbers." For Euclid, and for Greek philosophers more generally, unity was not a number but rather an object deprived of any mathematical function other than being the constituent of a number.[12]

---

[11] Billingsley (1570) f185v.

[12] See Acerbi (2010) p. 245. Mueller (1981) p. 58 goes even so far as to say that "the first of these [arithmetic first principles] defines the unit in a mathematically useless way."

This is made very clear in the first two definitions of Book VII:

Definition (VII.1) Unity is that, whereby everything that is, is said to be one.

Definition (VII.2) Number is a multitude composed of unities.[13]

If not for its mathematical meaning, the philosophical importance of the definition of unity is well reflected in the Chinese translation, where it is embedded in the Confucian doctrine of the Great Unity (*taiyi* 太一), the origin of all numbers and more generally of all objects in the physical world:

Definition (VII.1) Unity: Heaven, Earth and a Myriad Objects, there is none that does not stem from Unity.
第一界　一者天地萬物無不出乎一。

Definition (VII.2) Numbers: they are made by taking a multitude of unities and uniting them.
第二界　數者。以衆一合之而成。 [14]

The translations of the concepts of unity and number here are clearly not a faithful rendering of Greek notions, but a mediation between cultural values and mathematical notions necessary for the remaining statements in the number-theoretical books. A comparison of the Chinese terms used in the translation by Li and Wylie on the translation of the last seven books of Euclid's *Elements* reveals that Li Shanlan relies, in his own work on prime numbers, on precisely the terms adopted in the translation of the basic operations with numbers from Euclid's Book VII: part (*fen* 分), measuring (*du* 度) or exhausting (*jin* 盡) are just some examples systematically employed by Li to express mathematical operations with numbers.

3.1.1 Proposition IX.20

After having pointed out some of the conceptual flows in the process of transmission of early number theory to China, let me come back to prime numbers and Li Shanlan's project intended to complement Euclid's *Elements*. Although there is no explicit evidence for what exactly inspired Li Shanlan to develop methods for testing primality, I would conjecture that he was intrigued by Proposition IX.20 and the constructive nature of its proof. This operational approach fits well with the Chinese algorithmic and calculatory tradition within which Li Shanlan was accustomed to navigate. Although the standard historiography tells us that Euclid's proposition IX.20 is about the infinity of primes proven by a *reductio ad absurdum*

---

[13]Billingsley (1570) f184r.

[14]Chinese text quoted from Li and Wylie (1865) scroll 7, p. 1A.

scheme,[15] it is in fact a constructive approach showing that from any given set of prime numbers counting at least two elements, one can always construct another, larger prime number.[16]

An inverse question to constructing a prime number then would be, how, for a given integer number, can one know if it is prime? Or, in order to check the correctness of the construction, one might want to test by a certain algorithm, if one indeed has constructed a prime number. This is, I believe, where Li's idea might have emerged to develop algorithms for testing primality. My argument will become more detailed in the following exposition of IX.20.

Here is the proposition and its proof as given by Billingsley (Fig. 3.4):

Proposition (IX.20) Prime numbers being given how many soever, there may be given more prime numbers.

Suppose that the prime numbers given be A, B, C. Then I say, that there are yet more prime numbers besides A, B, C.

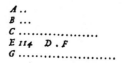

Take (by the 38. of the seventh)[17] the least number whom these numbers A, B, C do measure, and let the same be DE. And onto DE add unity DF. Now EF is either a prime number or not. First let it be a prime number, then are there found these prime numbers A, B, C and EF more in multitude than the prime numbers first given A, B, C. But now suppose that EF be not prime. Wherefore some prime number measures it (by the 24. of the seventh).[18] Let a prime number measure it, namely, G. Then I say, that G is none of these numbers A, B, C. For if G be one and the same with any of these A, B, C. But A,

---

[15]The standard account of the proof of IX.20 is as follows: Assuming that there exists a greatest prime N, since $N! + 1$ is not divisible by any of the numbers 2 to N, its smallest factor above unity must be a prime greater than N. This contradicts the hypothesis that N is the greatest prime. This is true but it is not what IX.20 proves; it cannot even be reduced to this.

[16]Clavius had already formulated this in a scholium as a suggestion to rewrite the proposition as a (construction) problem. See Clavius (1591) p. 77 shown in Fig. 3.6:

Poterat idem hoc theorema instar problematis hoc modo proponi. Primis numeris quotcunque propositis, invenire alium primum numerum ab illis diversum.

[17]See Billingsley (1570) f200r:

Proposition (VII.38) Three numbers being given, to find out the least number which they measure.

[18]See Billingsley (1570) f196r:

Proposition (VII.24) The least numbers that have one and the same proportion with them are prime the one to the other.

It seems that VII.24 is the wrong reference here. It should read "(by the 33. of the seventh)" instead. See Billingsley (1570) f198v:

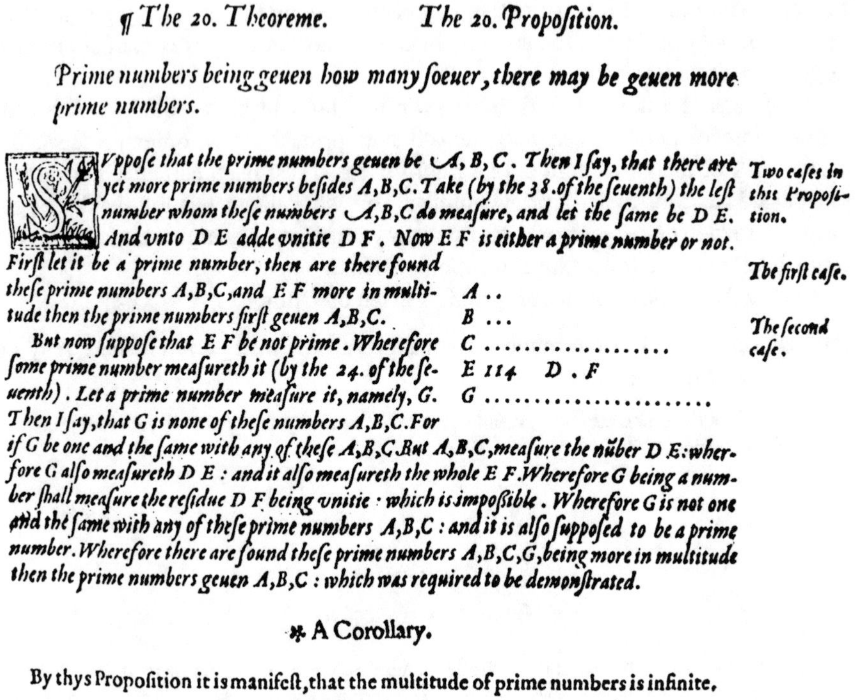

¶ *The 20. Theoreme.* *The 20. Propofition.*

Prime numbers being geuen how many foeuer, there may be geuen more prime numbers.

Vppofe that the prime numbers geuen be ⊿, B, C. Then I fay, that there are yet more prime numbers befides A,B,C. Take (by the 38. of the feuenth) the left number whom thefe numbers ⊿,B,C do meafure, and let the fame be D E. And vnto D E adde vnitie D F. Now E F is either a prime number or not. Firft let it be a prime number, then are there found thefe prime numbers A,B,C,and E F more in multitude then the prime numbers firft geuen A,B,C.

But now fuppofe that E F be not prime. Wherefore fome prime number meafureth it (by the 24. of the feuenth). Let a prime number meafure it, namely, G. Then I fay, that G is none of thefe numbers A,B,C. For if G be one and the fame with any of thefe A,B,C. But A,B,C, meafure the nüber D E: wherfore G alfo meafureth D E : and it alfo meafureth the whole E F. Wherefore G being a number fhall meafure the refidue D F being vnitie · which is impoffible . Wherefore G is not one and the fame with any of thefe prime numbers A,B,C : and it is alfo fuppofed to be a prime number. Wherefore there are found thefe prime numbers A,B,C,G, being more in multitude then the prime numbers geuen A,B,C : which was required to be demonftrated.

Two cafes in this Propofition.

The firft cafe.

The fecond cafe.

A ..
B ...
C ....
E 114   D . F
G ....

❊ A Corollary.

By thys Propofition it is manifeft, that the multitude of prime numbers is infinite.

Fig. 3.4 Billingsley (1570) Book IX Prop. 20

B, C, measure the number DE: wherefore G also measures DE: and it also measures the whole EF. Wherefore G being a number shall measure the residue DF being unity: which is impossible. Wherefore G is not one and the same with any of these prime numbers A, B, C: and it is also supposed to be a prime number. Wherefore there are found these prime numbers A, B, C, G, being more in multitude than the prime numbers given A, B, C: which was required to be demonstrated.[19]

And here is Proposition IX.20 in the Chinese translation of the latter books of Euclid's *Elements* (Fig. 3.5):

第二十題　任置若干數根。數根必不盡於此。

甲　　乙　　丙　　　　　解曰。任置甲乙丙等若干數根。題言數根必
二　　三　　五　　　　　不盡於此。
　　　　　　　　　　　　論曰。取甲乙丙所度之最小數 七卷三十八 為丁
戊　三十丁　一己　　　戊。
加丁己一。則戊己或為數根或不為數根。設為數根。則甲乙丙數根之外。又有戊己
數根。設不為數根。而為數根庚所度。七卷三十三 則庚與甲乙丙各不相等。若云庚與甲

Proposition (VII.33) Every composed number is measured by some prime number.

[19] Billingsley (1570) f232v.

Fig. 3.5 Chinese translation of Book IX Prop. 20 in Li and Wylie (1865)

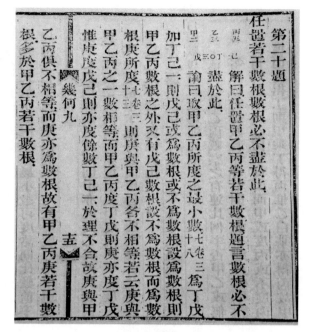

乙丙之一數相等。而甲乙丙度丁戊則庚亦度丁戊。惟庚度戊己。則亦度餘數丁己一。與理不合。故庚與甲乙丙俱不相等。而庚亦為數根。故有甲乙丙庚若干數根。多於甲乙丙若干數根。[20]

Proposition 20: No matter which and how many prime numbers one puts down, prime numbers are not exhausted by them.

A	B	C
2	3	5

E 30D 1 F

The explanation says: No matter which and how many prime numbers A, B, C, etc. one puts down, the proposition says that prime numbers are not exhausted by them.

The proof says: Take the smallest number that measures A, B, C. Book VII.38 Let it be DE. Add DF, 1. Then, EF is either prime or not prime. Let us suppose it is a prime number. Then, different from A, B and C, EF is another prime number. Let us suppose it [EF] is not a prime number and let it be measured by G. Book VII.34. Then, G and A, B, C are all different from each other. If I would say that G and one of the numbers A, B, C were the same and since A, B, C measure ED, then G would also measure ED. But it measures EF, therefore it also measures the remaining number DF, 1. This is not coherent with the principles. Thus, G and A, B, C are all different from each other and G is also a prime number. We thus have any prime numbers A, B, C, G, which are more than any prime numbers A, B, C.

The only numeric example given in the Chinese version is the smallest possible, in fact, with $A = 2$, $B = 3$, $C = 5$ and $DE = 30$. With the construction proposed in the proof of IX.20, it leads to $EF = 31$, which is itself prime and larger than A, B and C. The example is not taken from Billingsley, who shows $A = 2$, $B = 3$,

---

[20]Quoted from Li and Wylie (1865) scroll 9, p. 15A.

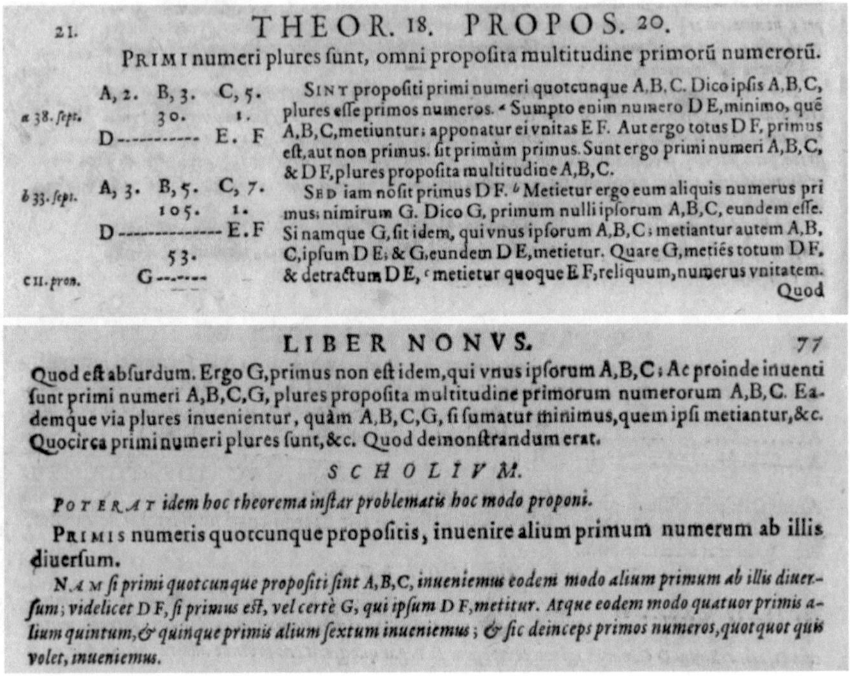

21.　　　　　THEOR. 18. PROPOS. 20.

PRIMI numeri plures funt, omni propofita multitudine primorū numerorū.

A, 2.　B, 3.　C, 5.　　SINT propofiti primi numeri quotcunque A,B,C. Dico ipfis A,B,C,
a 38. *fept.*　　30.　　1.　plures effe primos numeros. ⸿ Sumpto enim numero D E,minimo, quē
　　　D----------　E. F　A,B,C,metiuntur; apponatur ei vnitas E F. Aut ergo totus D F, primus
　　　　　　　　　　eft,aut non primus. fit primūm primus. Sunt ergo primi numeri A,B,C,
　　　　　　　　　　& D F,plures propofita multitudine A,B,C.
b 33. *fept.*　A, 3.　B, 5.　C, 7.　SED iam nōfit primus D F. ᵇ Metietur ergo eum aliquis numerus pri
　　　　　105.　　1.　mus; nimirum G. Dico G, primum nulli ipforum A,B,C, eundem effe.
　　　D----------　E.F　Si namque G,fit idem, qui vnus ipforum A,B,C; metiantur autem A,B,
　　　　53.　　　　C,ipfum D E; & G,eundem D E,metietur. Quare G,metiés totum D F,
c II. *pron.*　G--------　　& detractum D E, ᶜ metietur quoque E F,reliquum, numerus vnitatem.
　　　　　　　　　　　　　　　　　　　　　　　　　　Quod

LIBER NONVS.　　　　　　　　　　　　77

Quod eft abfurdum. Ergo G,primus non eft idem,qui vnus ipforum A,B,C; Ac proinde inuenti
funt primi numeri A,B,C,G, plures propofita multitudine primorum numerorum A,B,C. Ea-
demque via plures inuenientur, quàm A,B,C,G, fi fumatur minimus,quem ipfi metiantur,&c.
Quocirca primi numeri plures funt,&c. Quod demonftrandum erat.

SCHOLIVM.

POTERAT idem hoc theorema inftar problematis hoc modo proponi.

PRIMIS numeris quotcunque propofitis, inuenire alium primum numerum ab illis
diuerfum.

NAM fi primi quotcunque propofiti fint A,B,C, inueniemus eodem modo alium primum ab illis diuer-
fum; videlicet D F, fi primus eft, vel certè G, qui ipfum D F,metitur. Atque eodem modo quatuor primis a-
lium quintum,& quinque primis alium fextum inueniemus; & fic deinceps primos numeros,quotquot quis
volet, inueniemus.

Fig. 3.6 Clavius (1591) Book IX Prop. 20

$C = 19$, $DE = 114$ and $G = 23$. Yet the Chinese example corresponds to one of the two examples given in Clavius's marginal notes. Clavius additionally illustrates the situation where $DF = DE + 1$ is not prime but divisible by a prime greater than A, B and C, there $DF = 3 \cdot 5 \cdot 7 + 1 = 106 = 2 \cdot 53$ (see Fig. 3.6).

Where the Chinese text also differs from Billingsley's edition is that the corollary following proposition IX.20 is not translated. The corollary establishes the claim of theorem IX.20 as it is generally referred to by historians today:

> By thys Proposition it is manifest, that the multitude of prime numbers is infinite.[21]

Euclid himself had not operated with the notion of infinity in his formulation of the proposition. Instead he had devised a method that allowed one to construct from a finite set of prime numbers a new prime number that is larger than any of the given prime numbers. The notion of infinity is not necessary for his claims, since he only considers a limited number of objects that can be arbitrarily large. Later commentators on Euclid assume a deeper philosophical reason behind his choice, but in the absence of a systematic study of all the scholia, it is not possible at this stage to clarify what exactly the (anonymous) commentator here had in mind:

---

[21]Billingsley (1570) f232v.

Fig. 3.7 Scholium to Book IX Prop. 20, MS. *Dorvillianus* 301, fol. 171r

In this theorem he [Euclid] wants to prove that prime numbers are infinite; for if the prime are more than any assigned number, it is clear that the prime are infinite. But if this is the case, it seems to contradict a basic statement of the philosophers, for they say that what is prior must be circumscribed and less in number. What shall we say, then? "We shall say" that prime numbers are not a principle of numbers, but, if there is one, this is the unit—and this unit is compact and isolated. In this way, this "statement" holds also within numbers, namely, that the principle is not infinite, but circumscribed.[22]

In the remaining books of the *Elements* there are no further number-theoretical statements about prime numbers, regarding, for instance, their distribution or a method to find all prime numbers. Given the traditional interest in inverting procedures to check the result of an algorithm and to devise new mathematical techniques in the Chinese mathematical tradition—the so-called "back to the roots" problems (*huanyuan* 還源) which were systematic in some Yuan dynasty works[23]— Li Shanlan might have identified an opportunity to complement the *Elements* precisely for proposition IX.20. It is not only one of the rare occurrences of a constructive proof in Euclid's number theory and therefore might have pleased Li Shanlan who was interested in an algorithmic approach to number theory,[24] IX.20 also leaves unanswered the question of how to determine if a given number is prime or not. When wanting to apply IX.20 to the construction of ever-larger prime numbers, one first needs to be sure of having three numbers A, B and C which fulfil the condition of being prime before computing the resulting number $A \cdot B \cdot C + 1$, which is prime exactly because of the proof of IX.20. Here, Euclid's procedure has a conditional statement, it leads to a new prime number but contains no hint on how to decide practically upon the prime nature of any given number. This question, of course, does not arise in an Euclidean setting where the hypotheses made about A, B and C merely serve an argumentative purpose.

---

[22]Translated by Fabio Acerbi (personal communication) according to the Greek text in Heiberg's edition, vol. V, p. 407. The scholium belongs in the oldest layer of annotations to the *Elements*, since it can be read in MS. *Dorvillianus* 301, fol. 171r, dated 888 AD (Bodleian Library, Oxford) (see Fig. 3.7) and in MS. *Vaticanus Graecus* 190, fol. 132r (Biblioteca Apostolica Vaticana).

[23]See for example Bréard (1999) p. 194–212.

[24]See also Chap. 5 on his work in combinatorics.

So far, I have only mentioned Li Shanlan's acquaintance with a foreign, Euclidean tradition of defining and constructing prime numbers, and how certain conceptual and operational characteristics might have provided the inspiration for him to continue along the algorithmic line sketched out by Euclid. Li Shanlan's 1872 text *Methods for Testing Primality* nevertheless reveals that Euclid was not his only source. He also brings elements into play from the Chinese algebraic and modular arithmetic tradition, which I will describe in the following.

3.2   Primality in Chinese Sources

Li Shanlan's *Methods for Testing Primality* were first printed in 1872 in three instalments in the *Peking Magazine* (*Zhongxi wenjian lu* 中西聞見錄, lit. title *Records of Chinese and Western Knowledge*), a journal founded by the American Presbyterian missionary W. A. P. Martin (chin. Ding Weiliang 丁韙良, 1827–1916). By that time (from 1869 until 1895), Martin was the president of the Interpreter's College (*Tongwen guan* 同文館) in Beijing, and Li Shanlan its professor for mathematics (from 1869 until his death in 1882). A slightly abbreviated version of the text with less numeric examples was published posthumously in 1902 in the *Classified Edition of the Hunan Scholarly Journal* (*Xiangxue bao leibian* 湘學報類編), a tri-monthly newspaper founded by the reformer Huang Zunxian 黃遵憲 (1848–1905) as a medium to disseminate writings on reform, new ideas and information.

In late Qing China, both channels of publication, addressed to an emerging reading public interested in new kinds of knowledge, foreign and Chinese alike, were rather unusual for including mathematical novelties. At first glance, Li Shanlan's text certainly does not look revolutionary. Its purely rhetoric style, without recourse to formulaic expressions or diagrams, might as well be a page from the *Nine Chapters*. Yet concepts like "prime number" or "numbers prime to one another" had never occurred in traditional mathematical writings, not even in Qin Jiushao's *Mathematical Book in Nine Chapters* (*Shushu jiu zhang* 數書九章) from 1247, which was famous for containing methods to solve simultaneous integer congruences.[25]

Integer congruences, albeit not conceptualized, do nevertheless implicitly play an important role in Li Shanlan's methods. The first two examples from his text shall suffice here to illustrate this.[26] In order to test, for example, if the number 31 is prime, Li proceeds as follows:

---

[25] See Shen (1986).

[26] Textual and mathematical details can be found in Appendix B where I give a complete annotated translation of Li (1872a,b,c).

By choosing what he calls the "auxiliary number" equal to 2 and taking it to the fourth power,[27] one finds:

$$(31 - 2^4)^2 \equiv 2^3 \pmod{31}.$$

Since

$$2^{2 \cdot 4 - 3} = 2^5 \equiv 1 \pmod{31}$$

one has:

$$31 \equiv 1 \pmod 5.$$

Because 5 is odd and $N = 31$ is not divisible by $1 + 2 \cdot 5$, Li concludes according to the stated criteria that $N = 31$ is prime.

Another example provided by Li is 341. Here, one takes the ninth-power of the "auxiliary number" (again chosen as 2), which leads to the congruence:

$$(341 - 2^9)^2 \equiv 2^8 \pmod{341}.$$

Since

$$2^{2 \cdot 9 - 8} = 2^{10} \equiv 1 \pmod{341}$$

one has

$$341 \equiv 1 \pmod{10}.$$

Because 10 is even, but $N = 341$ is divisible by $1 + 10$, Li concludes that $N = 341$ is not prime.[28]

It has to be underlined that Li does not provide a general theory here linking prime numbers to congruence equations, nor does he give proofs to show the correctness of his algorithms. It seems that he only implicitly applies basic number-theoretical knowledge in his procedures. I have therefore chosen not to present his procedures in a general number-theoretical context, although one can, with a good capacity for reading between the lines, reconstruct a modern foundation to his algorithms with the help of the many examples he provides.[29] That such

---

[27] The "auxiliary number" is implicitly chosen as coprime to the number N that is to be tested. Since one immediately sees that a number N is not prime when it is even, 2 and an odd number N are always coprime.

[28] $N = 341$ is actually the smallest possible pseudoprime. See page 243.

[29] See, for example, Yan (1954a), Yan (1954b), Horng (1991) p. 420–424, Yang (2010) p. 5–11 and Born (2015) p. 44–50.

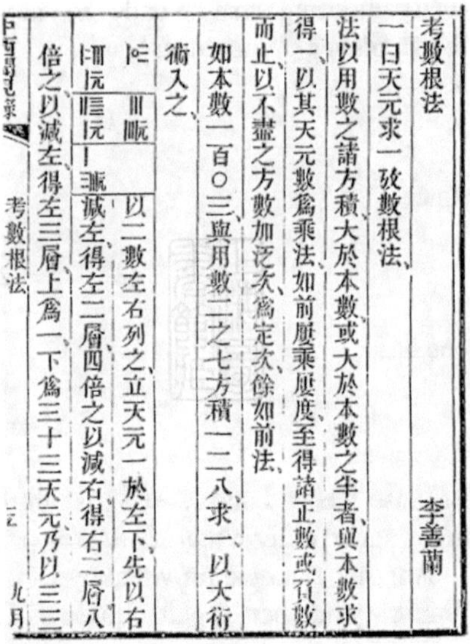
考數根法　李善蘭
一曰天元求一攷數根法、
法以用數之諸方積大於本數、或大於本數之半者、與本數求
得一、以其天元數爲乘法、如前屢乘屢度、至得諸正數或負數
而止、以不盡之方數加泛次、爲定次、餘如前法、
如本數一百〇三、與用數二之七方積一二八、求一、以大衍
術入之、
以二數左右列之、立天元一於左下、先以右
減左、得左二層、四倍之以減右、得右二層八
倍之以減左、得左三層上爲一、下爲三十三天元、乃以三三
考數根法

Fig. 3.8 "Method for testing primality by the celestial element [method] and reduction to One" in Li (1872b)

presentist interpretations can sometimes be misleading is best demonstrated by Wylie's announcement of a (false) "CHINESE THEOREM," which was quoted at the beginning of this chapter, and to which I will return in the next section.

What Li Shanlan makes explicit in his text, is his indebtedness to Song dynasty sources. In his second method, the "Method for testing primality by the celestial element [method] and reduction to One" (see Fig. 3.8) he refers to traditional Chinese algebraic methods and Qin Jiushao's technique for solving linear congruence equations, both dating from the Song dynasty (960–1279). In his fourth method for testing the primality of a rather large number N, he applies Qin's method to finding the prime factors of N, in case it has tested negative. The importance of the Chinese mathematical tradition for a late Qing mathematician like Li Shanlan is further amplified by yet another translator of Western science books, Hua Hengfang 華蘅芳 (1833–1902). Hua goes back in time to *Master Sun's Mathematical Classic* (*Sunzi suanjing* 孫子算經). This text, compiled approximately in the fourth century AD, contains a specific numeric example of three simultaneous linear congruences, and is solved for this one specific case but not for the general case. Nevertheless, referred to in a general formulation by modern historians and mathematicians, Master Sun's text has consequently become the *locus classicus* of the so-called "Chinese Remainder Theorem."[30] Hua picks up precisely the second of Li Shanlan's

[30]The *Sunzi suanjing* has been transmitted as part of the Tang dynasty collection of *Ten Books of Mathematical Classics* (*Suanjing shi shu* 算經十書), in which the *Nine Chapters* were equally included. For a translation and historiographic discussion of the problem on an unknown number of things and its transmission in versified form, see Bréard (2014) p. 170–173.

four methods, which in its title and approach goes back to the Song dynasty. In the preface to his *Prime Numbers Developed from the Past* (*Shugen yan gu* 數根演古), he emphasizes the benefits of reading ever more ancient sources for making new mathematical discoveries, which he believes to be easily accessible:

> Prime numbers are not easy to find. First, because there are many numbers that are not prime numbers, it is not easy to distinguish if [a number] is one or not. Another reason is that no prime numbers are connected to each other by a principle, it is not easy to find one from another. These are the reasons why today's mathematicians have not yet obtained prime numbers without recourse to searching factorizations.
>
> If one had a single method with which, from a number, of which one already knows that it is a prime number, by using methods of addition, subtraction, multiplication and division, one could find all the numbers of which one had not yet known that they are prime numbers, would that not be utterly convenient! But, I spared no effort to explore the yet undiscovered principles, and, inadvertently, understood all in a flash. I claim that this undoubtedly is an extremely easy affair! Its principle is definitely not utterly profound! Its method has also existed since ancient times. Among those who practiced mathematics, there was no one who did not know it. Only they could not yet apply it to prime numbers, that is all! From this, one can clearly see that the benefit of reviving ancient knowledge cannot be ignored.
>
> The "*Dayan* Reduction to One" (*Dayan qiu yi* 大衍求一) is that ancient method. In the problem about an unknown number of things from *Master Sun's* [*Mathematical Classic*] the numbers used, 3, 5 and 7 are all prime. "Counting by threes," "counting by fives," "counting by sevens," is just like measuring it [the given number] with prime numbers. The numbers that remain are just the numbers that are not exhausted when measuring. If odd numbers smaller than the "extended divisor" (*yan mu* 演母)[31] and 3, 5 and 7 all are not exhaustive when measuring, then [the given number] is not a prime number. So how could it not be that if there is a number that does precisely exhaust [the given number], there is not some prime number that is exhaustive when measuring? This means that it is not a multiplicative number [i.e. that the given number cannot be factorized]. So how could this be a principle that is difficult to understand?
>
> Now, according to the problem about an unknown number of things, I constructed problems to investigate the transformations thoroughly. Just by not using the problems in which there is a number that is precisely exhaustive, one can obtain forty-eight problems. I experimented to develop their calculations based on the ancient method as follows.[32]

Hua, one by one, goes through all the possible congruences to count how many numbers A are prime that satisfy $A \equiv r_1$ (mod 3), $A \equiv r_2$ (mod 5) and $A \equiv r_3$ (mod 7), where $r_1 = 1, 2$, $r_2 = 1, 2, 3, 4$ and $r_3 = 1, 2, 3, 4, 5, 6$. Among the 48 possible problems, he observes that half of the solutions give an odd number A and that all of them are prime. Not satisfied with the fact that his series of problems does produce even numbers, which obviously are not prime, he extends the series to another forty-eight problems that include a fourth congruence that is guaranteed to produce only odd results; that is $A \equiv 1$ (mod 2). Hua obtains again forty-eight numbers as the smallest possible solution to each problem, all of these numbers are odd and four of them are not prime. Adding the three prime numbers 3, 5 and 7 to the forty-four prime numbers, Hua has found altogether

---

[31]On the meaning of this technical term in Qin Jiushao's *Dayan* procedure to solve indeterminate problems, see Libbrecht (1973) p. 329n15.

[32]Translated from Hua (1893b) p. 1A–1B.

Fig. 3.9 List of all prime
numbers (here from 461 to
733) in Hua (1893a) p. 29B

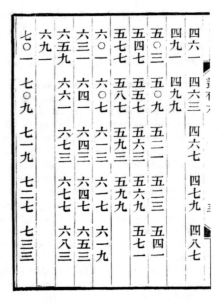

forty-seven prime numbers smaller than 200. But his goal was to find all prime
numbers below 1000; he thus sets out to list all the other solutions to the problems
that are not the smallest possible. This can easily be done by successively adding
210 (the product of 2, 3, 5 and 7) as long as the result remains below 1000.
Again, the resulting list of 228 numbers contains some numbers that are not prime
and need to be eliminated. Therefore, Hua enumerates all the products of prime
numbers from 11 to 31 (including their squares) that are smaller than 1000 and
obviously not prime. These sixty-three composite numbers are exactly the ones that
need to be eliminated from his list of 228. Hua states, without further comment,
that, by adding the prime numbers 3, 5 and 7, he thus has found all the 168
prime numbers smaller than 1000, for which he provides a complete list (see
Fig. 3.9).

The conclusion of Hua's list has a certain self-mockery. It is a conversation
between Hua, a guest, and an elderly neighbour, which goes as follows:

A guest who saw these calculations, could not help laughing. He said: If you successively
add the "extended divisor,"[33] the [quality of being a] prime number cannot be preserved.
How could it not be that one encounters numbers that are not prime? Is it not that for
each of the prime numbers, successively added to twice, the prime number turns into an
exhaustible number [i.e. divisible without remainder], and one thus has obtained a number
that is not prime? I said: Well, well. Then take a booklet and write down all the odd numbers
below 1000. First, start from 3. Doubling it makes 6. To 6 add 3, one obtains 9. Then,
below the character 9, mark a dot. To 9 add 6, one obtains 15. Then, below the character
15, mark a point. Like this, successively add and successively mark dots, until you arrive
at 999. One third [of the numbers] have already been reached. Take 5 and proceed the

---

[33] Chin. *yanmu* here the product of 2, 3, 5 and 7.

same way. Take 7, also proceed the same way. From 11 to 31, also proceed the same way. Altogether after 10 rounds, the numbers in the notebook that are not prime have already all been marked by a dot. Look for the ones that are not marked with a dot. By recording them one obtains the numbers smaller than 1000 that are not prime. In less than an entire dynasty, this affair can be terminated! Because we laughed together loudly, an old man from the neighbourhood asked us, saying: you two gentlemen who are talking about mathematics, who indeed is clever and who is clumsy? Neither the guest nor I dared to answer.[34]

I would suspect that Hua has recorded here a friendly, possibly fictional, conversation with a foreigner, who bursts out laughing when he sees Hua's method to find all the prime numbers between one and 1000. The alternative method that Hua proposes to his guest stems from the Greek mathematical tradition, which was well known as the sieve of Eratosthenes.[35] The neighbour, wondering if this method is any better or less clumsy than Hua Hengfang's (Chinese) method, acts as an arbiter between Western and Chinese science and asks the embarrassing question of superiority that Hua would neither dare to ask, nor to answer. At the end of the scene, both methods are depicted as laughable. Hua thus reveals himself as unsatisfied with what he had already mentioned in the preface; namely, that he could not find a general principle that would allow the deduction of one prime number from a previous one. But he also expresses a certain self-satisfaction by pointing to the obvious fact that the Greek method is comparably primitive and empirical, and, as Heath remarked in 1921, it "would be hopeless as a practical means of obtaining prime numbers of any considerable size."[36]

In a related essay on prime numbers, the *Explanation of the Procedures for Testing Primality* (*Shugen shujie* 數根術解), Hua Hengfang gives another example of how to use the ancients' writings to develop new procedures. In this case, it is again a rather cumbersome procedure, but one that is clearly inspired from Master Sun's congruences. The mechanism he suggests in order to find all the prime numbers smaller than 100 takes the product of the three one-digit primes, 3, 5 and 7, which comes up to 105. One then "only" needs to check whether any (odd) one- or two-digit number (which is not yet known to be prime) has a common factor with 105. If a common factor was found, the number is not prime; otherwise, it is a prime number. Hua admits that, although the procedure is extendible to three-, four- and more digit numbers, the test becomes increasingly unpractical.[37] For large numbers, he instead refers to Li Shanlan's method, which Hua describes as follows:

Master Li Renshu[38] has a quick procedure for testing primality which says: with the original number multiply the logarithm of 2 [$N \cdot \log 2$]. Find the integer number of the result [antilog ($N \cdot \log 2$)]. Subtract 2 and what remains measure it with the original

---

[34]Translated from Hua (1893a) p. 31A–31B.

[35]Described by Nicomachus in his *Introductio Arithmetica*, see Nicomachus <Gerasenus> (1866) bk. I, chap. XIII.2-8, p. 29–33 .

[36]*Idem.*

[37]Hua (1893a) p. 5A–5B.

[38]Li Renshu was the style name of Li Shanlan.

number. If one can measure exhaustively [i.e. if one can divide without a remainder],
the original number is a prime number. If one cannot measure exhaustively [i.e. if
one cannot divide without a remainder], the original number is not a prime number.[39]

This version of Li Shanlan's procedure does not resemble Li Shanlan's text,
which was published in August 1872 and discussed above.[40] In fact, it rather
strongly reminds one of Wylie's description of the (false) CHINESE THEOREM
quoted at the beginning of this chapter. I therefore assume that Hua Hengfang
must have either been familiar with Wylie's account, or directly with Li's earlier
ideas on how to test primality. In either case, this reminds us that Hua's pragmatic
discourse about the usefulness of the Chinese mathematical tradition needs to be
contextualized in a setting wider than that of the Chinese late Qing mathematical
community. In order to understand Hua's insistence on sole reliance on the ancients,
we need to take into account the presence of foreign mathematical knowledge
and foreign observers with whom Li and Hua interacted in their translation work
and to whom they reacted when it came to statements about scientific primarity.
Hua Hengfang, in his *Explanation of the Procedures for Testing Primality*, insists
that Li is not relying on methods other than those already present in the Chinese
mathematical tradition:

> This procedure, although it is a general method for testing primality, the supposed number, if
> it is too big, then multiplying with the logarithm of 2 will also lead to an enormous amount
> of positions. For logarithms of integer numbers one cannot search in tables to obtain the
> integer number [giving the antilog]. When using "the method for finding the integer number
> of a logarithm," the execution of the calculation to find it is not a simple affair either. That
> is why I still found it inconvenient to use this. I wanted to abandon logarithms and find
> another, more simple procedure. Gradually investigating the mysteries and profundities of
> prime number (*sushu* 素数) theory, I knew that Master Li's sources for establishing his
> procedures did not go beyond exponential series and continued proportions, and that was
> all.[41]

Hua dedicates the rest of his essay to relating prime numbers to the binomial theorem and the arithmetic triangle, in China known since Song dynasty
as the so-called Jia Xian triangle.[42] He also gives an algorithm for efficient
modular exponentiation to avoid the calculation of 2^N for large (odd) numbers N. The algorithm is quoted in 1871 in its original Chinese version and
translated into English by John Fryer in the *North China Herald and Supreme*

---

[39] Translated from Hua (1893a) p. 5B and Hua (1897) p. 2A.

[40] The preface to Hua (1893a) is written by Li Shanlan and dated Tongzhi 2/11 (= March 1872).
One can thus assume that Hua had not seen Li's texts (Li 1872a,b,c) containing the four prime
number tests. This assumption is confirmed by the content of Hua (1893a).

[41] Translated from Hua (1893a) p. 6A, found also in Hua (1897) p. 2A without textual variation.

[42] See Chap. 5, p. 120.

Court and Consular Gazette.[43] Fryer comes to Hua's defence, in a published dispute under the title "Chinese versus foreign mathematicians" by stressing, that:

> Mr. Hwa, however, went more carefully to work. After investigating the principle involved in Mr. Li's original formula, which translated from his notation reads $\frac{2^a-2}{a}$, he employed much of his leisure time in endeavouring to find out a reliable and satisfactory method of shortening the labour entailed in its use. Having discovered such a method he next endeavoured to write it in a form easily understood by Chinese who have not learnt the use of algebraical formulæ. That he has succeeded in his efforts, is a fact which any one can judge for himself by first taking the trouble to understand the rule and then working a few examples by it.

3.3   Fermat's Little Theorem

Fryer's remark is the first in what he refers to as "a harmless form of typhoon,"[44] a series of four polemic contributions to a controversy between himself and Baron Johannes von Gumpach (stands for an unkown date of birth–1875)[45] in the Correspondence section of the *North China Herald and Supreme Court and Consular Gazette* in 1871.[46] Fryer originally reacts to a note by von Gumpach, published 22 February 1870 in the *North China Daily News*, which again is a summary of another series of four contributions published in the *Notes and Queries on China and Japan* in response to Wylie's query on "A Chinese Theorem" quoted at the beginning of this chapter.[47] The "typhoon" that Li Shanlan created, through Wylie's note, among foreign actors was not really a dispute of "Chinese *versus* foreign mathematicians." Indeed, the Chinese didn't even have a direct voice in it. It was rather a dispute among foreigners about how they interpreted and the degree to which they esteemed the methods and results of Chinese native mathematicians.

At the beginning, the latent question posed by Wylie was whether anything similar to the "Chinese Theorem" was to be found in earlier European sources. This was a typical question where the competition between two nations' mathematics

---

[43] See Volume VI - Issue 214 (1871-06-09) p. 429–430. See also Han and Siu (2008) p. 947–948. On John Fryer's and Hua Hengfang's collaboration to translate Western scientific works into Chinese see p. 79.

[44] Fryer (1871a) p. 163.

[45] For von Gumpach's date of death, see Hart (1975) vol. 1, p. 203, Letter no. 135.

[46] All published under the title "Chinese versus foreign mathematicians." See Fryer (1871a), von Gumpach (1871a), Fryer (1871b) and von Gumpach (1871b).

[47] See von Gumpach (1870a) and von Gumpach (1869) where he points out p. 153 that Li's criterion is equivalent to testing if for a given integer number n, the quotient $(2^n - 2)/n$ "is found to be an integer." Two other responses to Wylie (1869) stem from Mr. W. McGregor of Amoy McGregor (1869) and R. A. Jamieson of Hankow Jamieson (1869), professor of Natural Philosophy and Mathematics in "The University of Peking" at Shanghai and not, as Han and Siu (2008) p. 946 believe, a British journalist. The final note in this series by von Gumpach is von Gumpach (1870b).

can be played out. Jamieson, who had identified, in a more general version, Fermat's Little (or Simple) Theorem as an equivalent,[48] dismisses Li's discovery as unimportant:

> your readers will be surprised to learn that Le's [sic] rule is merely a particular, very narrow, and imperfectly developed case of a theorem which is as old as the seventeenth century, and which (until I read Mr Von Gumpach's paper) I thought was known to every senior Schoolboy. It is referred to as "Fermat's Theorem," [...]. Le's rule is therefore merely reproduced from some elementary work on Algebra, and spoiled in the reproducing.[49]

"Oh, the naughty man, to hit me so hard," responded von Gumpach, who felt strongly Jamieson's rebuke that "with a senior school boy's ignorance" he had not identified Li's theorem as a special case of (actually the invalid inverse of) Fermat's statement.[50] This came at a time when von Gumpach's opinion of Li Shanlan was not at its best. Von Gumpach was a German-born, British-naturalized citizen, who had been engaged to teach mathematics and astronomy at the Interpreter's College in Beijing. In 1868 "von Gumpach turned out to be completely insatisfactory and was discharged."[51] Shortly after Li Shanlan was appointed as his successor, von Gumpach's first note was published in October 1869:

> It is manifest from Mr Le's mode of enunciation, that he has empirically deduced his rule from trials with some few low numbers; has not seized its principle; attaches an undue value to it; and was not justified in qualifying it as a theorem, without having demonstrated the mathematical necessity of its truth. Thus, if x represent the given number, Mr. Le's uncouth formula would be—
>
> $$\frac{\text{number}[\log(x.\log 2)] - 2}{x,}$$
>
> and I need hardly observe, that our Tables of Logarithms to seven and ten decimal places do not permit its application to any number exceeding 24 and 36 respectively.[52]

---

[48]In a letter to Bernard Frénicle de Bessy (ca. 1605–1675), Pierre Fermat (1607–1665) conjectured the following (Tannery and Henry 1894) p. 209:

> Tout nombre premier mesure infailliblement une des puissances −1 de quelque progression que ce soit, et l'exposant de la dite puissance est sous-multiple du nombre premier donné −1; et, après qu'on a trouvé la première puissance qui satisfait àla question, toutes celles dont les exposants sont multiples de l'exposant de la première satisfont tout de même à la question. [...]
> Et cette proposition est généralement vraie en toutes progression et en tous nombres premiers; de quoi je vous envoierois la démonstration, si je n'appréhendois d'être trop long.
> (Letter from Fermat to Frénicle de Bessy, dated 18 October, 1640)

[49]Jamieson (1869) p. 179.

[50]von Gumpach (1870b) p. 39.

[51]This turned into the famous "von Gumpach case," a *cause célèbre* because it was not clear whether the British Supreme Court or the Chinese Government had jurisdiction to deal with it. See Hart (1975) vol. 1, Introduction p. 14 and the many legal notes in the *North China Herald* between March 29 and May 5, 1870.

[52]von Gumpach (1869) p. 153.

Fig. 3.10 Table of
remainders in von Gumpach
(1869) p. 153

Given Number x.	Remainders. Bases, u.			Given Number x.	Remainders. Bases, u.		
	2	3	4		2	3	4
2	0	0	0	32	30	30	28
3	0	0	0	33	6	24	27
4	2	6	0*	34	2	6	12
5	0	0	0	35	16	9	5
6	2	0*	0*	36	34	6	14
7	0	0	0	37	0	0	0
8	6	6	4	38	2	6	12
9	6	6	6	39	6	24	21
10	2	6	2	40	14	38	16
11	0	0	0	41	0	0	0
12	2	6	0*	42	20	12	18
13	0	0	0	43	0	0	0
14	2	6	12	44	34	34	32
15	6	9	0*	45	15	18	25
16	14	14	12	46	2	6	12
17	0	0	0	47	0	0	0
18	14	6	6	48	14	30	12
19	0	0	0	49	0	0	14*
20	16	18	12	50	22	46	22
21	6	3	18	51	6	24	9
22	2	6	12	52	14	26	44
23	0	0	0	53	0	0	0
24	14	6	12	54	26	24	24
25	5	15	20	55	41	15	30
26	2	6	12	56	30	54	12
27	24	24	24	57	38	24	39
28	14	22	0*	58	2	42	12
29	0	0	0	59	0	0	0
30	2	6	12	60	14	18	12
31	0	0	0	61	0	0	0

Von Gumpach added a mathematical discussion of the fact that Li's rule is not
generally valid by proceeding empirically, and therefore rewriting Li's "formula"
into the simpler, but just as impractical, version for large x:

$$\frac{2^x - 2}{x}.$$

He then tests if for the more general version:

$$\frac{u^x - u}{x}$$

one also finds that when the quotient is an integer, x is prime. Listing all remainders
for $x = 2, 3, \ldots, 61$ and for $u = 2, 3, 4$ in a table, he finds several exceptions to the
rule (marked by an asterisk in Fig. 3.10). He concludes that he is no more able than
Li to give a formal number theoretical proof to this, other than by providing "counter
examples." Number 49 is one of them for $u = 2$, which von Gumpach believed was
based on a calculation error, error which McGregor in the following note on the
"Chinese Theorem" copied as proof for "the approximate accuracy of the rule."[53]
I shall not go through all the errors, reproaches, nonsense, misrepresentations,
muddled arguments and technical points of the discussion between the different
contributors here, but rather focus on Fryer's conciliatory approach to the issue.

---

[53]McGregor (1869) p. 167.

Fryer wanted to settle which of the foreign mathematicians reacting to Wylie's query was right in criticizing Li Shanlan's method. He did so by considering "how these criticisms appear from a Chinese point of view."[54] Very familiar with both Li and Hua, Fryer gave a voice to his Chinese mathematician friends who were neither subscribers of the journals in which the foreigners' debate took place, nor capable of reading English, except for Arab numerals. They were thus in "blissful ignorance" about the judgement of their work "at the hands of Foreign mathematicians."[55] Fryer recalls a conversation he had with Hua when he showed him von Gumpach's table of remainders, shown above in Fig. 3.10:

> The weak points in Mr. Von Gumpach's criticism have now to be noticed. A few days ago, it occurred to the writer to show it to a native mathematician, Hwa of Woo-se [=Hua Hengfang]. Mr. Hwa first looked at the table just mentioned, and seemed sorely puzzled. At last the light dawned upon his mind and he said "What reason can the Foreign Sien-shêng [=Sir] give for taking 3 and 4 as bases instead of 2? Anybody can see that he has been working in the dark. Of course there will be exceptions to any other number than 2 as base. Even if there were none, why use higher numbers if 2 answers all purposes?[56] Why waste so much time in building without any foundation? Why make up such a long table of mere possibilities that can be of no imaginable use to anybody? Truly you Foreigners are very profound!" [...]
>
> Mr. Hwa's remarks upon the above "uncouth" treatment of Mr Le's rule were very amusing. "You Foreigners," said he, "must certainly be very fond of making up such long and imposing rows of figures. No doubt to you they look very learned and grand, but to us Chinese they appear very ridiculous. Surely the Foreign Sien-shêng who took so much unnecessary trouble, must have done it on purpose to make fun of Mr. Le's rule! If you Foreigners do not know how to use figures better than that of what use is it for you to come to China to teach us?".[57]

This sounds like an echo of the dialogue Hua recorded in his own work, and which I suspected to be a conversation between a Chinese and a foreign mathematician.[58] In both cases, Hua underlines how unnecessary it is for Chinese mathematicians to look towards "ridiculous" Western mathematical knowledge, from which nothing can be learnt. What Fryer records of Li Shanlan is less polemic and more interesting from a heuristic point of view. Fryer tells the story of how Li discovered his rule. If Fryer's account is authentic, it confirms that Li (apart from the use of algebraic symbols) conjectured his rule on the basis of the Chinese mathematical tradition. It is also interesting for us, because it shows how Li proceeded through recognition of numeric patters in the arithmetic triangle[59]:

---

[54] See Fryer (1871a) p. 164.

[55] *Idem.*

[56] This passage shows that Hua had not understood that Li's rule is wrong, even for $u = 2$.

[57] Fryer (1871a) p. 164.

[58] See page 66.

[59] I will come back to this point in detail in Chap. 5. I would strongly disagree with Fryer's conviction that "Li could no doubt give a masterly demonstration of it if asked to do so" (Fryer 1871a) p. 164, at least in a normative sense of "demonstration."

Before leaving Shanghai to assume the duties of his Professorship at Peking, Mr. Le spent much of his leisure time in trying to discover a general and convenient rule for determining prime numbers. "Fermat's Theorem" he considered not to be depended upon, and "Wilson's Theorem" to be too laborious. At last he noticed that in the table of the expansions of $(1+1)^a$ where a successively $= 0, 1, 2, 3,$ &c., viz: —

$$
\begin{array}{ccccccc}
 & & & 1. & & & \\
 & & 1. & & 1. & & \\
 & & 1. & 2. & 1. & & \\
 & 1. & 3. & 3. & 1. & & \\
 1. & 4. & 10. & 4. & 1. & & \\
1. & 5. & 10. & 10. & 5. & 1. & \\
\&c., & & \&c., & & \&c. & &
\end{array}
$$

whatever value is given to a, the sum of the terms of the corresponding series must necessarily be equal to 2^a. Taking away the first and the last terms (viz: the two 1's) from any horizontal line in this table, the sum of the remaining terms $= 2^a - 2$. It is evident from the construction of the table that when a is a prime number, it must measure every term in its series except the first and last,[60] and must therefore measure their sum of $2^a - 2$. But if a be not a prime number it cannot measure their sum or $2^a - 2$.[61]

The mathematical storm of the "Chinese Theorem" does not end with the von Gumpach, Fryer et al. controversy. In fact, it reappears in different versions throughout the Western literature as a discovery attributed to the time of Confucius, first in a paper by J. H. Jeans (see Fig. 3.11), then in Giuseppe Peano's encyclopaedic collection of mathematical formulas,[62] but also in Dickson's nominal work on the *Theory of Numbers*.[63] That such historiographic myths are very hard to root out of the literature is well known. Indeed, references to an ancient "Chinese Theorem" still appear in contemporary sources in spite of the fact that its antiquity has been doubted since at least 1959, when it was questioned in Joseph Needham's *Science and Civilisation of China*.[64] What equally persists is the notion of a Theorem, not for Li Shanlan, but for Fermat, who did not provide any proof for what was a simple

---

[60]That this was not evident to a Chinese mathematician is shown in Hua (1893a) p. 13A, where Hua sets out to deduce that (in modern terms) for $k = 1, \ldots, p - 1$ the binomial coefficients $\binom{p}{k}$ are all divisible by p.

[61]Fryer (1871a) p. 164.

[62]Peano (1901) p. 96–97:

Les chinois ont connu cette P pour $b = 2$, dès le temps de *Confucius*. a. —550–477; cfr. Jeans Mm. a.1898 t.27 p. 171.

[63]Dickson (2002) vol. 1, p. 59:

The Chinese seem to have known as early as 500 B.C. that $2^p - 2$ is divisible by the prime p. This fact was rediscovered by P. de Fermat while investigating perfect numbers. Shortly afterwards, Fermat stated that he had a proof of the more general fact now known as Fermat's theorem: If p is any prime and x is any integer not divisible by p, then $x^{p-1} - 1$ is divisible by p.

[64]See Needham (1959) p. 54, n.d.

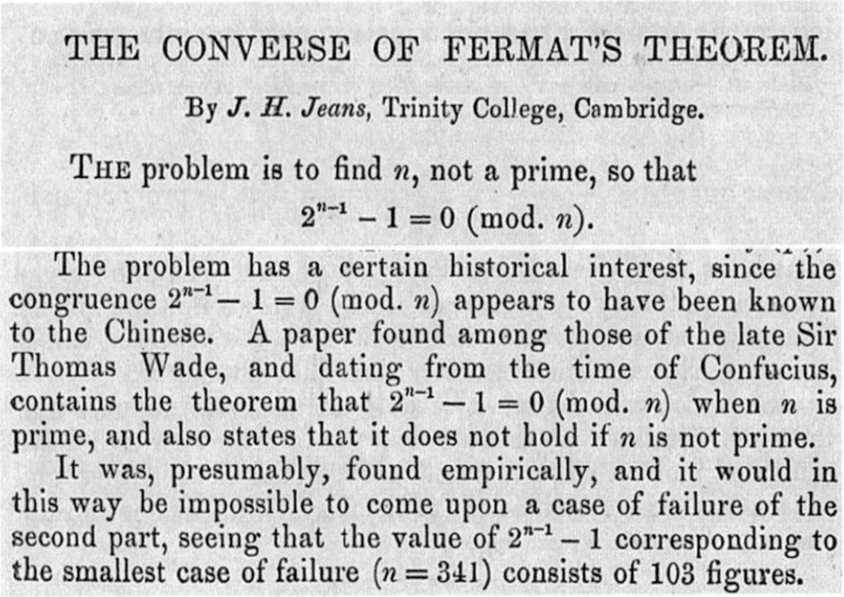

THE CONVERSE OF FERMAT'S THEOREM.

By *J. H. Jeans*, Trinity College, Cambridge.

THE problem is to find n, not a prime, so that

$$2^{n-1} - 1 = 0 \ (\text{mod. } n).$$

The problem has a certain historical interest, since the congruence $2^{n-1} - 1 = 0$ (mod. n) appears to have been known to the Chinese. A paper found among those of the late Sir Thomas Wade, and dating from the time of Confucius, contains the theorem that $2^{n-1} - 1 = 0$ (mod. n) when n is prime, and also states that it does not hold if n is not prime.

It was, presumably, found empirically, and it would in this way be impossible to come upon a case of failure of the second part, seeing that the value of $2^{n-1} - 1$ corresponding to the smallest case of failure ($n = 341$) consists of 103 figures.

Fig. 3.11 "The Converse of Fermat's Theorem" (Jeans 1898)

conjecture. Ribenboim probably comes closest to the historiographic entanglements of the "Chinese Theorem" by saying that:

the problem concerning the congruence $2^n \equiv 2 \ (\text{mod } n)$ [...] might be appropriately called, if not as a joke, the "pseudo-Chinese congruence on pseudoprimes."[65]

References

Acerbi, Fabio (2010). *Il silenzio delle sirene. La matematica greca antica*, Volume 59 of *Saggi*. Roma: Carocci editore.

Billingsley, Henry (1570). *The Elements of Geometrie of the Most Auncient Philosopher Euclide of Megara. Faithfully (now first) translated into the Englishe toung, by H. Billingsley, citizen of London. Whereunto are annexed certaine scholies, annotations, and inuentions, of the best mathematiciens, both of time past, and in this our age. With a very fruitfull præface made by M. I. Dee, specifying the chiefe mathematicall scie[n]ces, what they are, and wherunto commodious: where, also, are disclosed certaine new secrets mathematicall and mechanicall, vntill these our daies, greatly missed.* London: John Daye.

Born, Johannes (2015). Ausgewählte Primzahltests. Master's thesis, Johannes Gutenberg-Universität Mainz, Fachbereich Physik, Mathematik und Informatik, directed by Prof. Stefan Müller-Stach.

---

[65] See Ribenboim (2004) p. 88–89.

Bréard, Andrea (1999). *Re-Kreation eines mathematischen Konzeptes im chinesischen Diskurs: Reihen vom 1. bis zum 19. Jahrhundert*, Volume 42 of *Boethius*. Stuttgart: Steiner Verlag.

Bréard, Andrea (2014). On the Transmission of Mathematical Knowledge in Versified form in China. In Alain Bernard and Christine Proust (Eds.), *Scientific sources and Teaching Contexts Throughout History: Problems and Perspectives*, Volume 301 of *Boston Studies in the Philosophy and History of Science*, 155–185. Dordrecht: Kluver.

Clavius, Christophorus (1574). *Euclidis Elementorum Libri XV Accessit XVI de Solidorum Regularium Cuiuslibet Intra Quodlibet Comparatione, Omnes Perspicuis Demonstrationibus, Accuratisque Scholiis Illustrati, ac Multarum Rerum Accessione Locupletati*. Rome: V. Accoltum.

Clavius, Christophorus (1591). *Euclidis Elementorum Libri XV* (3rd ed.). Ciotti. first edition (Clavius 1574).

Dickson, Leonard E. (1999–2002). *History of the Theory of Numbers* (Repr. of the Washington, Carnegie Inst. of Washington, 1919–1923 ed.). Procidence, RI: AMS Chelsea Publishing.

Engelfriet, Peter M. (1998). *Euclid in China: the Genesis of the First Chinese Translation of Euclid's Elements Books I–VI (Jihe yuanben; Beijng, 1607) and its Reception up to 1723*, Volume 40 of *Sinica Leidensia*. Leiden: Brill.

Fryer, John (1871a, March 8). Chinese *versus* Foreign Mathematicians. *The North China Herald and Supreme Court and Consular Gazette 6*(201), 163–164.

Fryer, John (1871b, June 9). Chinese *versus* Foreign Mathematicians. *The North China Herald and Supreme Court and Consular Gazette 6*(214), 429–430.

Guo, Shuchun 郭書春 et al. (Eds.) (1993). *Zhongguo kexue jishu dianji tonghui: Shuxue juan* 中國科學技術典籍通彙: 數學卷 (Comprehensive Collection of Ancient Classics on Science and Technology in China: Mathematical Books), 5 vols. Zhengzhou: Henan jiaoyu chubanshe 河南教育出版社.

Han, Qi and Man-Keung Siu (2008). On the Myth of an Ancient Chinese Theorem about Primality. *Taiwanese Journal of Mathematics 12*(4), 941–949.

Hart, Sir, Robert (1975). *The I. G. in Peking: Letters of Robert Hart, Chinese Maritime Customs, 1868–1907*. Cambridge, Mass.: Belknap Press of Harvard University Press.

Heath, Thomas L. (1981). *A History of Greek Mathematics*, 2 vols. New York: Dover Publications. Unabridged republication of the 1921 edition, Oxford: Clarendon Pr.

Horng, Wann-Sheng (1991). *Li Shanlan: The Impact of Western Mathematics in China during the Late 19th Century*. Ph. D. thesis, Graduate Center, City University of New York.

Hua, Hengfang 華蘅芳 (1893a). Shugen shujie 數根術解 (Explanation of the Procedures for Testing Primality). In Hua Hengfang 華蘅芳 (Ed.), *Suancao congcun* 算草叢存 (Collection of Mathematical Workings) (Xingsu xuan suangao 行素軒算稿 1882 ed.), Volume 2. Wuchang. Preface by Li Shanlan dated Tongzhi 2/11 (March 1872).

Hua, Hengfang 華蘅芳 (1893b). Shugen yan gu 數根演古 (Prime Numbers Developed from the Past). In Hua Hengfang 華蘅芳 (Ed.), *Suancao congcun* 算草叢存 (*Collection of Mathematical Workings*) (Xingsu xuan suangao 行素軒算稿 1882 ed.), Volume 6. Wuchang.

Hua, Hengfang 華蘅芳 (1897). Shugen shujie 數根術解 (Explanation of the Procedures for Testing Primality). In Hua Hengfang 華蘅芳 (Ed.), *Huashi Zhongxi suanxue quanshu* 華氏中西算學全書 Master Hua's Complete Books of Western and Chinese Mathematics) (行素軒校本 Xingsu xuan ed.), Scroll 1, First Series (Chuji juan yi 初集 卷一), 1A–4B. Shenji shuzhuang 慎記書莊.

Jamieson, R. A. (1869, December). Chinese *versus* Foreign Mathematicians. *Notes and Queries on China and Japan 3*(12), 179.

Jeans, J. H. (1898). The Converse of Fermat's Theorem. *Messenger of Mathematics*, 174–174.

Li, Shanlan 李善蘭 (1872a). Kao shugen fa 考數根法 (Methods for Testing Primality). *Zhongxi wenjian lu* 中西聞見錄 (Peking Magazine) 2, 13A–17A.

Li, Shanlan 李善蘭 (1872b). Kao shugen fa 考數根法 (Methods for Testing Primality). *Zhongxi wenjian lu* 中西聞見錄 (Peking Magazine) 3, 15A–18B.

Li, Shanlan 李善蘭 (1872c). Kao shugen fa 考數根法 (Methods for Testing Primality). *Zhongxi wenjian lu* 中西聞見錄 (Peking Magazine) 4, 15A–20B.

Li, Shanlan 李善蘭 and Alexander Wylie 偉烈亞力 (1865). *Jihe yuanben* 幾何原本 *15* 卷 (The Elements), 15 Books (Jinling 金陵刻本 ed.). Nanjing. Includes the first six books (Ricci and Xu 1607), translation of Books VII through XV based on (Billingsley 1570), published separately under the patronage of Han Yingbi 韓應陛 in Shanghai in 1857. Reprint in (Guo et al. 1993) 5:1151–1500.

Li, Zhizao 李之藻 et al. (1965). *Tianxue chuhan* 天學初函 (First Collectanea of Heavenly Studies), 6 vols (Reprint ed.), Volume 23, First Series (Chubian 初編 卷二十三) of Wu Xiangxiang 吳相湘 (Ed.), *Zhongguo shixue congshu* 中國史學叢書. Taipei: Taiwan xuesheng shuju 臺灣學生書局.

Libbrecht, Ulrich (1973). *Chinese Mathematics in the Thirteenth Century. The Shu-shu chiu-chang of Ch'in Chiu-shao*. Cambridge (Mass.): MIT Press.

McGregor, W. (1869, November). A Chinese Theorem. *Notes and Queries on China and Japan* 3(11), 167–168.

Mueller, Ian (1981). *Philosophy of Mathematics and Deductive Structure in Euclid's* Elements. Mineola, NY: Dover.

Needham, Joseph (1959). *Mathematics and the Sciences of the Heavens and the Earth*, Volume 3 of *Science and Civilisation in China*. Cambridge: Cambridge University Press.

Nicomachus <Gerasenus> (1866). *Nikomachu Gerasēnu Pythagoreiku Arithmētikē eisagōgē*. Bibliotheca scriptorum Graecorum et Romanorum Teubneriana. Lipsiae: Teubner.

Peano, Giuseppe (1901). *Formulaire de Mathématiques*. Paris: Georges Carré et C. Naud.

Ribenboim, Paulo (2004). *The Little Book of Bigger Primes* (2nd ed.). New York, Berlin, Heidelberg: Springer.

Ricci, Matteo and Guangqi Xu (1607). *Jihe yuanben* 幾何原本 (The Elements). Beijing: [n.p.]. Translation of the first six books of (Clavius 1574), included in (Li et al. 1965).

Shen, Kangshen 沈康身 (1986). 《Shushu jiuzhang》 dayan lei suanti zhong de shulun mingti 《数书九章》大衍类算题中的数论命题 (=Analysis on Shushu Jiuzhang to Its Propositions about Theory of Numbers). *Hangzhou daxue xuebao (Ziran kexue ban)* 杭州大学学报 (自然科学版) (Journal of Hangzhou University, Natural Science ed.) *4*, 421–434.

Tannery, Paul and Charles Henry (Eds.) (1894). *Correspondance*, Volume II of *Oeuvres de Fermat*. Paris: Gauthier-Villars.

von Gumpach, Johannes (1869, October). A Chinese Theorem. *Notes and Queries on China and Japan* 3(10), 153–154.

von Gumpach, Johannes (1870a, February 22). A Chinese Theorem. To the Editor of the NORTH-CHINA HERALD. *The North-China Daily News*, 7059.

von Gumpach, Johannes (1870b, May). A Chinese Theorem. *Notes and Queries on China and Japan* 4(27), 39–42.

von Gumpach, Johannes (1871a, April 29). Chinese *versus* Foreign Mathematicians. To the Editor of the NORTH-CHINA HERALD. *The North-China Herald and Supreme Court and Consular Gazette* 6(208), 307–308.

von Gumpach, Johannes (1871b, June 23). Chinese *versus* Foreign Mathematicians. To the Editor of the NORTH-CHINA HERALD. *The North-China Herald and Supreme Court and Consular Gazette* 6(216), 469–471.

Wylie, Alexander (1869, May). A Chinese Theorem. *Notes and Queries on China and Japan* 3(5), 73.

Xu, Yibao (2005). The first Chinese Translation of the Last Nine Books of Euclid's Elements and its Source. *Historia Mathematica* 32(1), 4–32.

Yan, Dunjie 嚴敦傑 (1954a). Zhongsuanjia de sushulun 中算家的素數論 (The Chinese Mathematicians' Theories of Prime Numbers). *Shuxue tongbao* 數學通報 *4*, 6–10.

Yan, Dunjie 嚴敦傑 (1954b). Zhongsuanjia de sushulun 中算家的素數論 (The Chinese Mathematicians' Theories of Prime Numbers). *Shuxue tongbao* 數學通報 *5*, 12–15.

Yang, Li 杨丽 (2010). *Wan Qing shuxuejia guanyu sushu yanjiu de chengjiu yu bu zu* 晚清数学家关于素数研究的成就与不足 (About late Qing Mathematicians' Achievements and Insufficiencies in Research on Prime Numbers). Master's thesis, Tianjin shifan daxue 天津师范大学, Tianjin. Advisor: Li Zhaohua 李兆華.

Chapter 4
Negotiating a Linguistic Space In-Between

Contents

Since the early sixteenth century, the first Jesuits and sinologists defended the idea of Chinese characters as an ideographic writing system, a symbolic language disconnected from speech. Thus, Chinese characters were considered particularly suitable as a universal language, an "alphabet of human thought" based upon logical and combinatorial principles, or as Leibniz at one time believed, an alphabet which "could then be assigned symbols or numbers and manipulated in a way similar to arithmetic or geometry to attain new truths."[1] For Chinese mathematical rhetoric, such a conception would imply either that traditional algorithms can easily be translated into modern algebraic symbolism, or that, due to the formulaic nature of texts, symbolic transcription was not even considered necessary. Yet when looking at the history of the introduction of symbolic algebra during the Qing dynasty (1644–1911), one might easily notice that the process of the acceptance or rejection of mathematical signs did not correspond to either of these straightforward options.[2] The creation of a syncretistic symbolism, using elements from Chinese characters and maintaining a speakable format of symbols, was just one possible way of finding a compromise between two cultures of mathematical discourse.

---

[1] Mungello (1985) p. 192. For conflicting Western conceptions of Chinese characters as ideographic or logographic, see Harbsmeier (1998) p. 34–36.

[2] On the problem of a conceptual boundary between signs and names, see Bréard (2001) p. 312–313.

© Springer International Publishing AG, part of Springer Nature 2019
A. Bréard, *Nine Chapters on Mathematical Modernity*, Transcultural
Research – Heidelberg Studies on Asia and Europe in a Global Context,
https://doi.org/10.1007/978-3-319-93695-6_4

In this chapter I will analyse the transmission of algebra through the translations and paratexts accompanying the introduction of Western techniques against the background of Chinese Yuan dynasty algebraic techniques. Prominent translators and mathematicians, such as Li Shanlan 李善蘭 (1811–1882) and Hua Hengfang 華蘅芳 (1833–1902) in particular, maintained two different kinds of mathematical discourse in their commentatorial and translatory practices. But was this compartmentalization a purely conceptual or also a linguistic and political choice?

During a conference held in Oberwolfach in 1998, I gave a paper titled "On Mathematical Terminology: Culture Crossing in Nineteenth-Century China" in which I discussed the multi-layered issues involved in the history of mathematical terminologies at that time. In it, I portrayed "the introduction of Western mathematical notations into the Chinese consciousness of the late nineteenth century as a major signifying event."[3] I was young and probably thought that dramatic expressions were needed in order to make the topic interesting for the audience. But now I see things in a more nuanced light, and although I still believe that, from an epistemological point of view, the new algebraic symbolism invented in China in the mid-nineteenth century represented a significant transformation of mathematical discourse and practice in China, I will focus in this chapter on the perceptions of change in mathematical language by the Chinese and foreign scholarly community who were rather pragmatic about rejecting or adopting new modes of representing mathematical content. I will demonstrate how the writing of mathematics was not exempt from the ideological debates that unfolded along the central categories of tradition versus modernity, inferiority versus superiority, and authentic versus foreign knowledge.

In a first part, I will give some historical background to the translation enterprise undertaken during the late imperial period, before turning more concretely to examples of algebraic discourse, which was where traditional and Sinicized Western mathematical symbolism cohabited during the second half of the nineteenth century. In the end, I will try to identify the criteria that guided Chinese mathematicians and foreign translators in their choice of mathematical symbols and notations.

4.1   The Translation Enterprise

Following a first wave of translations of scientific texts by Jesuit missionaries, the end of the Opium Wars marked the beginning of a second stage in which missionaries continued to play a vital role. In Shanghai, during his 8 years (1852–1860) at The London Missionary Society Press (*Mohai shuguan* 墨海書館), Li Shanlan translated many scientific works with Protestant missionaries such as Alexander

---

[3] See Bréard (2001) p. 306.

Wylie and John Edkins.[4] In continuation of the Jesuits' efforts, Li and Wylie completed the translation of Euclid's *Elements*[5] but generally turned their attention to nineteenth-century sources. In spite of the many activities of missionaries in China, the most extensive translation of science texts in the second half of the nineteenth century was conducted within the context of a government institution, the Jiangnan Arsenal (*Jiangnan zhizao ju* 江南製造句) near Shanghai. In the 1860s, in the aftermath of the Second Opium War and the suppression of the Taiping Rebellion, China set in motion the first sustained attempt at "Self-Strengthening" (*ziqiang* 自強). Through the study and imitation of Western technology, Zeng Guofan 曾國藩 (1811–1872), a scholar-official and leader of the Self-Strengthening movement, hoped to prevent the collapse of the crumbling Qing dynasty. It was in this context that the Jiangnan Arsenal was founded in 1865 to become the paradigm of the Self-Strengthening enterprises that would catalyse technological innovation and develop military strength. Building modern steamships and guns depended much on the level of the fundamental sciences underlying firearms, engineering and machine technology. But no such knowledge was readily available to a Chinese audience. One of the Arsenal's central purposes, therefore, was the translation of scientific books related to military matters. "From 1871, when the first translation was completed and published at the Arsenal until 1912, when the Translation Department of the Arsenal came to its end, some 241 translations were undertaken, of which 193 were published."[6]

In mathematics, many books and book-length encyclopaedia entries were translated at the Arsenal. Chinese and native English language specialists were specifically employed for this purpose in its Translation Department. There, Hua Hengfang collaborated with John Fryer (1839–1928), who earlier in 1863 had come to the Beijing Interpreter's College (*Tongwen guan* 同文館) as an English teacher with the Church Missionary Society.[7] Hua and Fryer were responsible for the majority of the translations of mathematical texts produced at the Arsenal and they published

---

[4]See Chap. 2 for more details on Li Shanlan's translation of texts on calculus and conics. For the history of the secular publications of the Mission Press, see Su (1996) p. 327–331.

[5]Li and Wylie (1865).

[6]Quoted from Xu (2005) p. 152–153.

[7]On Fryer's early years in China, see Wright (1996) p. 4–5. Fryer headed for China after he received his teacher certificate from a London normal college and joined the newly established Interpreter's College in Peking to study Chinese and teach English. Between 1865 and 1867 he worked first as headmaster of the Anglo-Chinese School at Shanghai and then as editor of a Chinese-language newspaper. From 1868 to 1896 Fryer translated Western books into Chinese for the Jiangnan Arsenal, where he was responsible for the translation and publication of more than 100 works into Chinese. He was honorary secretary of the Chinese Polytechnic Institution in Shanghai from its inception in 1874, and from 1877 on he served as general editor of the School and Textbook Series Committee, which sponsored the publication of teaching materials for use in schools. In 1884 he organized his own science bookstore and publishing house in Shanghai, the Chinese Scientific Book Depot (*Gezhi shushe* 格致書室). His goal during this period was to introduce Western ideas and knowledge, with primary emphasis on Western science and technology, to China. See Bennett (1967).

valuable accounts of their work as translators.[8] The translation technique they employed was reminiscent of earlier Buddhist practices of working in tandem: the oral translation (*kouyi* 口譯) by a foreigner who mastered literary Chinese well enough to communicate the substance of a text, was taken down by a Chinese "brush-recorder" (*bishu* 筆述) who would transcribe, and sometimes retranslate and style-edit, the foreigner's words into classical Chinese:

> The foreign translator, having first mastered his subject, sits down with the Chinese writer and dictates to him sentence by sentence, consulting with him whenever a difficulty arises as to the way the ideas ought to be expressed in Chinese, or explaining to him any point that happens to be beyond his comprehension. The manuscript is then revised by the Chinese writer, and any errors in style, &c., are corrected by him. In a few cases the translations have been carefully gone over again with the foreign translator, but in most instances such an amount of trouble has been avoided by the native writers, who, as a rule, are able to detect errors of importance themselves, and who, it must be acknowledged, take great pains to make the style as clear and the information as accurate as possible.[9]

The power relation between the "dictator" and the "writer" that Fryer describes here was amplified when, 10 years later, he claimed in a defensive speech given at the *Missionary Society* that

> the responsibility for whatever undue haste or carelessness may characterise my work rests rather on my Chinese colleagues than myself.[10]

John Fryer certainly blamed—unfairly—the Chinese brush-recorder for possible mistakes, and denigrated the latter's important role in the translation process. The tasks of the Chinese scholars actually made great demands on both their literary and scientific skills. As Hua points out, "among the books of Western learning, there are books on chemistry, mathematics, medicine and others. Those who do not thoroughly understand these sciences, cannot serve as brush-recorders!"[11] It was, in fact, not unusual for the brush-recorders to be engaged in their own proper scientific research and scholarly writing in a continuation of the tradition of evidential research scholarship (*kaozheng xue* 考證學), a movement that was most prominent during the Qianlong (r. 1735–1796) and Jiaqing (r. 1796–1820) eras.[12] Interested in reviving China's authentic past, evidential scholars emphasized exacting research, rigorous analysis, and the collection of evidence drawn from ancient artefacts and historical documents and texts, including mathematical writings. Of particular interest to these scholars were those texts from the Yuan dynasty pertaining to an algebraic tradition that had been lost or were not well understood during the Ming dynasty. Due to its parallels with Western algebra, the rediscovery and re-edition of texts from this algebraic tradition challenged the Chinese mathematician's

---

[8]See Fu (1880) and Hua (1885) scroll 12, p. 16B–18B.

[9]Fryer (1880) p. 80, quoted in Wright (1998) p. 662–663.

[10]Fryer (1890) p. 536.

[11]Hua (1885) scroll 12, p. 16B.

[12]The group of scholars engaged in evidential research during this time is therefore also referred to as the Qian-Jia School (*Qian Jia xuepai* 乾嘉學派).

views of the uniqueness of foreign mathematical achievements.[13] Despite the many conceptual similarities, one immediate visual difference between both traditions was the symbolic language with which algebra was expressed:

> The Chinese method of "four elements" is just like the Western method of algebra. In the "four elements," the unknowns, their powers, their mutual products, are all distinguished by the order of their position [on a calculating surface], whereas in algebra they are differentiated by notation. Although the methods are different, their principles are the same.[14]

4.2   A Proto-Grammatical Symbolism

Li Shanlan's translations included mathematical books which were the first to (re-) introduce symbolical algebra to China. In 1712, the French Jesuit Jean-François Foucquet (1665–1741) tried to include algebraic symbols in his *New Method of Algebra* (*Aerrebala xinfa* 阿爾熱巴拉新法). This treatise, which was written for the Kangxi emperor, presented symbolical algebra for the first time to a Chinese audience using Descartes' notation. However, because of the emperor's negative attitude towards the "new" notation, the book exerted no influence on Chinese mathematics at the time. Unfortunately, the emperor did not understand the meaning of the multiplication of symbols: "Jia multiplied by Jia, Yi multiplied by Yi, do not give any number, and one does not know the value of the result," was the comment in a note duly transmitted to Foucquet: "It seems to me that [Foucquet's] algebra is very plain and insufficient. In a word, it is laughable."[15] For the period under consideration here, the second half of the nineteenth century, it is thus more appropriate to speak of a "reintroduction" of Western mathematical notations, which were not simply copied but transformed under local linguistic and cultural constraints.

In their translations, Li Shanlan, and slightly later Hua Hengfang, transcribed formulas with a kind of Siniciced notation, thus creating a new syncretistic algebraic symbolism that other late Qing mathematicians also applied in their own writings. Building on the Preface to the translation of Elias Loomis's *Elements of Analytical Geometry, and of Differential and Integral Calculus* (*Dai wei ji shi ji* 代微積拾級, 1859), Li and Wylie appended a table of the newly created Chinese correspondences for the original symbols and the English expressions as well as some explanations[16]:

---

[13] See Tian (1999).

[14] Preface by Li Shanlan in Li and Wylie (1859) p. 1A.

[15] See Horng (1991) p. 16–17.

[16] These equally applied to another book that Li and Wylie co-translated in the same year, the *Daishuxue* 代數學 (Li and Wylie 1872). It was a rendering of Augustus De Morgan's (1806–1871) *Elements of Algebra Preliminary to the Differential Calculus* (De Morgan 1835).

All the notations found in the present book have not been part of ancient mathematical books. Therefore, let us explain them in detail. "$\llcorner$" stands for positive, addition. "$\top$" stands for negative, subtraction. We subtract right from left. "$\div$" stands for division. Right divides left. We can also draw a "$-$." The divisor is above, the dividend below.[17]

Obviously, arranging the divisor above the dividend is inverse to what we can find in the original text. This means that the European way of writing fractions was inverted, so that, for example, three twelfths would be written as $\frac{12}{3}$. Nevertheless, if we consider the fact that in the nineteenth century, Chinese texts and mathematical formulae were still written in vertical columns, we can indeed speak of a proto-grammatical choice of layout. The arrangement $\frac{\equiv}{=}$ corresponds precisely to the rhetorical formulation in traditional mathematical language, i.e. "twelve parts, thereof three" (*shier fen zhi san* 十二分之三). It is thus "natural" for a Chinese reader to see the divisor above the dividend, with the horizontal line in between symbolizing the syntactical unit "parts thereof" (*fen zhi* 分之).

Another example of the syncretistic choice of speakable symbols presented in Li Shanlan's and Wylie's translations combines notations from thirteenth-century Chinese algebra and the Western alphabet. For the first twenty letters of the alphabet (*a, b, c,* etc.) the ten heavenly branches and the twelve earthen stems are used.[18] The four characters for "objects" (*wu* 物), "heaven" (*tian* 天), "earth" (*di* 地) and "men" (*ren* 人) correspond to the unknowns w, x, y and z in the source text. It is particularly interesting to note that in the "four elements" (*si yuan* 四元) algebra, as attested to in a 1303 Yuan dynasty treatise, the expressions "heaven," "earth," "men" and "objects" denote the first, second, third and fourth unknown in the algorithmic descriptions of solution procedures. Yet the nineteenth-century translators distorted the classical order from *tian, di, ren, wu* to *wu, tian, di, ren.* This seems to have been a deliberate choice, either in order to follow the (unwritten) convention that x, just like *tian*, represented the first unknown, or to obtain a phonetically more coherent sequence with the alphabetical subset w, x, y, z. Letters from the Greek alphabet were translated by simply using the characters of twenty-six out of the twenty-eight lunar mansions (*xiu* 宿) in Chinese astronomy.

There was a further mode of creating mathematical symbols that was related to the characteristics of the Chinese script. It consisted in composing and decomposing Chinese characters. For the respective capital letters of the Latin and Greek alphabets, a character-symbol was created by adding the mouth radical (*kou* 口) to the left of the corresponding character. For $\int$ (integral) and d (differential) the leftmost radicals 彳 (*chì*)[19] and 禾 (*hé*)[20] of the respective Chinese characters for

[17]Li and Wylie (1859) *Fanli* 凡例 (Guide to the Reader), p. 1A.

[18]In Li Ye's 李冶 (1192–1279) *Sea Mirror of Circle Measurements* (*Ceyuan haijing* 測圓海鏡, 1248) and earlier Jesuit translations on Euclidean geometry and trigonometry, the heavenly branches, the earthen stems and the lunar mansions were already used to designate points in geometrical figures. Cf. Jami (1990) p. 100–101.

[19]Literally meaning "step with left foot."

[20]Literally meaning "cereal" or "grain."

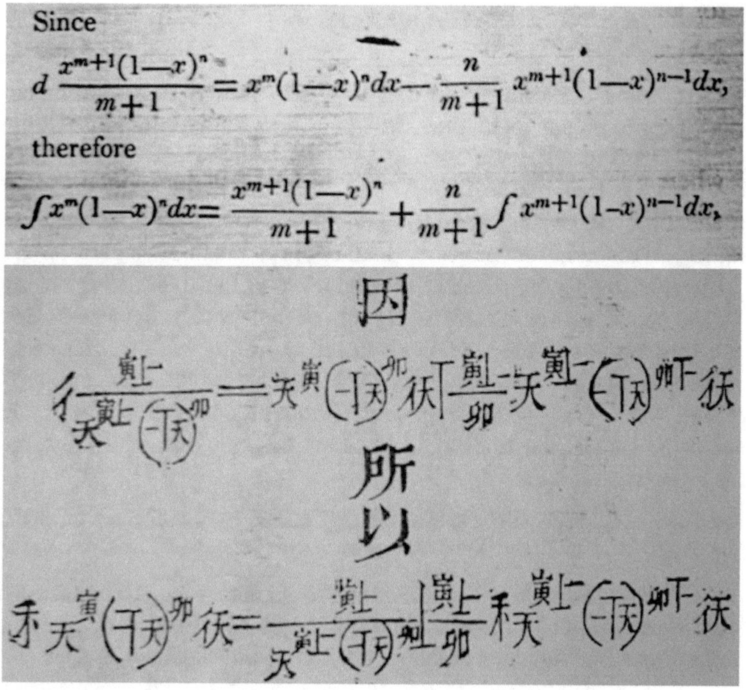

Since

$$d\,\frac{x^{m+1}(1-x)^n}{m+1} = x^m(1-x)^n dx - \frac{n}{m+1}\,x^{m+1}(1-x)^{n-1}dx,$$

therefore

$$\int x^m(1-x)^n dx = \frac{x^{m+1}(1-x)^n}{m+1} + \frac{n}{m+1}\int x^{m+1}(1-x)^{n-1}dx,$$

Fig. 4.1 Equations from Galloway (1839), translated in Fu and Hua (1896)

the concepts of "integral" (*wei* 微) and "differential" (*ji* 積) served as symbolic translations (Fig. 4.1).

As for the orientation of formulas and text, the page layout becomes problematic, and sometimes only one line of vertical text fits on a page, when a formula written horizontally occurs. But this also means that with these new conventions for two kinds of discourse, rhetoric and formulaic, we have two concurrent ways of representing numbers, i.e. numbers represented by Chinese characters, and used either for stating numerical values or for carrying out written calculations:

Method of noting numbers
Concerning the positional arrangement of the characters for notating numerals, there are two methods: one is in vertical layout, one is in horizontal layout.
Numbers laid out vertically are suitable in literary style, numbers laid out horizontally are suitable for calculations. Each method shall be used according to what is suitable. That is the reason why in books on mathematical methods numbers are noted vertically in explanations and problem statements, but in mathematical formulas horizontal layout must be used.[21]

---

[21] Hua (1885) vol. 1, p. 5A.

4.3 Western and Chinese Algebra

Following most of the terminology and symbolic transcriptions developed by Li Shanlan, Hua Hengfang took down in Chinese and translated with John Fryer (1839–1928) another text on algebra in 1873, the *Daishu shu* 代數術 (*The Art of Algebra*).[22] It was based on William Wallace's (1768–1843) article on *Algebra* for the eighth edition of the *Encyclopaedia Britannica*.[23]

With regard to the introduction of operational symbols, Hua's translation differs from Li Shanlan's only for division, since the use of the sign "÷" is not adopted. Instead, the same proto-grammatical layout for divisor and dividend is borrowed from the way fractions were already displayed in Li Shanlan's translations. In the Chinese preface to *The Art of Algebra*, Hua Hengfang discusses the utility of using the decimal system and algebra in general. Although reluctant to make an explicit judgement, he frames the translation in terms of a ranking of a Western and a Chinese algebraic tradition:

> Those who read the book will be able to know by themselves the difference or similarity, the superiority or inferiority of [Western] algebra and the "celestial element" method.[24]

Competition between Western and Chinese algebra was particularly fierce in nineteenth-century China following the rediscovery of several Song and Yuan dynasty texts where mathematical problems are solved with the much praised "celestial element" (*tian yuan* 天元) algebra.[25] In this context, the historical introduction from the original *Britannica* entry on algebra, which answers "a question much agitated, at what period and in what country [algebra] was invented"[26] by entirely ignoring the Chinese algebraic contributions from the Song and Yuan dynasties, might not have pleased Chinese scholars. Hua and Fryer have not made explicit whether this was the reason they chose not to translate the historical introduction, but it is interesting to note that they did not omit the historic part in their translation of the *Britannica* entry on probability theory.[27] This was a mathematical field that

---

[22] Fu and Hua (1873).

[23] Wallace (1842a). Fryer ordered a complete edition of the *Encyclopaedia Britannica* at Smith Elder and Company on 31 July 1868. According to Wright (1996) the relevant letter by Fryer is to be found in: Letter Journal No. 1, Fryer Papers, Bancroft Library. Several book length articles of the 8th edition of the *Encycloaedia Britannica* are subsequently translated into Chinese:

- Algebra (Fu and Hua 1873), translated from Wallace (1842a) or the identical essay in the 8th edition (Wallace 1853)
- Fluxions (Fu and Hua 1875), translated from Wallace (1842b) or Wallace (1855)
- Fortification (Fu and Xu 1880), translated from Portlock (1861)
- Probability (Fu and Hua 1896) (Fig. 4.2), translated from Galloway (1839) or the identical text in the 8th edition.

[24] Preface to Fu and Hua (1873) p. 2B.

[25] See Tian (1999).

[26] Wallace (1842a) p. 420.

[27] Fu and Hua (1896).

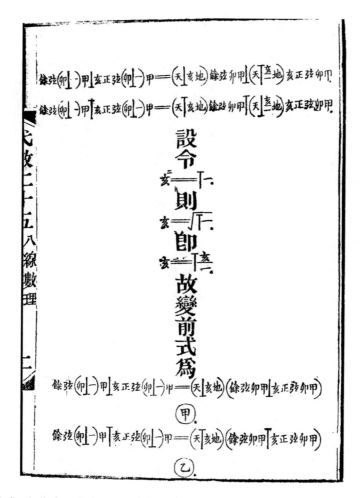

Fig. 4.2 Symbolic formula in Fu and Hua (1873), scroll 25

had no precedent in China, and historiographic incorrectness was thus less of a problem.

What the translators seem to have considered a crucial beginning for *The Art of Algebra* was a section on the "Notation and Explanation of the Signs" (Fig. 4.3). Hua and Fryer adapted it to the terminological choices of earlier translators by drawing clear boundaries between a Chinese and a Western symbolic system. When comparing the original English text with its translation, it is notable that the categories "China" and the "West" do not exist in the original English version[28]:

---

[28]The English original (Wallace 1842a) p. 427 and the Chinese translation (Fu and Hua 1873), scroll 1 卷首 釋號, p. 1A are quoted first, followed by my retranslation from the Chinese version back into English in italics. Emphases in red color are mine.

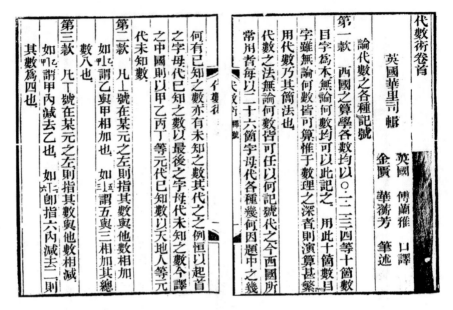

Fig. 4.3 "Notation and explanation of the signs" as translated in Fu and Hua (1873)

NOTATION AND EXPLANATION OF THE SIGNS.

論代數之各種記號

On all kinds of notations and signs in algebra

1. In arithmetic there are ten characters, which being variously combined, according to certain rules, serve to denote all magnitudes whatever. [...] In algebra quantities of every kind may be denoted by any characters whatever, but those commonly used are the letters of the alphabet; and as in every mathematical problem there are certain magnitudes given, in order to determine other magnitudes which are unknown, the first letters of the alphabet *a*, *b*, *c*, &c. are used to denote known quantities, while those to be found are represented by *v*, *x*, *y*, &c. the last letters of the alphabet.

第一款　西國之算學各數均以０一二三四等十箇數目字為本 無論何數均可以此記之　[...] 代數之法無論何數皆可任以何記號代之今西國所常用者每以二十六箇字母代各種幾何因題中之幾何有已知之數亦有未知之數其代之之例恆以起首之字母代已知之數以最候之字母代未知之數今譯之中國則以甲乙丙丁等元代已知數以天地人等元代未知數

In Western arithmetic all numbers take the 0, 1, 2, 3, 4, etc., the ten characters to denote numbers, as the basis. All numbers whatever can be noted with them. [...] In algebraic methods whatever numbers may be substituted by whatever notation. Nowadays, what in the West is commonly used are the twenty-six letters of the alphabet to substitute all kinds of magnitudes. As in [mathematical] problems there are magnitudes that are already known and also such that are not yet known, one always substitutes the already known numbers with the letters from the beginning of the alphabet and one substitutes the not yet known numbers with the letters from the end of the alphabet. Now translating this to China, one takes Jia, Yi, Bing, Ding, etc. as in Yuan dynasty for the already known numbers, and Tian, Di, Ren, etc. as in Yuan dynasty for the not yet known numbers.

What is interesting in this adaptation of the introductory section is that the choice in translating the symbols used for representing known magnitudes and unknown quantities explicitly points to the algebraic tradition of the Yuan dynasty. Li Ye 李冶 (1192–1279) and Zhu Shijie 朱世傑 (mid-thirteenth to early fourteenth century) had already employed in their writings the characters chosen by the translators as symbolic representations of givens and unknowns. In the following paragraph concerning positive and negative numbers, the emphasis on symbols understood as rhetorical devices in mathematical discourse is formulated in terms of "names" for these quantities. This linguistic aspect of mathematical symbols reveals a conceptualization of what I call speakable symbols, because the symbols' pronunciations are borrowed from natural language; quantities are given a "name" instead of a symbolic prefix:

2. The sign + (*plus*) denotes that the quantity before which it is placed is to be added to some other quantity. Thus $a + b$ denotes the sum of a and b; $3 + 5$ denotes the sum of 3 and 5, or 8.

The sign − (*minus*) signifies that the quantity before which it is placed is to be subtracted. Thus $a − b$ denotes the excess of a above b; $6 − 2$ is the excess of 6 above 2, or 4.

3. Quantities which have the sign + prefixed to them are called *positive* or *affirmative*; and such as have the sign − are called *negative*. [. . .] Quantities which have the same sign, either + or −, are said to have like signs. Thus, $+a$ and $+b$ have like signs, but $+a$ and $−c$ have unlike signs.

第二款　凡 ⊥ 號在某元之左。則指其數與他數相加。
如甲⊥乙。謂甲與乙相加也。　　如三⊥五。謂五與三相加其總數八也。
第三款　凡 丅 號在某元之左。則指其數與他數相減。
如甲丅乙。謂甲內減去乙也。　　如六丅二。謂六內減去二。則其數為四也。[. . .]
凡幾箇代數式。俱有 ⊥ 號。或俱有 丅 號者。謂之同號數。亦謂之同名數。
凡幾箇代數式。或有 ⊥ 號。或有 丅 號者。謂之不同號之數。亦謂之異名數。

2. In general, when the + sign is to the left of some unknown, then this means that this number and another number are added to each other. As in a + b, we say that a and b are added to each other. [Or] as in 5 + 3, we say that 5 and 3 are added to each other, their total is 8.

3. In general, when the − sign is to the left of some unknown, then this means that this number and another number are subtracted from each other. As in a − b, we say that b is subtracted from a. [Or] as in 6 − 2, we say that 2 is subtracted from 6. Thus its number makes 4. [. . .]

In general, when we have several algebraic expressions, of which all do carry the sign "+," or all do carry the sign "−," we call them numbers with like signs. We also call them numbers with equal names.

In general, when we have several algebraic expressions, of which some do carry the sign "+," and some do carry the sign "−," we call them numbers with unlike signs. We also call them numbers with different names.[29]

Here, symbols are thus understood as mathematically significant names given to algebraic objects. The "speakable" nature of the convention chosen for writing fractions with the divisor above the dividend has already been pointed out earlier in

---

[29]Chinese text in Fu and Hua (1873) scroll 1 卷首, p. 1B–2A. Original English text in Wallace (1842a) p. 427–428, followed by my retranslation from the Chinese version back into English in italics. Emphases in red color are mine.

relation to the Sinicized symbolism created by Li Shanlan and Wylie.[30] The same choice, following the cultural constraints of local reading and writing practices, is made in Hua Hengfang and John Fryer's translation:

> 7. The quotient arising from the division of one quantity by another is expressed by placing the *dividend* above a line, and the *divisor* below it. Thus, $\frac{12}{3}$ denotes the quotient arising from the division of 12 by 3, or 4; $\frac{b}{a}$ denotes the quotient arising from the division of b by a. This expression of a quotient is also called a fraction.
>
> 第七款　　凡幾何。以他幾何分之。記其約得之數其法作一線以界其法實。線之上為法。線之下為實。
>
> 如 $\frac{三}{十二}$。謂十二以三約之也。即謂其約得之四也。
>
> 如 $\frac{甲}{乙}$。謂置甲以乙約之。得乙分之甲也。此種之式。名之曰分數式。
>
> *7. In general, when a quantity is divided by a quantity, the rule for noting the number which one obtains by simplification is that one makes a line to delimit the divisor from the dividend. What is above the line makes the divisor, what is below the line makes the dividend.*
> *As in $\frac{3}{12}$, we say that 12 is simplified by three. That is to say that what results from simplification is four.*
> *[Or] as in $\frac{b}{a}$, we say that a is put down and simplified by b. One obtains b parts thereof a.*
> *This kind of expression, we call it the expression of a fraction.*[31]

Turning away from the kind of paratextual sections on mathematical symbolism in nineteenth-century translations, let me look more specifically at the issues at stake in the practice of algebra. Unlike conics, algebra was a mathematical field of inquiry that had its own history in China, where a tradition of solving polynomial equations in one to four unknowns is attested in sources since the Song dynasty.

Although Western algebra was transmitted to China from the seventeenth century onward, the traditional "celestial element" (*tianyuan* 天元) algebra was practised by mathematicians until the early twentieth century. It was then a topic of even greater interest because its past glory had just been rediscovered and reconstructed by Qian-Jia scholars during the second half of the nineteenth century, and it could be instrumentalized through informed argumentation in a Chinese–Western competition of scientific priority and ranking. Chinese concepts that were strikingly reminiscent of the imported algebra also stimulated further inquiries into the nature of the "celestial element" method and prevented traditional Chinese mathematics from suddenly disappearing and being replaced by modern Western science. The acceptance of a new system of algebraic notations in this field was thus a much more complex and multi-layered issue than was the case in mathematical domains like conics, calculus or probability theory, which were entirely new to Chinese scholars. Not only did the Chinese translator-mathematicians continue to do mathematics in both traditional and the new notational frameworks, but students also had to learn both styles.

---

[30] See page 81 and Li and Wylie (1859) *Fanli* 凡例 (Guide to the Reader), p. 1A.

[31] Wallace (1842a) p. 428. Chinese text in (Fu and Hua 1873) 卷首 p. 4A.

In 1880, Li Shanlan edited a collection of mathematical exercises together with the exam papers from eight of his students (*Suanxue keyi* 算學課藝), written when he was teaching mathematics at the Interpreter's College in Beijing. The institution was "founded in 1862 and—despite fierce opposition from conservative elements in the Qing court—included science teaching from 1867 on. Opponents feared that students would be corrupted by the Christian teachings of the foreign lecturers, most of whom were ex-missionaries who tended to see their educational work as part of a wider mission to Christianize the Celestial Empire."[32]

Some exercises in Book 2 were solved by the traditional algebraic "celestial element" method.[33] Yet this was clearly the concern of Book 3, which collected exercises drawn from Li Ye's 李冶 (1192–1279) *Sea Mirror of Circle Measurements* (*Ceyuan haijing* 測圓海鏡) from 1248. Li Shanlan used this work as a textbook in one of the two courses that he taught on traditional Chinese mathematics, entitled "Studying the *Sea Mirror of Circle Measurements*."[34] As late as 1876, when Li Shanlan went to Beijing as a mathematics instructor at the Interpreter's College, he acknowledged his debt to the Yuan dynasty algebraic method:

> Now I am teaching at the Interpreter's College using this book [the *Sea-mirror of the Circle Measurements*] as a textbook. My major concern is to synthesize Chinese and Western methods into one method by the way of demonstrating [the Western] algebra in the "celestial element" method.[35]

What Li also reveals here is his ultimate goal of merging the two algebraic traditions into a truly global discipline. But the chief instructor of the Interpreter's College, the American Presbyterian missionary W. A. P. Martin (chin. Ding Weiliang 丁韙良, 1827–1916), who obviously shares the ideal of an ecumenical science, relied solely upon Li Shanlan to make it become a reality. In his 1880 preface to the collection of mathematical exam papers, he says:

> It is all due to the strength of the teachings of Sir Li Renshu! Alas, synthesizing all procedures Chinese and Western, continuing the transmission from generation to generation of what the Ancients excelled in, letting [Chinese] mathematics become alive again in world mathematics, if not upon Renshu, whom could I count upon?[36]

In practice, the synthesis was not easy to form, and Li maintained separate discursive styles for solving mathematical problems in the courses he taught on algebra and on calculus. At least this is what we assume on the basis of the mathematical style adopted by his students in the published examination papers under the influence of Li's teachings. In an exercise asking to find the series expansion of $(a + x)^{1/2}$ the only solution given is in Western algebraic style using

---

[32] Quoted from Wright (1998) p. 661.

[33] Book 2 contained exercises about geometry, trigonometry, continued proportions and the summation of finite and infinite series.

[34] The other course was: Studying the *Nine Chapters*. Western mathematics were taught in the courses on Arithmetic, Algebra, and Differential and Integral Calculus.

[35] Quoted from Horng (1991) p. 61.

[36] Xi and Gui (1880) Preface, p. 2B–3A.

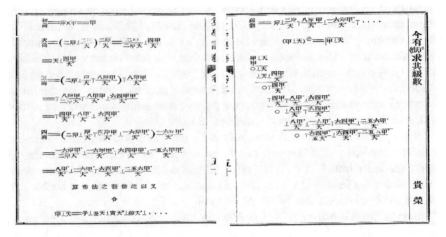

Fig. 4.4 Series expansion of $(a + x)^{1/2}$ in Xi and Gui (1880) vol. 2, p. 50A–50B

the syncretistic symbolism that Li Shanlan had developed as a translator of Western mathematical books earlier in his life (see Fig. 4.4). The solution given here is the correct infinite series:

$$\sqrt{a} + \frac{x}{2\sqrt{a}} - \frac{x^2}{8a\sqrt{a}} + \frac{x^3}{16a^2\sqrt{a}} - \cdots$$

Yet it seems that the student proceeded synthetically, showing that the series found is indeed the one required. First, he identifies the exponent $\frac{1}{2}$ of the binomial expression with the operation of root extraction, before showing that when squaring the series one indeed gets $a + x$. The consecutive products of partial sums and one term resulting from mutual multiplications of the terms are named first, second, third, etc. quotients (*chu shang* 初商, *ci shang* 次商, *san shang* 三商, etc.) as in traditional root extraction methods for polynomials and calculated on the left side of Fig. 4.4:

$$s_1 = \sqrt{a} \cdot \sqrt{a} = a$$

$$s_2 = \left(2\sqrt{a} + \frac{x}{2\sqrt{a}}\right)\frac{x}{2\sqrt{a}} = x + \frac{x^2}{4a}$$

$$s_3 = \left(2\sqrt{a} + \frac{x}{\sqrt{a}} - \frac{x^2}{8a\sqrt{a}}\right)\frac{-x^2}{8a\sqrt{a}} = -\frac{x^2}{4a} - \frac{x^3}{8a^2} + \frac{x^4}{64a^3}$$

$$s_4 = \left(2\sqrt{a} + \frac{x}{\sqrt{a}} - \frac{x^2}{4a\sqrt{a}} + \frac{x^3}{16a^2\sqrt{a}}\right)\frac{x^3}{16a^2\sqrt{a}} = \frac{x^3}{8a^2} + \frac{x^4}{16a^3} - \frac{x^5}{64a^4} + \frac{x^6}{256a^5}$$

The scheme on the right side of Fig. 4.4 is one of consecutive subtractions of these quotients from $a + x$ and shows implicitly by incomplete induction[37] that all terms are subsequently cancelled out, thus that $a + x - s_1 - s_2 - s_3 - s_4 - \cdots = 0$ if we continue the process to infinity:

$a + x$

a
_____

$0 + x$

$+x + \frac{x^2}{4a}$
_____

$0 - \frac{x^2}{4a}$

$-\frac{x^2}{4a} - \frac{x^3}{8a^2} + \frac{x^4}{64a^3}$
_____

$0 + \frac{x^3}{8a^2} - \frac{x^4}{64a^3}$

$+\frac{x^3}{8a^2} + \frac{x^4}{16a^3} - \frac{x^5}{64a^4} + \frac{x^6}{256a^5}$
_____

$0 - \frac{5x^4}{64a^3} + \frac{x^5}{64a^4} - \frac{x^6}{256a^5}$

$\cdots$

It is only at the end (lower left of Fig. 4.4), that the student remarks that one could also apply "the method of indeterminate coefficients" (*fan beishu fa* 泛倍數法)[38] to obtain the series:

$$\sqrt{a + x} = b + cx + dx^2 + ex^3 + \cdots$$

but he does not specify the algebraic form of the coefficients b, c, d, e, etc., nor how to obtain them.

For other kinds of problems, in particular those borrowed from Song and Yuan dynasty texts, two solutions—one using the "celestial element" method,[39] one using Western algebra—are published. Here is an example borrowed from Zhu Shijie's 朱世傑 *Jade Mirror of Four Elements* (*Si yuan yujian* 四元玉鑑, 1303):[40]

_____

[37]For a detailed discussion of this mode of implicit argumentation, see Chap. 5.

[38]It is interesting to note that this was the term used by Fryer and Hua Hengfang in their Chinese translation of Wallace (1853) p. 529 (see Fu and Hua 1873, scroll 7, p. 4B *et passim*), but not the term *weiding zhi xishu* 未定之係數 created by Li Shanlan and Wylie in their translation of De Morgan (1835) and which is the expression still in use today for indeterminate coefficients. See Li and Wylie (1872) scroll 8 *et passim*.

[39]This involves the use of rod numerals to note the coefficients of polynomials and to actually perform calculations. On how numbers were represented with counting rods, see for example (Martzloff 1997) p. 185–190.

[40]Problem II(10)5 in middle scroll 卷中, chap. "Number of calls according to images" (*Ru xiang zhao shu* 如象招數). For a complete translation of the problem in German, see Bréard (1999) p. 415–417, for a Chinese edition with English translation, see Zhu (2006) vol. 2, p. 449–455.

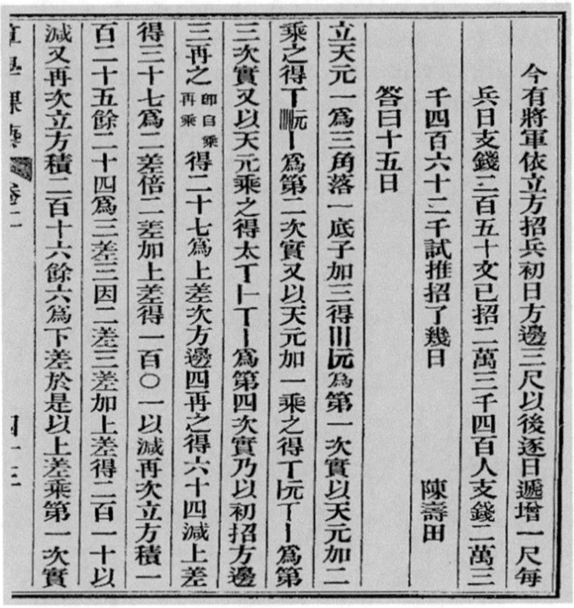

今有將軍依立方招兵初日方邊三尺以後逐日遞增一尺每
兵日支錢二百五十文已招二萬三千四百人支錢二萬三
千四百六十二千試推招了幾日　陳壽田
答曰十五日
立天元一爲三角落一底子加三得川元爲第一次實以天元加二
乘之得丅‖元丨爲第二次實又以天元加一乘之得丅元丅丨爲第
三次實又以天元乘之得太丅丄丅丨爲第四次實乃以初招方邊
三再之即自乘再乘得二十七爲上差次方邊四再之得六十四減上差
得三十七爲二差倍二差加上差得一百〇一以減再次立方積一
百二十五餘二十四爲三差三因二差三差加上差得二百一十以
減又再次立方積二百十六餘六爲下差於是以上差乘第一次實

Fig. 4.5 Solution with the traditional "celestial element" (*tianyuan* 天元) method in Xi and Gui (1880) vol. 2, p. 43A

Now we have a general, who calls for soldiers according to a cube. On the first day each side of the cube measures 3 *chi*. Afterwards day by day the side of the cube grows by one *chi*. Each of the soldiers is accorded a daily amount of 250 *wen* cash. 23.400 men have already been called; they have obtained 23.462 *qian* [= 23.462.000 *wen*] cash. Deduce for how many days [soldiers] were called.
Chen Shoutian
The answer says: 15 days.[41]

The solution given in traditional style (see Fig. 4.5) asks to establish the unknown, the "celestial element" (*tianyuan* 天元), as the number of days n corresponding to the same unknown in the following two finite series[42]:

$$\sum_{k=1}^{n} a_k = \sum_{k=1}^{n} [3+(k-1)]^3 = 3^3+4^3+\cdots+[3+(n-1)]^3 = 23400\ [men]$$

$$250\ [wen] \cdot \sum_{k=1}^{n} \left(\sum_{i=1}^{k} a_i \right) = 23462\ [guan]$$

$$= 250\ [wen] \cdot \left\{ 3^3+(3^3+4^3)+\cdots+\left(3^3+\cdots+[3+(n-1)]^3\right) \right\}$$

[41] Xi and Gui (1880) vol. 2, p. 32A.

[42] *Guan* and *wen* are monetary units for coins. 1 *guan*, lit. a string [of coins] corresponds to 1000 *wen* [cash].

With $m = n - 3$ the first statement about the total number of soldiers leads to the equation:

$$m^4 + 22 \cdot m^3 + 181 \cdot m^2 + 660 \cdot m - 92736 = 0$$

and the second statement related to the overall payment of the same soldiers gives:

$$3 \cdot m^5 + 90 \cdot m^4 + 1075 \cdot m^3 + 6390 \cdot m^2 + 18362 \cdot m - 5610840 = 0.$$

Of course, this is not what we find in the student's solution, who expresses rhetorically how he found the polynomial equation corresponding to the stated problem:

> The method to find the days: establish the "celestial element" (*tian yuan* 天元) as the seeds at the basis of the triangles, going one [level] down (*san jiao luo yi di zi* 三角落一底子).[43] [...] The method to find the days, [based on] the cash: establish the "celestial element" (*tian yuan* 天元) as the seeds at the basis of the triangles of scattered stars (*san jiao sa xing di zi* 三角撒星底子). Add four to obtain [4 + x]:
>
> ||||
> |元
>
> Multiply this with the "celestial element" augmented by three to obtain [$x^2 + 7x + 12$]:
>
> |=
> ‖元
> |
>
> as the first dividend. For the next dividend, multiply this with the celestial unknown augmented by two to obtain [$(x + 4) \cdot (x + 3) \cdot (x + 2) = 24 + 3x + 9x^2 + x^3$]:
>
> ‖ ≡
> ‖元
> ‖‖
> |
>
> as the second dividend. [...]

The procedure here actually corresponds to calculating four terms, multiplying these with the respective divided differences $\Delta^1 = 27$, $\Delta^2 = 37$, $\Delta^3 = 24$,

---

[43] This corresponds to the series:

$$n \cdot 1 + (n - 1) \cdot (1 + 2) + (n - 2) \cdot (1 + 2 + 3) + \cdots + 1 \cdot (1 + 2 + \cdots + n)$$

or the series of equal sum but of different factorization:

$$1 \cdot (1 + 2 + \cdots + n) + 2 \cdot (1 + 2 + \cdots + n - 1) + 3 \cdot (1 + 2 + \cdots + n - 2) + \cdots + n \cdot 1.$$

$\Delta^4 = 6$,[44] and then multiplying the cash by the smallest multiple of denominators of the four sums:

$$23462\,[guan] = 250\,[wen] \cdot \left\{ \frac{n(n+1)}{2!} \cdot \Delta^1 + \frac{(n-1)n(n+1)}{3!} \cdot \Delta^2 + \right.$$

$$\left. + \frac{(n-2)(n-1)n(n+1)}{4!} \cdot \Delta^3 + \frac{(n-3)(n-2)(n-1)n(n+1)}{5!} \cdot \Delta^4 \right\}$$

In the traditional solution this becomes with $m = n - 3$:

$$d_1 = (m+4)(m+3)$$
$$d_2 = d_1 \cdot (m+2)$$
$$d_3 = d_2 \cdot (m+1)$$
$$d_4 = d_3 \cdot m$$

$$d_1 \cdot \Delta^1 \cdot 30 + d_2 \cdot \Delta^2 \cdot 10 + d_3 \cdot \Delta^3 \cdot 10 \div 4 + d_4 \cdot \Delta^4 \div 2 = 60 \cdot 23462\,[guan] \div 250\,[wen].$$

In a second solution, operating with symbolic algebra (see Fig. 4.6), the student starts off with the equation in one unknown $x = n$:

$$\frac{23462000}{250} = \frac{x(x+1)\Delta^1}{2} + \frac{(x-1)x(x+1)\Delta^2}{6} +$$

$$+ \frac{(x-2)(x-1)x(x+1)\Delta^3}{24} + \frac{(x-3)(x-2)(x-1)x(x+1)\Delta^4}{120}$$

and ultimately obtains the following table of coefficients:

$$-22523520$$
$$+2168$$
$$+3060$$
$$+1060$$
$$+180$$
$$+12$$

---

[44]These can be obtained from the following equations:

$$1 \cdot \Delta^1 = a_1$$
$$1 \cdot \Delta^1 + 1 \cdot \Delta^2 = a_2$$
$$1 \cdot \Delta^1 + 2 \cdot \Delta^2 + 1 \cdot \Delta^3 = a_3$$
$$1 \cdot \Delta^1 + 3 \cdot \Delta^2 + 3 \cdot \Delta^3 + 1 \cdot \Delta^4 = a_4$$

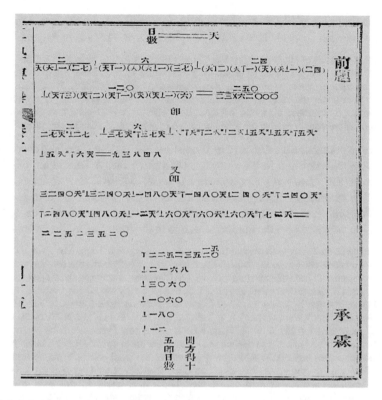

Fig. 4.6 Solution with algebraic methods in Xi and Gui (1880) vol. 2, p. 45A

which corresponds to the equation:

$$12x^5 + 180x^4 + 1060x^3 + 3060x^2 + 2168x - 22523520 = 0.$$

As in the traditional solution, there is no indication here either how the student then found the root $x = 15$ [days] to this equation. What is striking though is that both suggested methods result *in fine* in a tabular layout of the coefficients of the fifth-order equation, of which then a (positive) root needs to be extracted. The layout chosen was the vertical layout of coefficients of a polynomial equation proper to Chinese traditional algebraic techniques. It was the apparent similarity between the resulting equations obtained by the two algebraic traditions that led Chinese mathematicians to ask where precisely the differences between the newly imported and the traditional algebra were located. Hua Hengfang's personal position about apparent affinities becomes clear when one reads the jottings in his *Brush Talks on the Study of Mathematics* (*Xue suan bitan* 學算筆談, 1885). For Hua, mathematical formalism was the salient point: he claims that there were no major differences between traditional Chinese algebra and Western methods—except for notation. At least that is what he states in his essay "Explanation of Symbols in Algebra":

> Western methods of algebra are just like Chinese methods for the "celestial element" (*tian yuan*) and the "four elements" (*si yuan*). Only what is emphasized by the "celestial element" and the "four elements" is the layout in horizontal and vertical positions. Yet algebra does not bother about the order of layout in horizontal and vertical positions. Everything is elucidated by symbolic notations. All those who wish to study algebra, thus first have to memorize thoroughly each of the different symbolic notations and the nomenclature.[45]

When working as a translator of mathematical books, Hua Hengfang encountered much symbolic writing. In his essay "A Discussion of the Translation of Mathematical Books" (*Lun fanyi suanxue zhi shu* 論翻譯算學之書) we learn that the translation of symbols and figures was entirely in the hands of the Chinese brush-recorders, who had developed a rationalized logistic to speed up their work of translation and minimize the number of errors in the formulas, tables and diagrams:

> Since, when recording with the brush, when one comes upon a diagram or a formula one cannot translate one by one in detail, it should be marked clearly in the translated manuscript the occurrence of each diagram and formula. [. . .] The reason why one should proceed this way is that it is more convenient and rapid than to note one by one what is translated orally, and even, that one avoids errors. When I come across an algebraic formula in translating a mathematical book, I always mark it by a circle. The one who translates orally on the other hand encircles the formula in the Western book with a pen. Furthermore, he pronounces the word "circle" (*juan*). The one who notes with the brush draws equally a circle in the translated manuscript and should equally pronounce the word "circle." Among formulas one distinguishes big, small, long and short ones. One then should say it aloud and trace it by hand, which means to mark a long, short, big or small circle. All this facilitates their distinction. If one has a formula that is multiplied by another formula, then added to another formula, the one who translated orally should say: "there is a circle, this multiplied with a circle and added to a circle makes a circle." The one who writes notes: "there is a ○, this multiplied with a ○ and added to a ○ makes a ○."[46]

Hua Hengfang is very concerned here with the practical issues of localizing a foreign symbolic system into the Chinese cultural realm. He remains silent about the political motivations that underly such complex modes of knowledge transfer, and he gives no reason why he chose not to simply adopt mathematical symbolism as he found it in the foreign language originals he translated. This does not mean that there were no terminological debates that accompanied the translation enterprise of the late Qing, and I will dedicate the remaining part of this chapter to these debates. While, as we have seen thus far, linguistic issues seem to have played an important role in choosing a certain kind of symbolism for the newly translated mathematical texts, political issues eventually marked the end of traditional elements in mathematical discourse.

---

[45] See Hua (1885) vol. 8, p. 1A.
[46] See Hua (1885) vol. 12, p. 7A–18A.

4.3.1   A Pragmatic Choice?

The proto-grammatical system of notation which Li Shanlan had developed was used in all the Jiangnan Arsenal science texts for approximately half a century. John Fryer in 1880 defended this notation as being more likely to elicit approval among the supposedly conservative *literati*:

> What shall we say of those teachers of mathematics who insist on substituting the Arabic numerals for the Chinese throughout their text books? Is not any Chinese figure, ≡ "three," for instance, every bit as easy to read, write or print as the Arabic numeral 3? Is there any magic charm in the Arabic figures that we must drag them into our Chinese books to suit our hobbies, and to the perplexity of annoyance of the conservative Celestial mind?[47]

Fryer even justified the inversion of the Western convention for fractions, a convention to which he ascribed a faulty antiquity, since it was only during the nineteenth century that the denominator was placed above the numerator:

> Or still worse, what shall be said of those who would turn the mathematical world upside down, as far as lies in their power, by changing the time-honored and rational system of writing vulgar fractions with the denominator above and the numerator below? Is there any valid reason why a Chinese mathematician's mind should have to bear such a shock as to see fractions turned "topsy-turvy," just to suit the whim of a foreign professor? Is the practice of ages to be upset in this arbitrary manner? As well might all mathematical books be made to read from left to right because ours do, or to read from the bottom of the page to the top, so as to come to the numerators of Chinese fractions first![48]

Altogether, Fryer at that time, and in front of missionaries, defended a diplomatic strategy, respecting Chinese cultural traditions and aligning himself with the project of seventeenth-century humanists, such as Leibniz, to use Chinese as a universal language:

> The fact is that in all such trivial points we must be willing to sink our distinctive and conventional Western practices. We must carefully avoid standing in our own light if we want the Chinese to respect our Western learning. Our systems have no more right to universal use than the Chinese. Their ancient and wonderful language, which for some reasons is more suited to become the universal language of the world than any other, must not be tampered or trifled with by those who wish to introduce Western sciences.[49]

But the prevailing view among other translators and science educators was that China should not be deliberately isolated by a system of notation which was used nowhere else. Calvin Mateer (chin. Di Kaowen 狄考文, 1836–1908) , a pioneer Presbyterian missionary and founder of the first Christian college in Shandong province declared[50]:

---

[47] See Fryer's essay on *Scientific terminology: present discrepancies and means of securing uniformity* in Fryer (1890) p. 543.

[48] Ibid., p. 543.

[49] Protestant Missionaries of China (1890) p. 543.

[50] On Mateer's biography, see for example Hyatt (1971).

I differ *in toto* from Mr. Fryer in regard to the Arabic numerals and mathematical nomenclature. I consider that the effort to propagate in China a system of mathematical nomenclature, different from that which prevails in the whole civilized world outside of China, is to put a block in the way of progress, and greatly to retard the advancement of modern science in China. Those who invert fractions and introduce new signs into Chinese mathematics, are the theoretical men who sit in their studies and make books. They are not the practical men who are teaching mathematics in the school room. I have yet to hear that this system is used in a single school in China. [...] It takes thirteen strokes of the pen to write the Arabic numerals, and twenty-seven to write the Chinese numbers. In practice this makes an immense difference in the time consumed in performing mathematical operations, to say nothing of the endless confusion involved in writing Chinese numbers 一, 二, 三, in perpendicular columns over each other. I feel confident that the Chinese will never practically adopt our system of mathematics in their army and navy, and in their schools, save in connection with the figures and signs which form an integral part of it.[51]

Another missionary, the Methodist W.T.A. Barber (chin. Ba Xiuli 巴修理), brought forward more pragmatic arguments against Fryer's position. Barber insisted on the practical superiority of the Arabic numerals based upon the experience made by those who taught mathematics in China. Teachers observed that Chinese boys learned Arabic numerals faster, that it takes less strokes to write them down and that they are more distinct. Barber also advocated the use of Western symbols by questioning the notion of "the Chinese mathematician," who, in his eyes, represented a negligible portion of the Chinese population. Adepts of an insular mathematical practice using a syncretistic symbolism were an obstacle to the integration of China into the international scientific community through the universal language of mathematics:

Why perpetuate the barrier dividing the mental life of China from that of the rest of the world? If China is to join the intellectual comity of nations, let the junction be complete, and let a Chinaman, who sees a foreign mathematical book, whose words he cannot understand, yet understand without confusion the symbols common to the race.[52]

In his later years, Fryer seems to have slightly moved from his original position quoted above. When he was directing the Shanghai Polytechnic,[53] the *Curriculum of Western Studies in the Shanghai Polytechnic Institute* (*Gezhi shuyuan xixue kecheng* 格致書院西學課程) prescribed Arabic numerals and mathematical symbols for their practicality in teaching:

- In the present curriculum, one can obtain the answer to each mathematical problem regardless of the method used to solve it. But there are several problems which, when solved with Chinese methods, are not as simple as with Western methods. That is why it is appropriate to study jointly Western mathematical methods.
  本課諸數學題無論用何法演算能得答數相同為合然有數題用中法不如用西法簡便故宜兼習西法數學

---

[51] Intervention by Calvin Mateer, in Lewis et al. (1890) p. 550.

[52] Intervention by W. T. A. Barber, in Lewis et al. (1890) p. 551.

[53] The Shanghai Polytechnic opened in 1876 under the direction of John Fryer who expressed its aims "as bringing 'the Sciences, Arts, and Manufactures of the Western Nations in the most practical manner possible before the Chinese'." Quoted in Wright (1996) p. 9.

Fig. 4.7 Arabic numerals explained in the *Curriculum of Western Studies in the Shanghai Polytechnic Institute* (Fu 1895)

- In mathematics, for [writing] numbers one can use both, either Western or Chinese characters. But for calculations, Chinese characters alone are complicated. They do not equal the simple writing of Western methods. That is why, when teaching calculations, we still use Western characters for the numbers. Their form is as below [...] [see Fig. 4.7] 數學各數目字用華字西字均可演算惟華字繁果不如西法便寫故教算時仍用西字數 目其式如下 [...][54]

In spite of these explicit recommendations, we do find Chinese number words in the statements of the exercises printed in the Polytechnique's collection of mathematical problems. Figure 4.8 shows a fractional expression, with the numerators above and the denominators below, adopting Western conventions for fractions but using Chinese number words instead of Arabic numerals. There, it is asked to simplify the following expression:

$$\frac{5\frac{5}{8} + 5\frac{1}{6}}{5\frac{5}{8} - 5\frac{1}{6}} \div \frac{\left(4\frac{2}{3} - 4\frac{7}{12}\right) \cdot \frac{24}{33}}{\left(4\frac{2}{3} + 4\frac{7}{12}\right) \cdot \frac{24}{33}}$$

---

[54] Fu (1895) p. 10A.

Fig. 4.8 An expression involving fractions in Fu (1895) p. 12A

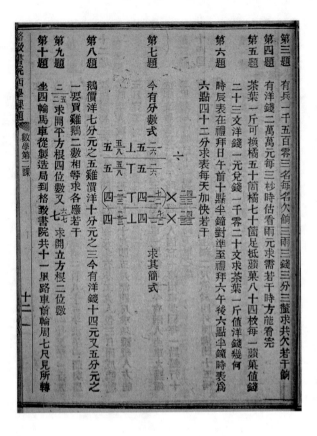

It was only at the turn of the twentieth century, after the abolition of the state examination system and the establishment of a Ministry of Education (*Xuebu* 學部) in 1905, that a decisive move was made towards universal education with mathematics forming an integral part, and that official textbooks started to promote the use of Arabic numerals more systematically (see Fig. 4.9). During the educational and political reforms of the late Guangxu reign, Chinese students were sent abroad in large numbers, mainly to Japan, for their studies. They were the new generation of translators of scientific books and government bureaucrats, who became increasingly alienated from China's traditional past. The local Education Departments in the provinces were eager for textbooks that could serve in the newly established nationwide school system after 1904 when regulations for a national school system had been drawn up. It was stipulated that "the goal was to regulate the education system and advocate the academic atmosphere from abroad."[55] Adopting

---

[55]"意在整齊國內之學制提倡海外之學風." Quoted from a September 1910 memorial of the Ministry of Education on relevant matters for preparing the constitution (*Xuebu zou yubei lixian di san nian shang jie chouban shiyi zhe* 學部奏預備立憲第三年上屆籌辦事宜折), see *XTXF* (1911).

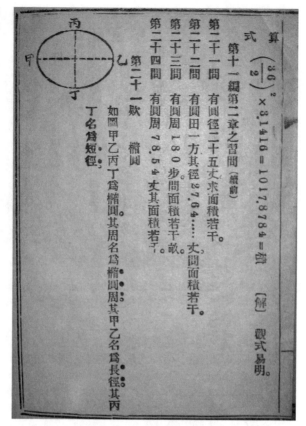

Fig. 4.9 Textbook, compiled by the Education Department (Xuewuchu 1907)

a foreign symbolic system was clearly a political choice and went well with the rejection of the old literary style that was partly made responsible for what Chinese reformers saw as a scientific backwardness in comparison with foreign nations. Thus Li Shanlan's project to synthesize Western and Chinese mathematics came to an end, a project that had been successful in forming the language in which mathematics was written for about half a century. Li's conventions for writing algebraic formulae were followed by other translators, including those who translated Li Shanlan's translations into Japanese,[56] but after the abolishment of the centuries-old civil service examinations linguistic authority in mathematics was gradually shifted from China to the West.

The criteria in choosing a syncretistic symbolism for transcribing or translating formulas from Western mathematical writings during the late Qing were, above all, linguistic criteria. The retained formalism revealed close links to natural language where mathematical symbols were pronounceable and inserted into discursive text

---

[56]Such was the case for the *Differential and Integral Calculus*. See Lee (2004).

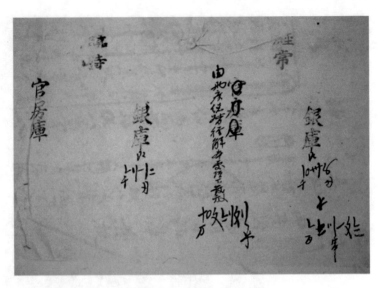

Fig. 4.10 Double checking the numbers in statistical tables from the fur and leather storage (*Piku tongji biao* 皮庫統計表), Archives of the Imperial Household Department (1909)

as part of a normal rhetoric and a syntaxically correct phrase. The newly created symbols were thus not much different from a Chinese character word: non-phonetic and ideographic in nature they were nevertheless pronounceable, speakable signs. Strictly speaking in Saussurian or Peircean terms, one cannot characterize all of them as "symbols." Some are signs whose relation to the conceptual object to which they refer is not entirely arbitrary. Instead it is motivated by association with a character denoting a mathematical concept or the idea of an operation.[57]

The nineteenth-century project of merging algebraic symbolism with Chinese characters, thereby rejecting a blind adoption of a foreign "language" before accepting it in the early twentieth century, was nevertheless a scholarly one, a project of cultural loyalty for textbooks and translations. I have not considered here those mathematical practices where the resistance against standardization to Arabic numerals continued to persist until recently. In the imperial and Republican administration, the use of Suzhou numerals in accounting books is widespread (see Fig. 4.10). This phenomenon seems like a singular linguistic continuum between the last dynasty, with its emphasis on a culture defined through its written language, and the new culture, which is expressed in vernacular language and foreign words.

---

[57] Such was the case of the characters for *shang* 上 (going up) and *xia* 下 (going down) degenerated into 丄 (plus) and 丅 (minus).

References

Bennett, Adrian Arthur (1967). *John Fryer: The Introduction of Western Science and Technology into Nineteenth-Century China*, Volume 24 of *Harvard East Asian Monographs*. Cambridge, Mass.: Harvard University Press.

Billingsley, Henry (1570). *The Elements of Geometrie of the Most Auncient Philosopher Euclide of Megara. Faithfully (now first) translated into the Englishe toung, by H. Billingsley, citizen of London. Whereunto are annexed certaine scholies, annotations, and inuentions, of the best mathematiciens, both of time past, and in this our age. With a very fruitfull præface made by M. I. Dee, specifying the chiefe mathematicall scie[n]ces, what they are, and wherunto commodious: where, also, are disclosed certaine new secrets mathematicall and mechanicall, vntill these our daies, greatly missed.* London: John Daye.

Bréard, Andrea (1999). *Re-Kreation eines mathematischen Konzeptes im chinesischen Diskurs: Reihen vom 1. bis zum 19. Jahrhundert*, Volume 42 of *Boethius*. Stuttgart: Steiner Verlag.

Bréard, Andrea (2001). On Mathematical Terminology—Culture Crossing in 19th Century China. In Michael Lackner, Iwo Amelung, and Joachim Kurtz (Eds.), *New Terms for New Ideas: Western Knowledge & Lexical Change in Late Imperial China*, Volume 52 of *Sinica Leidensia*, 305–326. Leiden, Boston, Köln: Brill.

Clavius, Christophorus (1574). *Euclidis Elementorum Libri XV Accessit XVI de Solidorum Regularium Cuiuslibet Intra Quodlibet Comparatione, Omnes Perspicuis Demonstrationibus, Accuratisque Scholiis Illustrati, ac Multarum Rerum Accessione Locupletati.* Rome: V. Accoltum.

De Morgan, Augustus (1835). *Elements of Algebra Preliminary to the Differential Calculus.* London: Taylor and Walton.

Fryer, John (1880, 29 Jan.). Account of the Department for the Translation of Foreign Books. *North China Herald*, 77–81.

Fryer, John (1890). Scientific Terminology: Present Discrepancies and Means of Secure Uniformity. *Records of the General Conference of the Protestant missionaries of China held at Shanghai May 7–20, 1890*, 531–549.

Fu, Lanya 傅蘭雅 (Fryer, John) (1880). Jiangnan zhizao zongju fanyi xishu shilüe 江南製造總局翻譯西書事略 (A brief Account of the Translation Enterprise of Western Books at the Jiangnan Arsenal). *Gezhi huibian* 格致彙編 *(The Chinese Scientific and Industrial Magazine) III*(5), 10A–12B. Continued in vols. III (6), 9A–11B and III (7), 9A–11B.

Fu, Lanya 傅蘭雅 (Fryer, John) (Ed.) (1895). *Gezhi shuyuan xixue kecheng* 格致書院西學課程 (Curriculum of Western Studies in the Shanghai Polytechnic Institute). Shanghai: [n.p.].

Fu, Lanya 傅蘭雅 (Fryer, John) and Hua Hengfang 華蘅芳 (1873). *Daishu shu* 代數術 (Algebra). Shanghai: Jiangnan jiqi zhizao zongju 江南機器製造總局. Original: (Wallace 1853), or probably the earlier edition (Wallace 1842a) which was identical.

Fu, Lanya 傅蘭雅 (Fryer, John) and Hua Hengfang 華蘅芳 (1875). *Weiji suyuan* 微積溯源 (Fluxions), 8 vols. Shanghai: Jiangnan jiqi zhizao zongju 江南機器製造總局. Original: (Wallace 1855).

Fu, Lanya 傅蘭雅 (Fryer, John) and Hua Hengfang 華蘅芳 (1896). *Jueyi shuxue* 決疑數學 (A Treatise on Probability) (周氏刊本 Zhou ed.). China: [n.p.]. Original: (Galloway 1839).

Fu, Lanya 傅蘭雅 (Fryer, John) and Xu Shou 徐壽 (ca. 1880). *Yingcheng jieyao* 營城揭要 (Fortification), 2 vols. 卷, *with illustrations appended* 附圖. Shanghai: Jiangnan jiqi zhizao zongju 江南機器製造總局. Translated from (Portlock 1861).

Galloway, Thomas (1839). *A Treatise on Probability: Forming the Article under that Head in the 7th Edition of the Encyclopaedia Britannica* (Reprint ed.). Edinburgh: Adam and Charles Black.

Guo, Shuchun 郭書春 et al. (Eds.) (1993). *Zhongguo kexue jishu dianji tonghui: Shuxue juan* 中國科學技術典籍通彙: 數學卷 (Comprehensive Collection of Ancient Classics on Science and Technology in China: Mathematical Books), 5 vols. Zhengzhou: Henan jiaoyu chubanshe 河南教育出版社.

Harbsmeier, Christoph (1998). *Language and Logic*, Volume 7 part I of *Science and Civilisation in China*. Cambridge [Eng.]; New York; Melbourne: Cambridge University Press.

Horng, Wann-Sheng (1991). *Li Shanlan: The Impact of Western Mathematics in China during the Late 19th Century*. Ph. D. thesis, Graduate Center, City University of New York.

Hua, Hengfang 華蘅芳 (1885). [*Zhong xi* 中西] *Xue suan bitan* 學算筆談 (Brush Talks on the Study of [Chinese and Western] Mathematics), 12 scrolls 卷, 4 vols. 冊 (Liangxi Hua shi zangban 梁谿華氏藏版 ed.), Volume 5 of *Xingsu xuan suan gao* 行素軒算稿. [Reprint: Suanxue guan qianyin ben 算學館鉛印本, 1902].

Hyatt, Irwin T. (1971). The Missionary as Entrepreneur: Calvin Mateer in Shantung. *Journal of Presbyterian History (1962–1985)* 49(4), 303–327.

Index (1910). *Da Qing Guangxu Xuantong xin faling fenlei zongmu* 大清光緒宣統新法令分類總目 (Classified Index to New Administrative Methods and Regulations of the Guangxu and Xuantong Reign, Qing Dynasty). Shanghai: *Shangwu yinshuguan* 商務印書館.

Jami, Catherine (1990). *Les méthodes rapides pour la trigonométrie et le rapport précis du cercle (1774)*. Mémoires des Hautes Études Chinoises. Paris: Collège de France.

Lee, Chia-Hua 李佳嬅 (2004). Jukyu seiki tō ajia ni okeru seiō sugaku no denba 19世紀東アジアにおける西欧数学の伝播 (The Transmission of Western Mathematics to East Asia in the Nineteenth Century). *Kagakushi · Kagakutetsugaku* 科学史・科学哲学 *18*, 41–54.

Lewis, W. J., W. T. A. Barber, and J. R. Hykes (Eds.) (1890). *Records of the General Conference of the Protestant Missionaries of China, held at Shanghai, May 7–20, 1890*. Shanghai: American Presbyterian Mission Press.

Li, Shanlan 李善蘭 and Alexander Wylie 偉烈亞力 (1859). *Dai weiji shiji* 代微積拾級 (Elements of Analytical Geometry and of the Differential and Integral Calculus) 18 卷. Shanghai: Mohai shuguan 墨海書館. Original by Elias Loomis 羅密士 (Loomis 1851).

Li, Shanlan 李善蘭 and Alexander Wylie 偉烈亞力 (1865). *Jihe yuanben* 幾何原本 *15* 卷 (The Elements), 15 Books (Jinling 金陵刻本 ed.). Nanjing. Includes the first six books (Ricci and Xu 1607), translation of Books VII through XV based on (Billingsley 1570), published separately under the patronage of Han Yingbi 韓應陛 in Shanghai in 1857. Reprint in (Guo et al. 1993) 5:1151–1500.

Li, Shanlan 李善蘭 and Alexander Wylie 偉烈亞力 (1872). *Daishuxue* 代數學 (Algebra) (Japanese ed.). Tokyo: Tsukamato Neikai. Original: (De Morgan 1835).

Li, Zhizao 李之藻 et al. (1965). *Tianxue chuhan* 天學初函 (First Collectanea of Heavenly Studies), 6 vols (Reprint ed.), Volume 23, First Series (Chubian 初編 卷二十三) of Wu Xiangxiang 吳相湘 (Ed.), Zhongguo 中國史學叢書. Taipei: Taiwan xuesheng shuju 臺灣學生書局.

Loomis, Elias (1851). *Elements of Analytical Geometry and of the Differential and Integral Calculus*. New York: Harper & Brothers.

Martzloff, Jean-Claude (1987). *Histoire des mathématiques chinoises; préf. de J. Gernet et J. Dhombres*. Paris [etc.]: Masson.

Martzloff, Jean-Claude (1997). *A History of Chinese Mathematics*. Berlin, Heidelberg [etc.]: Springer. Originally published in French (Martzloff 1987).

Mungello, David E. (1985). *Curious Land: Jesuit Accommodation and the Origins of Sinology*, Volume 25 of *Studia Leibnitiana: Supplementa*. Stuttgart: Steiner.

Portlock, Joseph Ellison (1861). Fortification. In *The Encyclopædia Britannica or Dictionary of Arts, Sciences, and General Literature* (8th ed.), Volume IX. Edinburgh: Adam and Charles Black.

Protestant Missionaries of China (Ed.) (1890). *Records of the General Conference of the Protestant Missionaries of China held at Shanghai, May 7–20, 1890*. Shanghai: American Presbyterian Mission Press.

Ricci, Matteo and Guangqi Xu (1607). *Jihe yuanben* 幾何原本 (The Elements). Beijing: [n.p.]. Translation of the first six books of (Clavius 1574), included in (Li et al. 1965).

Su, Ching (1996, September). *The Printing Presses of the London Missionary Society among the Chinese*. Doctor of philosophy, The School of Library, Archive and Information Studies, University College London, The University of London.

GXXF (1909). *Da Qing Guangxu xin faling* 大清光緒新法令 (New Administrative Methods and Regulations of the Guangxu Reign, Qing Dynasty).. Shanghai: Shangwu yinshuguan 商務印書館. 20 vols., indexed in (Index 1910).

Tian, Miao (1999). Jiegenfang, Tianyuan, and Daishu: Algebra in Qing China. *Historia Scientiarum 9*(1), 101–119.

Wallace, William (1842a). Algebra. In *The Encyclopædia Britannica or Dictionary of Arts, Sciences, and General Literature* (7th ed.), Volume II, 420–502. Edinburgh: Adam and Charles Black.

Wallace, William (1842b). Fluxions. In *The Encyclopædia Britannica or Dictionary of Arts, Sciences, and General Literature* (7th ed.), Volume IX, 631–708. Edinburgh: Adam and Charles Black.

Wallace, William (1853). Algebra. In *The Encyclopædia Britannica or Dictionary of Arts, Sciences, and General Literature* (8th ed.), Volume II, 482–564. Edinburgh: Adam and Charles Black.

Wallace, William (1855). Fluxions. In *The Encyclopædia Britannica or Dictionary of Arts, Sciences, and General Literature* (8th ed.), Volume IX, 670–747. Edinburgh: Adam and Charles Black.

Wright, David (1996). John Fryer and the Shanghai Polytechnic: Making Space for Science in Nineteenth-Century China. *British Journal of the History of Science 29*, 1–16.

Wright, David (1998). The Translation of Modern Western Science in Nineteenth-Century China, 1840–1895. *Isis 89*(4), 653–673.

Xi, Gan 席淦 and Gui Rong 貴榮 (Eds.) (1880). *Suanxue keyi* 算學課藝 Mathematical Examination Problems), 4 vols. 四卷. Beijing: Tongwen guan 同文館.

XTXF (1911). *Da Qing Xuantong xin faling* 大清宣統新法令 (New Administrative Methods and Regulations of the Xuantong reign, Qing Dynasty). Shanghai: Shangwu yinshuguan 商務印書館. 35 vols. Indexed in (Index 1910).

Xu, Yibao (2005). *Concepts of Infinity in Chinese Mathematics*. Doctor of philosophy, Ph.D. Program in History, City University of New York.

Xuewuchu 學務處 (Ed.) (1907). *Shuxue jiaoke shu* 數學教科書 (A Teaching Manual for Mathematics) (2nd ed.). Originally published 光緒 30 (1904).

Zhu, Shijie 朱世杰 (2006). *Si yuan yujian: Han Ying duizhao* 四元玉鑒 汉英对照 (Jade Mirror of the Four Unknowns. Chinese–English), 2 vols. Library of Chinese Classics 大中华文库. Shenyang: Liaoning jiaoyu chubanshe 辽宁教育出版社. Translated into Modern Chinese by Guo Shuchun 郭书春, translated into English by Chen Zaixin 陈在新, revised and supplemented by Guo Jinhai 郭金海.

Chapter 5
Discourse Transformed: Changing Modes of Argumentation

Contents

In modern manuals on number theory or combinatorics one finds the following equality that holds for any natural number n:

$$\sum_{j=0}^{k} \binom{k}{j}^2 \binom{n+2k-j}{2k} = \binom{n+k}{k}^2 \tag{5.1}$$

It is the so-called "Li Shanlan Identity,"[1] supposedly found in Li Shanlan's *Comparable Categories of Discrete Accumulations* (*Duoji bilei* 垛積比類) from 1867, a collection of procedures and diagrams related to the arithmetic triangle and similar forms derived from it. But the above formula is not the style in which Li Shanlan had stated the identity. It is an anachronistic introduction of present-day formalism into interpretations of the past. When analysing the representation of the "Li Shanlan Identity" in modern algebraic symbolism from a historical point of view, we find that it is the result of translingual practices and global circulations of mathematical knowledge and formalism.

---

[1] Also: "Li Renshu Identity." Renshu was the style name of Li Shanlan. See Horng (1991) p. 206.

© Springer International Publishing AG, part of Springer Nature 2019
A. Bréard, *Nine Chapters on Mathematical Modernity*, Transcultural
Research – Heidelberg Studies on Asia and Europe in a Global Context,
https://doi.org/10.1007/978-3-319-93695-6_5

Fig. 5.1 Josef Kaucký,
Remarks on a work by Turán,
*Matematicko-Fyzikálny
Časopis* (1962)

MATEMATICKO-FYZIKÁLNY ČASOPIS SAV, 12. 3. 1962

POZNÁMKA K JEDNOMU ČLÁNKU P. TURÁNA

JOSEF KAUCKÝ, Bratislava

1. P. Turán v článku [1] dokazuje kombinatorickou identitu

$$(1) \qquad \sum_{j=0}^{k} \binom{k}{j}^2 \binom{n + 2k - j}{2k} = \binom{n + k}{k}^2,$$

kterou v roce 1867 bez důkazu uveřejnil čínsky matematik Le-Jen Shoo.

In 1937, the Hungarian mathematician György Szekeres took refuge in Shanghai in the hope of escaping the rising Nazi threat. There, he happened to meet Zhang Yong 章用(1911–1939), a young polyglot Chinese mathematician who was born in Aberdeen, Scotland and who had studied mathematics at Göttingen University in Germany. Influenced by Otto Neugebauer (1899–1990), the eminent historian of mathematics and astronomy, the young Chinese mathematician took up Chinese pre-modern mathematics and reported his findings to his Hungarian friend. Szekeres was particularly impressed by the then unheard of combinatorial formula.[2] Several mathematicians provided proofs for it: Zhang Yong was first,[3] and *in fine* the identity found its way into contemporary combinatorial manuals.

The Hungarian number theorist Pál Turán (1910–1976), for example, proved the identity through the properties of Legendre polynomials in a 1954 article (in Hungarian), and Josef Kaucký (1895–1982), a Czechoslovakian mathematician, gave an elementary proof using a simple substitution of variables and by taking only the binomial theorem into account (Fig. 5.1). But why was it that the "Li Shanlan Identity" needed to be complemented by a proof? Did Li not provide any proof or, if he did, was it not communicated to the European mathematicians? Actually, Li's treatise did not contain any proofs that correspond to Western standards of rigour, nor to the formal language of algebra.

What Li gave instead, rhetorically, are the explicit cases for the sum of the first diagonals of a squared arithmetic triangle, of which he provides a diagram, before stating that the general sum of any diagonal "can be induced by analogy" (lit. *ke leitui* 可類推). Figure 5.2 shows the squared arithmetic triangle both in its original and its transcribed version. For example, the diagonal which I coloured in red shows the squares of the natural numbers, $1^2, 2^2, 3^2, \ldots$, and the diagonal in blue depicts the squares of the triangular numbers $1^2, 3^2, 6^2, 10^2, 15^2, \ldots$. Li Shanlan's identity is therefore a statement about the partial sums in any such a diagonal.

In addition to this squared arithmetic triangle, Li Shanlan illustrates the numbers in its cells constructively in the form of figured numbers, or pebble diagrams where

---

[2]See Martzloff (2006) p. 341–342.
[3]See Zhang (1939).

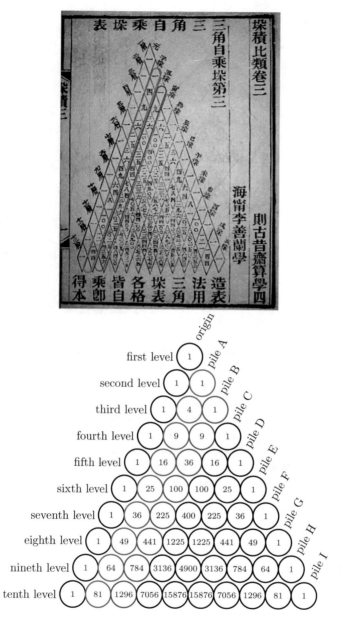

Fig. 5.2 Squared Arithmetic Triangle in Li (1867), scroll 3, where each cell is $(C_k^n)^2$

Fig. 5.3 Pebble diagrams related to the squared arithmetic triangle in Li (1867)

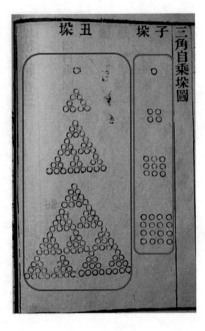

each circle represents a numerical unit (Fig. 5.3).[4] The quadratic numbers 1, 4, 9 and 16 in the red diagonal[5] are depicted on the right as squares with 1, 2, 3 and 4 pebbles as their side. The first four squared triangular numbers in the blue diagonal, 1^2, 3^2, 6^2 and 10^2 are shown as triangles of triangles on the left.[6]

The statement of the so-called "Li Shanlan Identity" is provided on the pages following these diagrams. Figure 5.4 shows what the "Identity" looks like in Li's 1867 book where the reader is invited to recognize a regular pattern after having been given rhetorically for the sum of the first n cells in a diagonal something equivalent to these four consecutive instances:

$$\sum_{k=1}^{n} \binom{k}{1}^2 = 1 \cdot \binom{n+2}{3} + 1 \cdot \binom{n+1}{3}$$

$$\sum_{k=2}^{n+1} \binom{k}{2}^2 = 1 \cdot \binom{n+4}{5} + 4 \cdot \binom{n+3}{5} + 1 \cdot \binom{n+2}{5}$$

---

[4]For an extensive discussion of the strands of the Chinese tradition of considering piles of discrete objects in different geometric shapes as figured numbers, see Bréard (1999a). On figured numbers (and in extension, polygonal numbers) in the Greek tradition, see Diofanto (2011) p. 39–40.

[5]These correspond to the coefficients $(C_k^{n+k})^2$ for the values $k = 1$ and $n = 0, 1, 2$.

[6]The squared triangular numbers correspond to the binomial coefficients $(C_k^{n+k})^2$ for the values $k = 2$ and $n = 0, 1, 2, 3$.

Fig. 5.4 Formulation of the "Li Shanlan Identity" in Li (1867)

三角自乘垛者三角垛逐層皆自乘也子垛爲一乘垛逐
層自乘之其積丑垛爲二乘垛逐層自乘之其積寅垛爲
三乘垛逐層自乘之其積卯垛以下可類推
三角自乘垛有層求積術
子垛有方一隅一方以層爲高隅以層減一爲高各以三
角二乘垛求積術入之
丑垛有方一廉四隅一方以層爲高廉以層減一爲高各以三角四乘垛求積術入之
以層減二爲高各以三
寅垛有方一甲廉九乙廉九隅一方以層爲高甲廉以層
減一爲高乙廉以層減二爲高隅以層減三爲高各以三

$$\sum_{k=3}^{n+2}\binom{k}{3}^2 = 1\cdot\binom{n+6}{7} + 9\cdot\binom{n+5}{7} + 9\cdot\binom{n+4}{7} + 1\cdot\binom{n+3}{7}$$

$$\sum_{k=4}^{n+3}\binom{k}{4}^2 = 1\cdot\binom{n+8}{9} + 16\cdot\binom{n+7}{9} + 36\cdot\binom{n+6}{9} + 16\cdot\binom{n+5}{9} + 1\cdot\binom{n+4}{9}$$

$$\sum_{k=j}^{n+j-1}\binom{k}{j}^2 = ?$$

By colouring the coefficients that correspond to the red and blue diagonals shown in Fig. 5.2 I indicated the relation to the numbers found in the diagonals of the squared arithmetic triangle. A regular pattern appears and can easily be recognized to conjecture the general expression for the partial sum of any diagonal in the triangle. Li's invitation to think the problem through to the general case would certainly have pleased George Pólya (1887–1985). Li's challenge suits Pólya's ideas about teaching the art of plausible reasoning in mathematics: good guesses that precede rigorous mathematical proofs. Had he known about the "Li Shanlan Identity" in its authentic version, he would certainly have included it as an example of pattern recognition as a heuristic in number theory in his renowned book on *Induction and Analogy* from 1954.[7]

From Polya's point of view, as from the point of view of contemporary mathematics, Li's approach only precedes a rigorous mathematical proof, but is in itself not a proof. I will argue in the following that this was not the case in the eyes of late Qing Chinese mathematicians. Their standards of mathematical validity had a different format: modes of inductive argumentation were both visual and rhetorical, and the authority of these modes of argumentation was built upon both Chinese philosophical and foreign elements. In my analysis, I will focus on analogy and the role it played in the Chinese mathematical tradition. From a comparative and transcultural perspective, I will investigate the *Comparable Categories of Discrete Accumulations* by Li Shanlan, a Chinese mathematician who was familiar with foreign mathematical discourse through his translation work and who systematically proceeded in his number-theoretical writings by "analogical extension" (*leitui* 類 推), to infer from the particular to the universal. To conclude, I will discuss the historiographic problem of claims about a specifically Western scientific rigour of inductive proofs in Blaise Pascal's *Traité du Triangle Arithmétique* (1665).[8]

5.1   The Concept(s) of "Comparable Categories"

Many Chinese mathematical writings contain the expression *bilei* 比類 in their title. Generally translated as "analogy" in the mathematical context,[9] it literally designates "comparable categories" or arrangements "according to the classes" and can be found in early Chinese texts such as *The Classic of Rites* (*Liji* 禮記, Warring States 475–221 BCE). In the chapter "Proceedings of Government in the Different Months" (*Yue ling* 月令), for example, one finds the following reference to quantitative similarity between different classes:

> In this month, orders are given to the officers of slaughter and prayer to go round among the victims for sacrifice, seeing that they are entire and complete, examining their fodder and grain, inspecting their condition as fat or thin, and judging of their looks. They must arrange them according to their classes. In measuring their size, and looking at the length (of their horns), they must have them according to the (assigned) measures. When all these points are as they ought to be, God will accept the sacrifices.

---

[7]Pólya (1954).

[8]The treatise was composed towards the end of 1654 and published posthumously, together with others connected to it, in 1665 under the complete title *Traité du Triangle arithmétique, avec quelques autres petits traités sur la même matière*.

[9]See, for example Martzloff (2006) p. 20, 158 and 342 and Martzloff (1993) p. 165.

是月也，乃命宰祝，循行犧牲，視全具，案芻豢，瞻肥瘠，察物色。必比類，量小大，視長短，皆中度。五者備當，上帝其饗。[10]

Among many other contexts, *bilei* is also found as an abstract method for diagnosis in medicine. The medical classic, the *Inner Canon of the Yellow Emperor* (*Huangdi Neijing* 黃帝內經) warns the practitioner that if one does not know how to compare categories (*bilei*), situations will not be clarified, which will ultimately lead to internal disruptions.[11] In pre-modern mathematics in China, *bilei* is more specifically a mode of structuring knowledge, where procedures or mathematical objects are grouped together following an operational similarity.[12] Although not explicitly theorized by the Chinese authors themselves as a mode of argumentation for applying the same algorithms to similar mathematical objects, I will show in the first part of this chapter that the concept of *bilei*, in its various epistemological variations from the thirteenth to the nineteenth century, has become an important instrument of plausible inference and thus for producing valid mathematical knowledge.

A word needs to be said here on the notion of "analogy" in order to clarify the conceptual differences and eventual influences between foreign and native mathematics in China. One of the meanings of the Greek word ἀναλογία, from which the word "analogy" originates, is "proportion." In fact, the system of the two numbers 6 and 9 is analogous to the system of the two numbers 10 and 15 in so far as the two systems agree in the ratio of their corresponding terms,

$$6 : 9 = 10 : 15.$$

Proportionality, or agreement in the ratios of corresponding parts, which we may see intuitively in geometrically similar figures, is a very suggestive case of analogy.[13]

Cases of such kind of analogy can be found in China after the introduction of the Euclidean theory of proportions, for example in the *Essence of Numbers and their Principles* (*Shuli jingyun* 數理精蘊, 1723).[14] In a chapter on "Curvilinear figures"

---

[10]Translation from Legge (1885) Section III. Part II. 9. Another example can be found in the chap. "Record of Music" (*Yue ji* 樂記). See Legge (1885) sec. II.14.

[11]See Wang (2012) p. 85–93, here p. 92.

[12]For the earlier significance of "analogy" in mathematics in China as a form of reasoning based on examples, see Volkov (1992).

[13]Schironi (2007) shows that the originally mathematical concept of analogy was later taken over by Greek grammarians.

[14]On the genesis of the *Shuli jingyun*, which extends from 1690 to 1723, see Jami (1998) p. 118–119: Between 1690 and appr. 1695 Jesuits—in particular by Antoine Thomas (1644–1709) and Jean-François Gerbillon (1654–1707)—wrote manuals for the emperor, essentially in three fields: arithmetic, algebra and geometry. From 1713 on, these texts were revised and rewritten by Chinese scholars working in the Academy of Mathematics (*Suanxue guan* 算學館), an institution created especially at the *Mengyang zhai* 蒙養齋—a Bureau where scholarly books were compiled under imperial patronage—for the compilation of the *Origins of musical harmony and the calendar (Lüli*

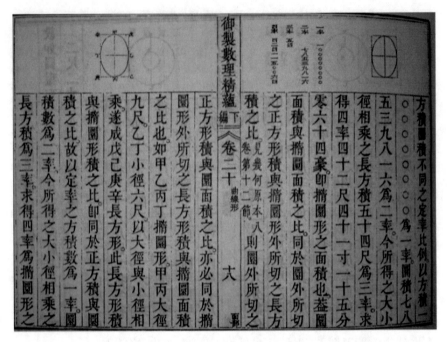

Fig. 5.5 *Collected Essentials of Mathematical Principles* (*Shuli jingyun* 數理精蘊) Yunzhi 允祉 (1723)

(*quxian xing* 曲線形) the surface of the ellipse is calculated following a proportional argument equalizing the ratio between the surface of an ellipse and its circumscribed rectangle with the ratio of a circle and its circumscribed square (Fig. 5.5):

> In general, the ratio (*bi* 比) between the surface of a circle and the surface of an ellipse is equal to the ratio between the surface of a square circumscribed to the circle and the surface of a rectangle circumscribed to an ellipse.
> 盖圜面積與橢圜面積之比。同於圜外所切之正方形積與橢圜形外所切之長方積之 比。[15]

But in traditional Chinese mathematics one does not typically find this kind of proportional analogy. Objects are rather classified or explained by being likened or assimilated to one or another, and the related problems are then solved with similar or even identical procedures. Yang Hui's 楊輝 (ca. 1238–1298) *Detailed Explanations of the Nine Chapters on Mathematical Methods* (*Xiangjie jiu zhang suanfa* 詳解九章算法) printed in 1261 contains mathematical problems from the

---

yuanyuan 律曆淵源), of which the *Essence of Numbers and their Principles* were the mathematical part.
[15]Translated from Yunzhi 允祉 (1723) vol. 20, p. 18A and already quoted on page 38. I provide the quote again here because my focus is on argumentation rather than on the mathematical aspects of the ellipse.

Fig. 5.6 Yang Hui's *Detailed Explanations of the Nine Chapters on Mathematical Methods* Yang (1261)

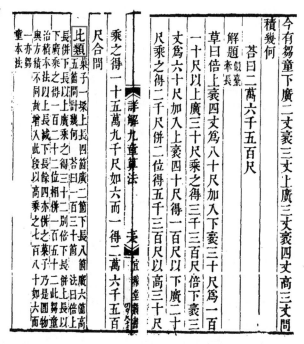

Nine Chapters that are rearranged and complemented by problems of "comparable categories" (see Fig. 5.6). The following example illustrates the procedural similarity between the new problem of "comparable category" and its counterpart from the canonical book. The *Nine Chapters* give the procedure for calculating the volume of a truncated pyramid with a rectangular base, a so-called *chutong* 芻童 (lit. a haystack). Yang Hui adds a problem of "comparable category" where the same geometric shape (turned upside down) is not a solid volume but built out of discrete elements. The question then is not about the volume, but rather about the total number of countable unitary objects that are piled up in the shape of a truncated pyramid with a rectangular base. Yang Hui even makes explicit, in the solution procedure, which sequences of operations remain unchanged, and which operations modify the original method:[16]

> Let us suppose that we have a *chutong*. Its lower breadth is 2 *zhang*, its width 3 *zhang*, its upper breadth is 3 *zhang*, its width 4 *zhang*, its height is 3 *zhang*. The question is: what is its [volume-] product?
>
> The answer says: 26500 [cube] *chi*.[17]
>
> The explanation of the problem: it is similar (*si* 似) to an observation platform, extended in length.

---

[16]Translated from Yang (1261) p. 78A–78B.

[17]Measure of length, where 1 *zhang*=10 *chi*.

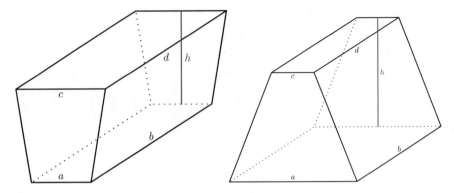

Fig. 5.7 Two *chutong* of comparable category in Yang (1261)

The calculation says: Double the upper width of 4 *zhang*, it makes 80 *chi*. Add the lower width 30 *chi*, it makes 110 *chi*. With the upper breadth 30 *chi* multiply this, one obtains 3300 *chi*. Double the lower width 3 *zhang*, it makes 60 *chi*. Add the upper width 40 *chi*, one obtains 100 *chi*. With the lower breadth 20 *chi* multiply this, one obtains 2000 *chi*. Add together the two positions, one obtains 5300 *chi*. With the height 30 *chi* multiply this, one obtains 159000 *chi*. When dividing by 6, one obtains 26500 *chi*. This corresponds to what was asked for.[18]

The comparable category (*bilei* 比類)[19]: A pile of seeds. An upper length of 4, a breadth of 2, a lower length of 8, a breadth of 6, a height of 5. The question is: how many are there altogether?   The answer says: 130.   The method says: Double the upper length, add the lower length. With the upper breadth multiply this, one obtains 32. Furthermore, double the lower length, add the upper length. With the lower breadth multiply this, one obtains 120. The two positions added together, 152. This is the original method that governs the [volume-] product of a *chutong*. With the upper length reduce the lower length, the remainder 4 is also added. A pile of seeds then is not equal to a circular object, nor to a rectilinear [volume-] product. That is why one has to add this section. With the height multiply this, 780. Dividing by 6 is also the original method of a *chutong*.[20]

This kind of textual organization, where ancient mathematical problems are rearranged based upon procedural and structural similarities, is more systematically applied by Yang Hui in his book *Fast methods of multiplication and division compared to [various] categories of fields and [their] measures (Tianmu bilei chengchu jiefa* 田畝比類乘除捷法) of 1275 and reprinted in Korea in 1433.

---

[18]The procedure here corresponds to the following calculation (for the various dimensions, see Fig. 5.7 to the left):

$$A = [(2d + b) \cdot c + (2b + d) \cdot a] \cdot h \div 6.$$

[19]Text framed in red in Fig. 5.6.

[20]The procedure of "comparable category" here corresponds to the following calculation (for the various dimensions, see Fig. 5.7 to the right):

$$[A = (2d + b) \cdot c + (2b + d) \cdot a + (b - d)] \cdot h \div 6.$$

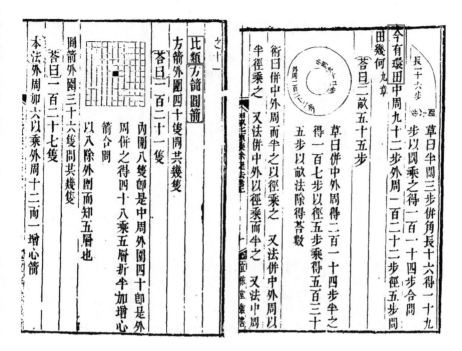

Fig. 5.8 Yang Hui's *Fast methods of multiplication and division compared to [various] categories of fields and [their] measures* Yang (1842)

Figure 5.8, for example, shows a "ring field" (*huan tian* 環田) accompanied by the "comparable categories" of "square arrows" and "round arrows" (*fang jian yuan jian* 方箭圓箭) bundled together. Here, the procedure of the continuous geometric surface needs no major modification, but requires rather a calculation trick. Yang Hui considers, for example, the cross-section of a bundle of round arrows without an arrow at its centre as comparable to a ring. The inner diameter of such a bundle then corresponds to the number of arrows around the missing, central one. He adds it to the outer diameter, i.e. the given number of arrows along the outer edge of the bundle, divides by two and multiplies with the number of layers. In the end he simply adds the central arrow.

Yang Hui's undertaking of reorganizing ancient mathematical knowledge and associating it with new objects created a new mathematical discipline in China, which remained a hot topic until the late nineteenth century, the so-called domain of "discrete accumulations."[21] It evolved out of analogies to the fifth chapter of the *Nine Chapters* dealing with the volumes of different geometric solids. A look at the table of contents in Wu Jing's 吳敬 (fl. 1450) *Great Compendium of Comparable*

---

[21]For an extensive discussion of Yang Hui's contributions see Bréard (1999a) chap. 3.2, p. 119–158.

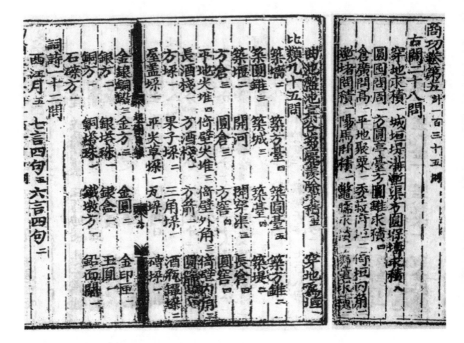

Fig. 5.9 TOC of Wu Jing's *Great compendium of Comparable Categories to the Nine Chapters on Mathematical Methods* Wu (1450)

Categories to the Nine Chapters on Mathematical Methods (*Jiu zhang suanfa bilei daquan* 九章算法比類大全) shows the many extensions of the twenty-eight canonical problems of the fifth chapter from the *Nine Chapters* (see Fig. 5.9): ninety-five problems of "comparable categories" are added, most of them interested in the total number of pebbles laid out or piled up in a certain two- or three-dimensional geometric form.

Since I am mainly concerned in this book with the development of mathematics in China in late imperial times, before Western mathematics entirely superseded the local traditions, I will skip the development of "discrete accumulations" until its almost final stage and come back to Li Shanlan's treatise from 1867 introduced at the beginning of this chapter.[22]

5.2 Li Shanlan's "Comparable Categories"

In this second part I will investigate the late nineteenth-century number-theoretical writing by a Chinese mathematician that I began this chapter with, Li Shanlan's *Comparable Categories of Discrete Accumulations* (*Duoji bilei* 垛積比類) written

---

[22]The interested reader might consult Bréard (1999a) chap. 4 and 5 or Tian (2003).

around 1850.[23] It was praised by many modern mathematicians for its high achievements in combinatorial results and contained in particular the remarkable identity already mentioned at the beginning of this chapter. Li was familiar with foreign mathematical discourse through his work as a translator of Euclid's *Elements* and other more contemporary foreign mathematical treatises. Although he praised the clarity of Euclidean style, Li Shanlan did not provide any explicit discourse on his demonstrative scheme in his own work, neither did he use any algebraic formalism at all to develop his formulas. Instead, he systematically proceeded by "analogical extension" (*leitui* 類推), to infer from the particular to the universal, the finite to the infinite, for the different kinds of "discrete accumulations." As the title of his work in four scrolls indicates, he arranged them in "comparable categories."

To be more precise, Li expressed all the stated combinatorial identities entirely in traditional algorithmic and rhetorical language, complemented by diagrams and tables. Rhetorical and visual argumentation are combined into a consistent textual structure, which makes his book one of the aesthetic masterpieces of Chinese mathematics, in my eyes. But the format of visual tools and the type of inductive argument set out to justify the passage from several particular (consecutive) situations to the general numerical case were clearly modelled upon Wang Lai's 汪萊 (1768–1813) essay *The Mathematical Principles of Sequential Combinations* (*Dijian shuli* 遞兼數理).[24] Yet with Li Shanlan the status of such an argumentative scheme changed: the kind of inductive argumentative practice found in Wang Lai for recursive procedures,[25] and to a certain extent even earlier in Chen Houyao's 陳厚耀 (1648–1722) *Meaning of the Methods of Combination and Alternance* (*Cuozong fayi* 錯綜法義), was now considered a valid model by Li Shanlan, upon which he could build an entire set of complex combinatorial results without a second order discourse on how to justify them.

In Li Shanlan's *Comparable Categories of Discrete Accumulations*, tables (in particular arithmetic triangles), diagrams and methods are systematically presented and a variety of number-theoretical relations in arithmetic triangles are stated. I have chosen the text here as a case study because Li's book features two crucial transcultural elements: First, because his book is structured deductively, it logically builds up results that he refers back to permanently, and, not unlike Euclid's *Elements*, it is self-contained.[26] Second, because Li himself considers his work as

---

[23]Martzloff translates the title as *Accumulated "Heaps" Studied from an Analogical Point of View.* See Selin (2008) p. 1225.

[24]By relying on diagrams with separate explanations, Wang Lai uses (incomplete) induction to prove rhetorically the correctness of a recursive algorithm that he represents graphically up to a certain finite step. For details on the inductive pattern and the visual tools in Wang Lai's text, see Bréard (2019).

[25]I use the terms "algorithm" and "procedure" interchangeably, the latter being the literal translation of *shu* 術, which in the Chinese mathematical tradition designates the list of operations to follow for calculating the numerical solution of a mathematical problem.

[26]See Mueller (1981).

an extension of China's mathematical tradition, thus reacting against the monopoly of Western science.

5.2.1   A Deductive Structure

Without any meta-discourse on modes of argumentation to move from case n to the general, Li applies a normative discursive structure to a vast amount of mathematical procedures related to natural numbers. He gives explicit credit to Wang Lai, praising his "book as being methodical" (*you tiaoli* 有條理).[27] Li adopts the same linguistic and textual structures as Wang Lai did for his procedures of successive combinations and triangular piles to induce generality. The fact that Li is much less explicit than Wang Lai and does not give any arguments to prove the inductive step in any of his many combinatorial results underlines the possibility of characterizing combinatorics as a science of patterns, where valid results can be conjectured through pattern recognition. However, it is not my concern here to speculate about Li Shanlan's heuristics nor to defend the idea that Li might have had a solid proof scheme. The interesting feature of his work on *Discrete Accumulations* is rather that quite complex, mathematically valid results are presented in a homogenous style of writing mathematics, a style that has precedents and that proved fruitful for the production of mathematical knowledge.

Li's book is a deductive construction of generalized arithmetic triangles, starting from what in the history of mathematics is known as the "Pascal Triangle" or the "Arithmetic Triangle" (see Fig. 5.10). In the West, Pascal equally applies the arithmetic triangle[28] to the theory of combinations, the powers of binomial quantities and the problem of points—the division of stakes in an interrupted game of chance.[29] In pre-modern China, we only know of its important use in the context of interpolation techniques, the resolution of polynomial equations and the construction of finite arithmetic series in particular through an early fourteenth-century treatise, the *Jade Mirror of Four Elements* (*Si yuan yujian* 四元玉鑑, 1303) by Zhu Shijie 朱世傑, who places the diagram at the beginning of his book. Although Zhu does not give a combinatorial interpretation of the diagram, his interest in combinatorial practices played out on another level: the terminology used for the finite series was constructed out of combinations of binomial expressions

---

[27]Li (1867) p. 1a. Li qualifies identically Dong Youcheng's 董佑誠 (style name Fangli, 1791–1823) work on quadrangular piles, but since Li does not reproduce discursive elements from Dong's text, I do not analyse it here with respect to Li Shanlan's schemes of justification.

[28]Depicted in the *Traité du Triangle Arithmétique* "apparently printed in 1654 (though circulated in 1665)." See Daston (1988) p. 9.

[29]Todhunter (1865) chap. 2 as well as many other historians of probability theory considers the Problem of Points, which prompted the seminal correspondence between Pascal and Pierre Fermat in 1654, as the beginning of the theory of probability. See also Todhunter (1865) chap. 9 on the history of the arithmetic triangle.

Fig. 5.10 Arithmetic Triangle, where each cell corresponds to the combinations C_k^n

in the linguistic sense. A mathematical meaning was attached to each prefix or suffix, reflecting the factorization of each term in the arithmetic series. Inverse problems—i.e. how many terms for a given sum—were solved with interpolation techniques using the coefficients from the triangle.[30]

In Li's book, the arithmetic triangle is followed by diagrams representing constructively the cells in its first five diagonals with unitary pebbles. These illustrations are nearly identical to those found slightly earlier in Wang Lai's text (cf. Figs. 5.12 and 5.11) where they are endowed explicitly with a combinatorial interpretation. In one procedure, Wang calculates individually the possible outcomes of drawing k objects out of a set of n objects, the combinations C_k^n which correspond to the values of the cells of the k-th diagonal in the triangle ($k = 0, \ldots, n$).[31] Li, in his explanation (*jie* 解) of the diagrams, gives the rationale of the recursive construction of the first to the fourth diagonal of the arithmetic triangle, what he calls the first to the fourth-order piles: the first-order pile is a pile of layers of unitary piles, the second-order pile is a pile of layers of first-order piles, etc.[32]

---

[30]See Bréard (1999a) chap. 4.2 and Bréard (1999b).

[31]On the meaning of the procedures and diagrams in Wang's text, see Bréard (2015).

[32]Recursive algorithms calculate the solution to a problem depending on solutions to previously calculated instances of the same problem. It is thus possible to define a mathematical object for any instance by a finite statement.

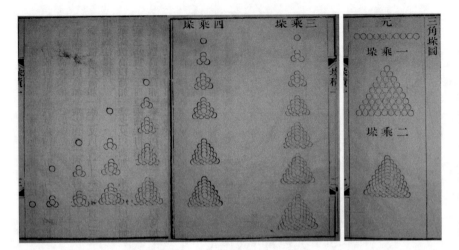

Fig. 5.11 Li Shanlan, *Comparable Categories of Discrete Accumulations* (1867), scroll 1

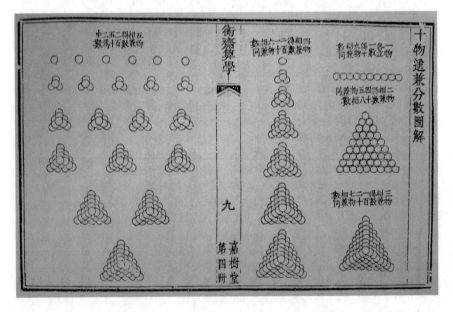

Fig. 5.12 "Diagrammatic explanation of the partial number of sequential combinations for ten objects" in Wang Lai (1854)

Mathematically, this corresponds to calculating the sum of an arithmetic sequence recursively:

$$C_k^n = \sum_{i=k-1}^{n-1} C_{k-1}^i$$

Fig. 5.13 Illustration of $C_1^{10} = C_9^{10} = 1 + 1 + 1 + \cdots + 1 = 10$

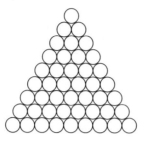

Fig. 5.14 Illustration of $C_2^{10} = C_8^{10} = 1 + 2 + 3 + 4 + \cdots + 9$

Concretely, Wang and Li give drawings for the case $n = 10$, illustrating the number of possible combinations C_k^{10} through surfaces and piles of unit pebbles for $k = 1, \ldots, 5$. Both authors are aware of the symmetry $C_k^{10} = C_{10-k}^{10}$, which explains why they do not illustrate the cases C_k^{10} for $k = 6, \ldots, 10$, but only the cases where sequentially one, two, three, four or five objects are drawn from a set of ten objects (see Fig. 5.12 from top right to left).

The illustrations visually suggest the recursive patterns of formation of all terms of the arithmetic series that constitute the C_k^n. Starting with a row of unit pebbles, where each stands for one possible combination, the following figure of a triangle is composed of rows, in the next step, layers of triangles form a regular triangular pyramid, followed by a sequence of such pyramids, and leading finally to a sequence of sequences of such pyramids. In detail, the number of combinations for drawing one object (or nine objects) out of ten is represented as a row of ten unitary pebbles (see Fig. 5.13).

When in successive rows one to nine pebbles are piled up from top to bottom, one obtains a plane triangle, whose total number of pebbles represents the number of combinations for drawing two (or eight) objects out of ten (see Fig. 5.14).[33]

The next picture in the series is then a regular pyramid, where each layer is a plane triangle, each composed of 1, 3, 6, $\ldots$, and 36 unitary elements. Expressed in terms of their respective rows, each triangle has 1, 1+2, 1+2+3, $\ldots$, and 1+2+3+$\cdots$+8 pebbles, and altogether their sum corresponds to the number of combinations when drawing three (or seven) out of ten objects (see Fig. 5.15).[34]

---

[33]Expressed formally, the recursive character of Wang's construction becomes even more apparent:
$C_2^{10} = \sum_{i=1}^{n-1} C_1^i = C_1^1 + C_1^2 + \cdots + C_1^9.$
[34]Idem: $C_3^{10} = \sum_{i=2}^{n-1} C_2^i = C_2^2 + C_2^3 + \cdots + C_2^9.$

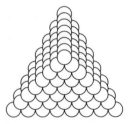

Fig. 5.15 Illustration of $C_3^{10} = C_7^{10} = 1 + 3 + 6 + 10 + \cdots + 36$

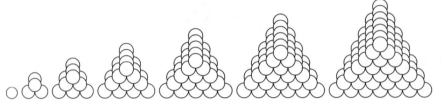

Fig. 5.16 Illustration of $C_4^{10} = C_6^{10} = 1 + (1+3) + (1+3+6) + \cdots + (1+3+6+10+15+21+28)$

Then, seven such triangular pyramids are shown with one to seven layers (see Fig. 5.16), where the total number of pebbles of all the seven pyramids corresponds to the number of combinations when drawing four (or six) objects out of ten. Finally, an ensemble of twenty-one pyramids (see left half of Fig. 5.12) represents C_5^{10} as a series of a series of pyramids with one to six layers.[35]

Verbally, Li Shanlan indicates up to the "fifth-order pile" the way in which the cells of the arithmetic triangle are found by combination of previous cells, and concludes simply, as if the general pattern had naturally become apparent through "analogy":

> Beyond the fifth-order pile, one can induce by analogy (*leitui* 類推).[36]

Based on this argumentative scheme for passing from the specific to the general and giving throughout his book the same or a similar statement as the one explained in detail above, many complex identities are deduced in the following sections of Li's book. The summation of the squares of natural numbers in the second scroll, for example, is deduced as the sum of multiples of sections from the third diagonal of

---

[35]It can be grouped and ordered in two ways. A horizontal reading from top to bottom gives the terms:

$$C_5^{10} = 6 \cdot 1 + 5 \cdot 4 + 4 \cdot 10 + 3 \cdot 20 + 2 \cdot 35 + 1 \cdot 56$$

whereas a diagonal reading produces different terms, yielding the same sum:

$$C_5^{10} = 1 + (1 + 4) + (1 + 4 + 10) + (1 + 4 + 10 + 20) +$$

$$+ (1 + 4 + 10 + 20 + 35) + (1 + 4 + 10 + 20 + 35 + 56).$$

[36]Li (1867) vol. 1, p. 3B.

the "Arithmetic Triangle" in the first scroll. The cells in this diagonal constitute the so-called "second-order triangular pile" (see the cells coloured in blue in Fig. 5.10):

> The first order squared pile has a square [coefficient] of 1 and a corner [coefficient] of 1. For the square [coefficient] take the number of layers as the height [n], for the corner [coefficient] take the number of layers reduced by one as the height [$n-1$]. Each proceeds by using the procedure to find the accumulation for second-order triangular piles.
> 一乘方垛有方一，隅一。方以層數為高，隅以層數減一為高。各以三角二乘垛求積術入之。[37]

$$\underbrace{\sum_{k=1}^{n} k^2}_{-乘方垛} = \underbrace{1}_{方} \cdot \underbrace{\sum_{k=1}^{n} C_2^{k+1}}_{三角二乘垛} + \underbrace{1}_{隅} \cdot \underbrace{\sum_{k=1}^{n-1} C_2^{k+1}}_{三角二乘垛}$$

The two terms of the sum on the right above are both "second-order triangular piles" (*sanjiao ercheng duo* 三角二乘垛), the former having n layers (*ceng* 層), the latter $n-1$ layers:

$$\sum_{k=1}^{n} C_2^{k+1} + \sum_{k=1}^{n-1} C_2^{k+1} = \sum_{k=2}^{n} \binom{k+1}{2} + \sum_{k=2}^{n-1} \binom{k+1}{2} = \sum_{k=2}^{n} \frac{k(k+1)}{2} + \sum_{k=2}^{n-1} \frac{k(k+1)}{2}$$

Two more examples shall suffice to illustrate the deductive structure in Li's text.[38] First, in the second scroll the summation formula for the cubes of natural numbers follows the above formula for squares of natural numbers. It is deduced as the sum of multiples of sections from the fourth diagonal of the "Arithmetic Triangle," the cells in the diagonal corresponding to the "third-order triangular pile" in the first scroll (see the cells coloured in red in Fig. 5.10). It reads, word for word, like an extension of the previous procedure carried to the next higher numerical order:

> The second order squared pile has a square [coefficient] of 1, an edge [coefficient] of 4 and a corner [coefficient] of 1. For the square [coefficient] take the number of layers as the height [n], for the edge [coefficient] take the number of layers reduced by one as the height [$n-1$] and for the corner [coefficient] take the number of layers reduced by two [$n-2$] as the height. Each proceeds by using the procedure to find the accumulation for third-order triangular piles.
> 二乘方垛有方一，廉四，隅一。方以層數為高，廉以層數減一為高，隅以層數減二為高。各以三角乘垛求積術入之。[39]

$$\sum_{k=1}^{n} k^3 = 1 \cdot \sum_{k=1}^{n} C_3^{k+2} + 4 \cdot \sum_{k=1}^{n-1} C_3^{k+2} + 1 \cdot \sum_{k=1}^{n-2} C_3^{k+2}.$$

---

[37]Translated from Li (1867) scroll 2, p. 2B-3A. Punctuation in the Chinese quote is mine.

[38]For more details, see my introduction to Li (2019).

[39]Translated from Li (1867) scroll 2, p. 3A. Punctuation in the Chinese quote is mine.

Li continues, stating explicitly the procedures up to the fourth-order squared pile, before stating in general terms how one can find further procedures from the fifth-order squared pile onwards:

> From the fifth-order squared pile onwards, one successively adds an edge [coefficient]. The numbers of the edge [coefficients] are detailed in the table to the left. The remaining methods can be induced by analogy (*ke leitui* 可類推).

The table mentioned is yet another arithmetic triangle, constructed in a specific way and displaying the correct coefficients—without justification—for calculating the "squared piles":

> Method to construct the table: Each cell is formed by looking at the above left and above right cell. For the cell to the left it is the line in which it lies diagonally to the left, for the cell to the right it is the line in which it lies diagonally to the right.[40] With these numbers multiply each and add them together. This makes the basis for the cells of the table.[41]

Combining the coefficients found in the above table (Fig. 5.17) and extending further analogically the four explicitly stated procedures for squared piles, one could then, for example, conjecture the procedure for the sixth-order squared pile. With the coefficients from the sixth line (in green) in Fig. 5.17 it would correspond to:

$$\underbrace{\sum_{k=1}^{n} k^6}_{\text{六乘方垛}} = 1 \cdot \sum_{k=1}^{n} C_6^{k+5} + 57 \cdot \sum_{k=1}^{n-1} C_6^{k+5} + 302 \cdot \sum_{k=1}^{n-2} C_6^{k+5}$$

$$+302 \cdot \sum_{k=1}^{n-3} C_6^{k+5} + 57 \cdot \sum_{k=1}^{n-4} C_6^{k+5} + 1 \cdot \sum_{k=1}^{n-5} C_6^{k+5}$$

Ever more complex, the following summation formula from the third scroll, in which we also find the "Li Shanlan Identity," is then the partial sum of cells in the fourth diagonal of the squared arithmetic triangle shown in Fig. 5.2:

> Pile C has a square [coefficient] of 1, an edge [coefficient] α of 9, an edge [coefficient] β of 9, and a corner [coefficient] of 1. For the square [coefficient] take the number of layers as the height [n], for the edge [coefficient] α take the number of layers reduced by one as the height [$n-1$], for the edge [coefficient] β take the number of layers reduced by two as the height [$n-2$], and for the corner [coefficient] take the number of layers reduced by three [$n-3$] as the height. Each proceeds by using the procedure to find the accumulation for sixth-order triangular piles.

---

[40]For example, the number 57 in the second cell from the left in the green line in Fig. 5.17 is the sum of $5 \cdot 1$ and $2 \cdot 26$, since the 1 on its upper left is in the fifth position of its line that goes diagonally down to the left and 26 on its upper right is in the second position in the line going diagonally down to the right.

[41]Translated from Li (1867) vol. 2, p. 3B–4A.

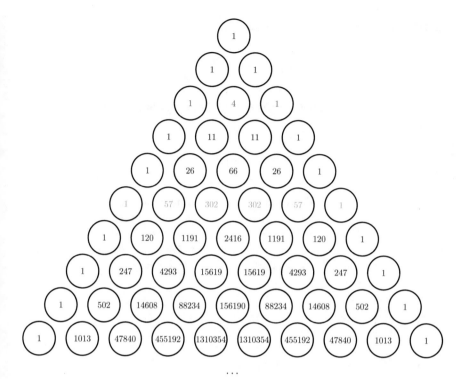

Fig. 5.17 Coefficient table for squared piles

寅垛有方一，甲廉九，乙廉九，隅一。方以層為高，甲廉以層減一為高，乙廉以層
數減二為高，隅以層減三為高。各以三角六乘垛求積術入之。[42]

$$\underbrace{\sum_{k=1}^{n}(C_3^{k+2})^2}_{\substack{三角自乘垛寅垛}} = \underbrace{1}_{方} \cdot \underbrace{\sum_{k=1}^{n}C_6^{k+5}}_{甲廉\,三角六乘垛} + \underbrace{9}_{甲廉} \cdot \underbrace{\sum_{k=1}^{n-1}C_6^{k+5}}_{三角六乘垛} + \underbrace{9}_{乙廉} \cdot \underbrace{\sum_{k=1}^{n-2}C_6^{k+5}}_{三角六乘垛} + \underbrace{1}_{隅} \cdot \underbrace{\sum_{k=1}^{n-3}C_6^{k+5}}_{三角六乘垛}$$

Here, Li Shanlan uses the identity for the sixth-order triangular pile from the first
scroll to calculate the sums in each term. It is not given explicitly there and can only
be deduced by "analogical extension" from the stated procedures for the first- to
fourth-order triangular piles.

In the case of the "Li Shanlan Identity" (5.1), Li gives a generalized arithmetic
triangle, setting out the square binomial coefficients $(C_n^{m+n})^2 = \binom{m+n}{n}^2$. Thus, the
second diagonal from the left (see Fig. 5.2 red encircling) shows for $m = 0, \ldots, 11$

---

[42] See Li (1867) vol. 3, p. 3A–3B.

diagonally the values (from top to bottom):

$$(C_1^{m+1})^2 = \binom{m+1}{1}^2,$$

the third diagonal from the left (blue encircling) depicts for $m = 0, \ldots, 10$ the values:

$$(C_2^{m+2})^2 = \binom{m+2}{2}^2.$$

In the same way as for the standard arithmetic triangle and in Wang Lai's manner of drawing figured numbers, Li Shanlan illustrates these numbers constructively with unitary pebbles (see Fig. 5.3). The quadratic numbers or the coefficients $(C_n^{m+n})^2$ for the values $n = 1$ and $m = 0, 1, 2$ are depicted as squares on the right, and the coefficients $(C_n^{m+n})^2$ for the values $n = 2$ and $m = 0, 1, 2, 3$ as triangles of triangles on the left:

$$(C_2^2)^2 = \binom{2}{2}^2 = 1^2 = 1 \cdot 1$$

$$(C_2^3)^2 = \binom{3}{2}^2 = (1+2)^2 = (1+2) \cdot 1 + (1+2) \cdot 2$$

$$(C_2^4)^2 = \binom{4}{2}^2 = (1+2+3)^2 = (1+2+3) \cdot 1 + (1+2+3) \cdot 2 + (1+2+3) \cdot 3$$

$$(C_2^5)^2 = \binom{5}{2}^2 = (1+2+3+4)^2 = (1+2+3+4) \cdot 1 + \cdots + (1+2+3+4) \cdot 4$$

The sums of the first n cells in the red[43] and blue diagonals of Fig. 5.2 are given as:

$$\sum_{k=1}^{n} \binom{k}{1}^2 = 1 \cdot \sum_{k=1}^{n} \binom{k+1}{2} + 1 \cdot \sum_{k=1}^{n-1} \binom{k+1}{2}$$

$$\sum_{k=1}^{n} \binom{k+1}{2}^2 = 1 \cdot \sum_{k=1}^{n} \binom{k+3}{4} + 4 \cdot \sum_{k=1}^{n-1} \binom{k+3}{4} + 1 \cdot \sum_{k=1}^{n-2} \binom{k+3}{4}$$

---

[43] A familiar formulation is in terms of the sum of squares of natural numbers:

$$\sum_{k=1}^{n} k^2 = \frac{n(n+1)(2n+1)}{6}.$$

Furthermore, Li also states the following two cases explicitly for the first n cells in the fourth and fifth diagonal of the square arithmetic triangle shown in Fig. 5.2:

$$\sum_{k=1}^{n}\binom{k+2}{3}^2 = 1\cdot\sum_{k=1}^{n}\binom{k+5}{6} + 9\cdot\sum_{k=1}^{n-1}\binom{k+5}{6}$$
$$+9\cdot\sum_{k=1}^{n-2}\binom{k+5}{6} + 1\cdot\sum_{k=1}^{n-3}\binom{k+5}{6}$$
$$\sum_{k=1}^{n}\binom{k+3}{4}^2 = 1\cdot\sum_{k=1}^{n}\binom{k+7}{8} + 16\cdot\sum_{k=1}^{n-1}\binom{k+7}{8}$$
$$+36\cdot\sum_{k=1}^{n-2}\binom{k+7}{8} + 16\cdot\sum_{k=1}^{n-3}\binom{k+7}{8} + 1\cdot\sum_{k=1}^{n-4}\binom{k+7}{8}$$

before concluding the paragraph by saying:

> From the fifth pile on, one can induce by analogy (*ke lei tui* 可類推). All the cells in every horizontal line in the present table are the coefficients of each pile.

The pattern that one can observe from the above four formulaic instances can indeed lead to a general identity as shown in 5.2. To determine the sum of the first n cells in the j-th diagonal of the squared arithmetic triangle, one adds up the products of the numbers in the j-th horizontal line of the very same triangle and the partial sums of certain diagonals in the "Arithmetic Triangle":

$$\sum_{k=1}^{n}\binom{k+j-1}{j}^2 = \sum_{i=1}^{j+1}\left[\binom{j}{i-1}^2\cdot\sum_{k=1}^{n+1-i}\binom{k+2j-1}{2j}\right] \qquad (5.2)$$

The identity 5.2 is thus contained implicitly in the statement on the recursive construction of the cells of the generalized triangle in Fig. 5.3 as the sum of cells in triangles treated in an earlier chapter. Again, Li relies on the explicit cases for the first values of n and states the possibility of "analogical extension" (*ke lei tui* 可類推) to the general case, something he does with remarkable linguistic and structural stability throughout his book. A survey of the kind of expressions used to point to the general procedure after having stated explicitly the procedures for a small number of instances for specific subsequent values of n is provided in Figs. 5.18 and 5.19. The tables list for all the types of arithmetic triangles (lit. "tables" *biao* 表), the number of specific cases spelled out explicitly, and the wording used for the passage to the general case for each of the structural elements in Li's text:

- the (triangular) table: the number of rows depicted and the method of construction to fill its cells;

		Construction of table 表法		Construction of piles		Table 表	Illustration 圖
reference	Type of 'accumulation'	# specific cases stated explicitly	any number of rows 層	# specific cases stated explicitly	any # of rows 層	# rows 層	# piles depicted
juan 1:1B–5B	三角堆	-	stated in general	4（一乘堆…四乘堆）	五乘堆以上可類推	13	5（元＋4）
juan 1:6A–10A	一乘支堆	1（第二層）	三層以下如三角堆表法	3（方堆、第一、第二堆）	第三堆以下仿此	13	5（方堆＋4）
juan 1:10B–15B	二乘支堆	1（第二層）	三層以下如三角堆表法	5（方堆、甲堆、第一、二、三堆）	第四堆以下可類推	13	5（方堆、甲堆＋3）
juan 1:15B–20A	三乘支堆	1（第二層）	餘如三角堆表法	-	-	-	6（方堆、甲堆、乙堆＋3）
juan 1:13A	三乘支堆						
	以下理俱同						
juan 2:1A–7B	乘方堆	-	stated in general	5（太堆、元堆、一、二、三乘方堆）	四乘方堆以上可類推	13	5（太堆、元堆＋3）
juan 2:7B–14B	二乘方支堆	1（左邊）	餘如三角堆表法	5（方堆、甲堆、第一、二、三堆）	第四堆以下可類推	13	5（方堆、甲堆＋3）
5.18	三乘方支堆	1（左邊）	餘如三角堆表法	5（方堆、甲堆、乙堆、第一、二堆）	第三堆以下可類推	13	6（方堆、甲堆、乙堆＋3）
juan 3:1A–6A	三角自乘堆	-	stated in general	3（子堆、丑堆、寅堆）	卯堆以下可類推 + general rule	13	4（子堆、丑堆、黃堆、卯堆）
juan 3:6B	子支堆			其各堆俱與一乘支堆同			
juan 3:6B	丑支堆						
juan 3:7A–19B	寅支堆	1（左邊）	餘法如三角堆表	7（方堆、甲堆、乙堆、丙堆、丁堆、第一、二堆）	第三堆以下可類推	13	8（方堆、甲堆、乙堆、丙堆、丁堆＋3）
juan 3:20A–36B	卯支堆	1（左邊）	餘法如三角堆表	3（方堆、甲堆、乙堆）	丙堆以下類推 辰支堆以下理皆如是	13	9（方堆、甲堆、乙堆、丙堆、丁堆、戊堆、己堆、第一、二堆）
juan 4:1A–7B	三角變堆	-	stated in general	5（第一堆…第五堆）	第六堆已[sic]下可類推	13	5（第一堆…第五堆）
juan 4:8A–15B	三角再變堆	-	stated in general	4（第一堆…第四堆）	第五堆以下可類推	13	5（一堆…五堆）
juan 4:16A–23B	三角三變堆	-	stated in general	4（第一堆…第四堆）	第五堆以下可類推	13	4（一堆…四堆）
juan 4:23B	四變以下諸堆今不復演也						
juan 4:23B–24A	變堆皆有支堆一變諸支堆已附見三角變堆後一卷中一變三角支堆今不復演者自能類反也						

Fig. 5.18 Technical vocabulary used for deductive patterns in Li Shanlan's *Comparable Categories of Discrete Accumulations* (Li 1867)

reference	Type of 'accumulation'	procedure to find the sum for a given *n* 有高[or: 層(數)]求積術		有積求高[or: 層(數)]求術		Additional tables	
		# cases explicitly given for specific *n*	procedure for general *n*	# cases for *n*	general *n*	# rows	name
juan 1:1B-5B	三角垛	4 (一乘垛 … 四乘垛)	procedure explicitly stated	4 (一乘垛 … 四乘垛)	五乘垛以上以天元仿此推之	10	三角垛有積求高開方廉隅裵
juan 1:6A-10A	一乘支垛	5 (方垛 + 4)	第五垛以上可類推	5 (方垛 + 4)	-	-	-
juan 1:10B-15B	二乘支垛	5 (方垛 · 甲垛 + 3)	第四垛以下可類推	5 (方垛 · 甲垛 + 3)	-	-	-
juan 1:15B-20B	三乘支垛	6 (方垛 · 甲垛 · 乙垛 + 3)	第四垛以下可類推	5 (方垛 · 甲垛 · 乙垛 + 2)	-	-	-
juan 1:13A	三乘支垛	-	-	-	-	-	-
	以下理俱同						
juan 2:1A-7B	乘方垛	6 (太垛 · 元垛 + 4)	五乘方垛以上一虆增一廉之虆詳左裵除法可類推	6 (元垛 + 5)	六乘方垛以上以天元仿此推之	13	乘方垛各廉裵
juan 2:7B-14B	三乘方支垛	7 (方垛 · 甲垛 + 5)	第六垛以下可類推	5 (方垛 · 甲垛 + 3)	-	-	-
juan 2:15A-26A	三乘方支垛	9 (方垛 · 甲垛 · 丙垛 + 6)	第七垛以下可類推	6 (方垛 · 甲垛 · 乙垛 + 3)	-	-	-
juan 3:1A-6A	三角自乘垛	4 (子垛 · 丑垛 · 寅垛 · 卯垛)	辰垛以下可類推	4 (子垛 · 丑垛 · 寅垛 · 卯垛)	-	-	-
juan 3:6B	子支垛	-	-	-	-	-	-
juan 3:6B	丑支垛	-	-	-	-	-	-
juan 3:7A-19B	寅支垛	8 (方垛 · 甲垛 · 乙垛 · 丙垛 · 丁垛 + 3)	第四垛以下可類推	8 (方垛 · 甲垛 · 乙垛 · 丙垛 · 丁垛 + 3)	-	-	-
juan 3:20A-36B	卯支垛	9 (方垛 · 甲垛 · 乙垛 · 丙垛 · 丁垛 · 戊垛 · 己垛 · 第一 · 二垛)	第三垛以下可類推	9 (方垛 · 甲垛 · 乙垛 · 丙垛 · 丁垛 · 戊垛 · 己垛 · 第一 · 二垛)	-	-	-
juan 4:1A-7B	三角變垛	5 (第一垛 … 第五垛)	第六垛以下可類推	5 (第一垛 … 第五垛)	-	-	-
juan 4:8A-15B	三角再變垛	5 (第一垛 … 第五垛)	第六垛以下可類推	5 (第一垛 … 第五垛)	-	-	-
juan 4:16A-23B	三角三變垛	5 (第一垛 … 第五垛)	第六垛以下可類推	5 (第一垛 … 第五垛)	-	-	-
juan 4:23B	四變以下諸垛今不復演學者自能隅反也						
juan 4:23B-24A	變垛皆有支垛一變諸支垛借作三角支垛已詳上卷中一變支垛今不復演學者分不復演學者亦能隅反也						

Fig. 5.19 Technical vocabulary used for deductive patterns in Li Shanlan's *Comparable Categories of Discrete Accumulations* (Li 1867) (cont.)

- the piles: the number of piles depicted and the method of construction to find the number of unit pebbles in each layer of the subsequent piles;
- the procedures for finding the sums for a given n: the number of procedures formulated explicitly and the kind of expression used to induce the general case;
- the inverse procedures, i.e. finding n for a given sum: these are never stated in general, they have to be deduced algebraically one by one following the model for previous instances as pointed out by Li at two instances.[44]

The frequently used expression *leitui* 類推 in Li's technical vocabulary for making inductive arguments is not entirely new in the Chinese lexicon. It had already played a prominent role in Neo-Confucian epistemology in the "investigation of things" (*gewu* 格物), as a key "method of 'extending' (*tui* 推) the knowledge of a thing or an event to that of other things or events belonging the same 'kind' (*lei* 類)."[45] The most famous Song dynasty Neo-Confucian philosopher Zhu Xi 朱 熹 (1130–1200) explained it as follows:

> It is not necessary to jump higher and look farther. It is not to roam about advancing and stopping, either. It simply is to take up those subjects that are here near at hand and are understood and to keep going. It is like [the case in which,] having understood this one event thoroughly, one proceeds to infer from [the knowledge of] this event to work on that event, and knows that this lantern has much light, one then proceeds to infer from [the knowledge about] this lantern to know that that candle is similarly bright. It is like climbing up stairs. Having climbed the first level, one bases oneself on this first level and advances to the second level. Also, basing oneself on the second level, one advances to the third level. One simply keeps proceeding like this. [...] If, [standing on] the first level, one wants to jump to the third level, the step will have to be too wide, and one will [end up] wasting one's effort.[46]

Whether Li Shanlan was familiar with the meaning of the term *leitui* 類推 in Zhu Xi's epistemology is unclear. But he certainly understood it as a heuristic principle of analogical reasoning when he employed the term in the context of a specific argumentative scheme that he encountered in his work as a translator. In the Chinese version of Elias Loomis's book on the differential and integral calculus, the expression *leitui* 類推 stands for "etc. etc." Loomis first indicates the equations for the quadratic, cubic and biquadratical parabolas, followed by "etc. etc." after which he gives the general form: "All of these parabolas are included in the equation $y^n = ax$."[47] The mathematical context is thus structurally equivalent to the one found in Li's own work on *Comparable Categories of Discrete Accumulations*, where the expression is systematically used to indicate that the general procedure can be found by combining observations of the first instances and the subsequent analogies.

---

[44] See scroll 1, p. 5B and scroll 2, p. 7B: "By the Celestial Element [method] induce this accordingly." (*Yi tianyuan fang ci tui zhi* 以天元仿此推之).

[45] See Kim (2004) p. 41.

[46] Quote from Zhu Xi's *Classified Dialogues of Master Zhu* (*Zhuzi yulei* 朱子語類, compiled in 1270), translated in Kim (2004) p. 57.

[47] See Li and Wylie (1859) scroll 8, p. 8A and Loomis (1868) p. 105.

5.2.2 A Pillar Different from the Nine Chapters (and Euclid)

> I want those who learn mathematics to know that the procedures for discrete accumulations erect another flag beyond the *Nine Chapters*. Their theory has begun with Li Shanlan.
> 欲令習算家知垛積之術於九章外別立一幟其説自善蘭始[48]

This is what Li Shanlan says in his own preface to the *Comparable Categories of Discrete Accumulations*. He sees himself as a revolutionary, as someone who has established a new mathematical theory outside of the canonical disciplines of Chinese mathematics and, although not explicitly mentioned, a domain of inquiry that also lies outside of the Euclidean tradition, which he was well aware of. That "the *Nine Chapters* have entirely provided all [mathematical] methods, but lack any type of method for alternations and combinations,"[49] was already noticed by Chen Houyao. Yet designating his theory as a new pillar of mathematical knowledge is a much stronger claim. It strengthens the scope of the Chinese mathematical tradition and underlines the fact that Li felt that this was lacking not only in Chinese antiquity but also in the Greek one.

However, not everyone approved of Li Shanlan's ingenious achievements and original contributions. One particularly traditionalist mathematician, Chen Song 陳崧 (style name Mengshi 夢石), criticized him for not having followed the Yuan dynasty nomenclature for the different kinds of piles:

> I read the four scrolls of the *Comparable Categories of Discrete Accumulations*. When having finished all the chapters, I understood that Renshu boasts in vain with many procedures but he could not illuminate the principles of *discrete accumulations*. The ancients had no reason not to speak about *discrete accumulations*. As for example, in the *Mathematical Methods of Ding Ju* 丁巨算法, there is a song to calculate piles; in the *Detailed Workings Revealing a Screen* 透簾細草 one also finds all the piles of aquatic grass, of triangles and squares. And then in Zhu Hanqing's 朱漢卿 [courtesy name of Zhu Shijie] *Jade Mirror of Four Elements*, we have the chapters entirely devoted to the subtleties of accumulations of piles. I ever say that among the ancients, only Hanqing is the most profound as for what concerns the principles of accumulated piles. Even if Luo Mingxiang [style name of Luo Shilin 羅士琳 (1774–1853)] meticulously developed the workings, there are many of which he has not understood their purpose. Therefore, one knows that Hanqing's principles of piles are very profound. Then, Renshu wanted to surpass the ancients and had no fear of extending and broadening each of the piles. The way he applied his mind can truly be called diligent. It is only regrettable that from the names of the piles in the ancients' writings he could not verify and illustrate their shapes and thoroughly understand their meanings. Instead, for all the names of the piles in the *Jade Mirror*, he [Li Shanlan] followed the principle of cutting them out. He merely used "first multiplication," "second multiplication," up to "*n*-th multiplication" to name them and "first pile," "second pile," up to "*n*-th pile" to order them. He either called the piles Jia, Yi, Bing, Ding 甲乙丙丁 or he called them Zi, Chou, Yin, Mao 子丑寅卯.[50] All piles' ramifications have further branches. Therefore, by showing the gloomy and obscure, the readers are dazzled by the

---

[48]Li (1867) scroll 1, p. 1A–1B.

[49]Chen (17th) p. 685.

[50]*Jia* 甲, *Yi* 乙, *Bing* 丙 and *Ding* 丁 are the first four of the Ten Heavenly or Celestial Stems (*tiangan* 天干) which, together with the Twelve Earthly Branches (*dizhi* 地支), beginning with *Zi* 子, *Chou* 丑, *Yin* 寅 and *Mao* 卯, form China's traditional calendrical system. They were also

complicated situation. It is impossible that things are bright and clear at one glance. That such is the case, no one can have a doubt![51]

Chen continues to complain of inconsistencies in Li Shanlan's textual structure, where the number of procedures given for the first instances of n varies, or the number of diagonals in the arithmetic triangles illustrated with unitary pebbles differs from case to case, etc. In spite of these critiques, there were several other authors during the late Qing who continued to develop the domain of "discrete accumulations," often in relation to Western algebraic techniques, and it certainly was one of the hot topics of mathematical research at the time.[52]

5.3 "Comparable Categories" in the West

In Li Shanlan's *Comparable Categories of Discrete Accumulations*, no explicit mention is made of foreign influence. There nevertheless is a structural affinity with the Euclidean canon. Given the original format of Li's book with respect to the Chinese mathematical tradition, where texts usually followed the canonical scheme of a topically arranged collection of problem-answer-procedure units, one can assume a certain influence from the *Elements*. Li Shanlan builds his work on a deductive pattern, proceeding logically from simple to more complex results. Yet he does not provide any explicit proofs for them,[53] neither does his text show the characteristics that were regarded as essential parts of argumentative practices in the Greek mathematical tradition.

In the epistemic environment of late imperial China more generally, we cannot point to an accepted canon of explicit standards of validity. In particular, Chinese mathematicians had not developed an axiomatic-deductive theory of proof. Nevertheless, this does not mean that correct mathematical statements were developed on an *ad hoc* basis, neither that they would have been accepted by the mathematical community when formulated by arbitrary linguistic or visual tools. Often in commentarial form to the classics, aspects of proof were closely intertwined with the steps of the solution algorithm to perform. Later authors, such as Wang Lai 汪萊 (1768–1813), rather sought to make the underlying "principles" of generally formulated procedures apparent, as indicated in the preface and title of his text, *Mathematical Principles of Sequential Combinations* (*Dijian shuli* 遞兼數理). By relying on diagrams with separate explanations, Wang Lai uses (incomplete) induction to prove rhetorically the correctness of a recursive algorithm that he

---

used since the first translation of the first six books of Euclid's *Elements* (Ricci and Xu 1607) for representing the alphabetic letters in the geometric diagrams.

[51] Translated from Chen (1904) p. 1A–1B.

[52] See Tian (2003) p. 58–69.

[53] By "proof" I mean an argument very much in the style of Greek proofs, considered rigorous by contemporary authors and later commentators.

represents graphically up to a certain finite step.[54] With respect to applying inductive arguments, number theory, or more particularly combinatorics, is of course a rather ideal mathematical domain, since recursive algorithms and inference from case n to case $n + 1$, with n being any natural positive number, play a prominent role.

In the history of mathematics, there are many examples of mathematical or logical principles that are used before they are explicitly formulated and formalized.[55] Induction is one such case, and its earliest occurrence or even just its possibility as an unformalized scheme of proof has been much debated for the history of Western mathematics. For the ancient mathematical corpus, the survival of one inductive proof is attested in Plato's *Parmenides*.[56] For early modern mathematics in Europe, it is usually assumed that Blaise Pascal's *Traité du Triangle Arithmétique* (1665) contains the first explicit application of the principle of induction. Pascal applies this pattern of proof, with almost identical wording, to prove four number-theoretical identities related to the Pascal Triangle. Although not explicitly formalized, one could thus argue that Pascal (and his subsequent readers) did recognize induction as a general demonstrative scheme, even if he did not formulate it as a general principle of inference. But, as the following short discussion of some of Pascal's theorems will show, it is certainly not right to claim that combinatorial results outside of China were justified by an explicit and solid proof scheme, something that has often been put forward as contrasting Chinese "intuitive" from "deductive" Western science.

Part I of the *Traité du Triangle Arithmétique* amounts to a systematic development of all the main results then known about the properties of the numbers in the arithmetic triangle (Fig. 5.20). One of them, for example, is that the sum of entries in the n-th row (in Pascal's illustration actually a diagonal) equals 2^n. This is Pascal's corollary 8 and can be proved by induction. The main point in the argument is that each entry in row n, say C_k^n is added to two entries below: once to form C_k^{n+1} and once to form C_{k+1}^{n+1} which follows from the construction of the arithmetic triangle

---

[54] As an application of the recursive algorithm, he only gives one example drawn from divination with hexagrams. In this case, the algorithm allows for the calculation of the total number of possible mutations of a diagram of six lines, with two possibilities for each line. In Bréard (2012) I have shown that in late imperial China, among cultural practices like gaming or divination, hexagrams became the paradigmatic model for observing and analysing stable combinatorial patterns. Bréard (2019) contains a complete translation of Wang Lai's text.

[55] For examples from the Babylonian and Greek mathematical traditions, see Høyrup (2012) and Mueller (2012).

[56] At 149A-C. See Acerbi (2000). The text "displays a series of phrases, adverbs, and syntactical constructs which enable him to word in a very refined way the explicitly iterative character of the proof." Acerbi (2003) p. 477. More generally, see *idem* p. 476–481 for an overview of recursive or quasi-inductive proofs in Greek mathematical writings. Outlining "the composition of the combinatorial *humus* in which calculations *must* have grown out" (p. 466), Acerbi shows, for example through a passage of Plutarch's *De Stoicorum repugnantiis*, that, even without an explicitly formulated proof scheme, an astonishing result related to Schröder numbers could be obtained.

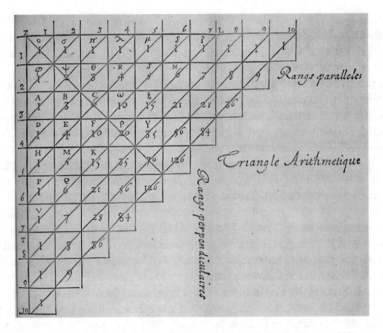

Fig. 5.20 The Arithmetic Triangle in Pascal's *Traité du Triangle Arithmétique* (1665)

given by Pascal:

$$C_k^{n+1} = C_{k-1}^n + C_k^n,$$
$$C_{k+1}^{n+1} = C_k^n + C_{k+1}^n.$$

For this reason, the sum of entries in row $n + 1$ is twice the sum of entries in row n. This is what Pascal states in corollary 7:

> In any arithmetic triangle, the sum of the cells in each base [i.e. diagonal], is twice that of the preceding base.[57]

As a consequence, in corollary 8 (Fig. 5.21), he finds, that "in any arithmetic triangle, the sum of the cells in each base is a number of the double progression that begins with unity, and of which the power is that of the base."[58]

---

[57] Pascal (1665) p. 5:

> En tout Triangle Arithmetique, la somme des cellules de chaque base, est double de celles de la base precedente.

[58] Pascal (1665) p. 5:

> En tout Triangle Arithmetique, la somme des cellules de chaque base, est un nombre de la progression double, qui commence par l'unité, dont l'exposant est le mesme que celuy de la base.

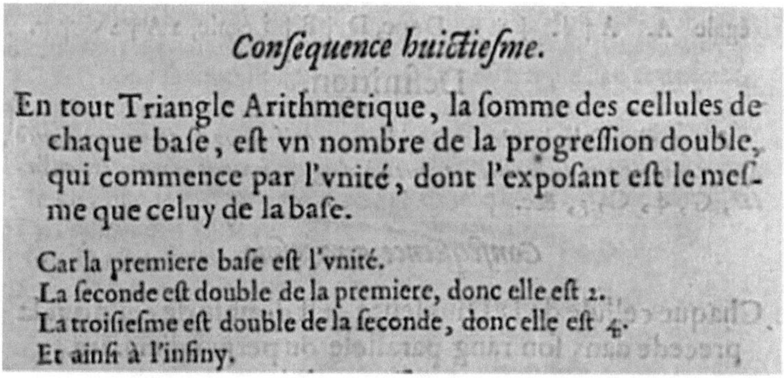

Confequence huiEtiefme.

En toutTriangle Arithmetique, la fomme des cellules de chaque bafe, eft vn nombre de la progreffion double, qui commence par l'vnité, dont l'expofant eft le mef-me que celuy de la bafe.

Car la premiere bafe eft l'vnité.
La feconde eft double de la premiere, donc elle eft 2.
La troifiefme eft double de la feconde, donc elle eft 4.
Et ainfi à l'infiny,

Fig. 5.21 Corollary 8 in the *Traité du Triangle Arithmétique* (1665)

But Pascal's induction to prove these two corollaries is incomplete. He only shows for particular cases that the statements hold, before concluding, not unlike Li Shanlan, upon their generality: "the same thing can be shown in the same way in all other cases," "and so on until infinity."[59]

It is a generally accepted opinion that the first instance of conscious use of complete induction as a proof method is contained precisely in the twelfth corollary of the *Traité du Triangle Arithmétique* (see Fig. 5.22). Pascal proves by explicit use of mathematical induction, that (in our notation):[60]

$$k \cdot C_k^n = (n - k + 1) \cdot C_{k-1}^n.$$

The problem is to find C_k^n as a function of n and k, which Pascal does by applying the twelfth corollary recursively and by splitting his proof into two parts:

Even if this proposition has an infinity of cases, I will give a very brief proof here, by supposing two lemmata.
[Quoy que cette proposition ait une infinité de cas, j'en donneray une démonstration bien courte, en supposant 2 lemme.][61]

The first lemma concerns the cases $n = 2$, $k = 1, 2$ and, as Pascal points out, is self-evident[62]:

$$1 \cdot C_1^2 = 2 \cdot C_0^2$$
$$2 \cdot C_2^2 = 1 \cdot C_1^2$$

---

[59] *Idem.* Final phrases to corollary 7 and 8 respectively: "La mesme chose se demonstre de mesme de toutes les autres" and "Et ainsi à l'infiny."

[60] Pascal expresses this relation proportionally. For more details, see Edwards (1987) p. 63–65.

[61] *Idem* p. 91 in the Latin version: "Quamvis infiniti sint hujus propositionis casus, [...], breviter tamen demonstrabo, positis duobus assumptis."

[62] Pascal (1665) p. 7:

Le 1. qui est évident de soy-mesme, que cette proportion se rencontre dans la seconde base; car il est bien visible, que f est às comme 1 est à 1.

Conſequence douziéſme.

En tout Triangle Arithmetique, deux cellules contigues
   eſtant dans vne meſme baſe, la ſuperieure eſt à l'inferieure, comme la multitude des cellules depuis la ſuperieure iuſques au haut de la baſe, à la multitude de cel-
   les, depuis l'inferieure iuſques en bas incluſiuement.

Soient deux cellules contigues quelconques d'vne meſme baſe, E, C,
   Ie dis que   E   eſt à C   comme 2   à   3

                        ⌄   ⌄

        inferieure, ſuperieure, parce qu'il y a deux   parce qu'il y a trois
                                    cellules depuis E   cellules depuis C
                                    iuſques en bas,   iuſques en haut,
                                    ſçauoir, E, H ;   ſçauoir C, R, μ.

Quoy que cette propoſition ait vne infinité de cas, i'en donneray
vne demonſtration bien courte, en ſuppoſant 2 lemme.

Le 1. qui eſt euident de ſoy-meſme, que cette proportion ſe rencontre dans la ſeconde baſe ; car il eſt bien viſible que σ eſt à σ comme 1, à 1.

Le 2. que ſi cette proportion ſe trouue dans vne baſe quelconque,
elle ſe trouuera neceſſairement dans la baſe ſuiuante.

D'où il ſe voit, qu'elle eſt neceſſairement dans toutes les baſes : car,
elle eſt dans la ſeconde baſe, par le premier lemme, donc par le ſecond
elle eſt dans la troiſieſme baſe, donc dans la quatrieſme, & à l'infiny.

Il faut donc ſeulement demonſtrer le ſecond lemme, en cette ſorte.
Si cette proportion ſe rencontre en vne baſe quelconque, comme en
la quatrieſme D λ, c'eſt à dire, ſi D eſt à B comme 1 à 3. Et B, à θ
comme 2 à 2. Et θ à λ comme 3 à 1. &c.

Ie dis, que la meſme proportion ſe trouuera dans la baſe ſuiuante,
H μ, & que par exemple E eſt à C comme 2 à 3.

Car D eſt à B comme 1 à 3. par l'hypotheſe.
Donc D † B eſt à B comme 1 † 3 à 3.
             ⌣⌣

        E   à B comme   4   à 3.
De meſme, B eſt à θ comme 2 à 2   par l'hypotheſe.
Donc, B † θ à B comme, 2 † 2 à 2.
          ⌣⌣   ⌣⌣

        C   à B comme   4   à 2
Mais   B   à E comme   3   à 4 comme il eſt monſtré.
Donc par la proportion troublée, C eſt à E comme 3 à 2.
        Ce qu'il falloit demonſtrer.

Fig. 5.22 Corollary 12 in the *Traité du Triangle Arithmétique* (1665)

The second lemma, if proved generally, would then be the necessary inductive step,
allowing a declaration of the corollary's validity:

> The second [lemma] is that this proportion holds in whatever base, it will necessarily hold
> in the next base.

[Le 2. que si cette proportion se trouve dans une base quelconque, elle se trouvera nécessairement dans la base suivante.]

From this can be seen that it is necessarily in all the bases, since it is in the second base by the first lemma; thus by the second [lemma] it is in the third base, thus in the fourth, and so on to infinity.

[D'où il se voit qu'elle est nécessairement dans toutes les bases: car elle est dans la seconde base par le premier lemme; donc par le second elle est dans la troisième base, donc dans la quatrième, & à l'infiny.][63]

The problem is that Pascal deals in the proof of this second lemma only with a particular case, the fourth and fifth base of the triangle. He claims quite cleverly, that this proof held "regardless for which base, for example the fourth." He nevertheless concludes:

We will show this in the same way for all the rest, since this proof is based only upon the fact that this proportion holds for the previous base and that each cell is equal to the previous one, plus the following one, that which is true everywhere.

[On le monstrera de mesme dans tout le reste, puisque cette preuve n'est fondée que sur ce que cette proportion se retrouve dans la base précédente, & que chaque cellule est égale à la precedente, plus à la superieure, ce qui est vray par tout.][64]

Pascal's inference certainly resembles what modern logicians like Giuseppe Peano (1858–1932) understood by "induction," since Pascal concludes that a universal generalization is true because each of its instances are also true. The deficiency of logical rigour is thus not a flaw in the form of Pascal's argument, but rather in the stated premises that do not adequately support his conclusion. By appealing to an argument by incomplete induction, his demonstrative scheme is not generally validated, a weakness I have equally identified in Li Shanlan's set of procedures.

One critical historiographic conclusion from the comparison of Li Shanlan's and Pascal's discursive argumentative features is that the formalization of Li's rhetorical procedures was misleading. It led to a whole historiography crediting Li Shanlan with the formulation of a ready-made general statement that is not only not in his text but can also only be reconstructed after several manipulations using the results of modern calculus. Furthermore, this presentist approach does not reveal anything about the conjectural and argumentative strategies in Li's work. These can only be understood by a careful analysis of the structure, technical vocabulary and visual tools that allowed late Qing mathematicians like Li Shanlan to recognize general number-theoretical patterns. This process of abstract thinking, where the perceptual and rational element of mathematical thinking intersect, obviously played a crucial role in what historians often characterized as an intuitional experience.

From a transcultural perspective, Li's text reveals the mutual influences of Western and Chinese epistemological traditions. Structured deductively as the Euclidean canon, Li's highly systematic work on discrete accumulations claims at

---

[63] Quoted from *Conséquence Douziesme* in Pascal (1665) p. 7.

[64] Translated from *Conséquence Douziesme* in Pascal (1665) p. 8.

the same time in its title an affinity to a Chinese tradition of arranging mathematical objects by "comparable categories" (*bilei*). This was a way to establish relationships between the procedures to calculate the surface or volume of certain geometric objects of similar shape. For Li, "comparing categories" (*bilei*) is THE heuristic technique. As a source of hypotheses in the context of number-theoretical methods, it allowed him to conjecture and discover many combinatorial identities. He expressed the action of combining observations and following analogies with the term "analogical extension" (*leitui* 類推), a term that relates to analogical reasoning in Chinese philosophical contexts. For the great Neo-Confucian scholar Zhu Xi, "analogical extension" was a method to reach the underlying "principles" (*li* 理) by extending the knowledge of a thing or an event to that of other things or events belonging to the same kind.[65] Li equally applies the expression (*leitui* 類推) in order to translate a pattern of incomplete induction in Western mathematical works.

The remarkable degree of standardization in the formulation of an (incomplete) inductive argument displayed in the linguistic side of Li Shanlan's text confirms the full extent to which new kinds of argumentative modes were recognized as legitimate standards of mathematical discourse and applied systematically to make the underlying "principles" of generally formulated mathematical procedures apparent.

References

Acerbi, Fabio (2000). Plato: *Parmenides* 149a7–c3. A Proof by Complete Induction? *Archive for History of Exact Sciences 55*, 57–76.

Acerbi, Fabio (2003). On the Shoulders of Hipparchus. A Reappraisal of Ancient Greek Combinatorics. *Archive for History of Exact Sciences 57*, 465–502.

Bréard, Andrea (1999a). *Re-Kreation eines mathematischen Konzeptes im chinesischen Diskurs: Reihen vom 1. bis zum 19. Jahrhundert*, Volume 42 of *Boethius*. Stuttgart: Steiner Verlag.

Bréard, Andrea (1999b). The Reading of Zhu Shijie. In Sik Kim Yung and Francesca Bray (Eds.), *Current Perspectives in the History of Science in East Asia*, 291–306. Seoul National University Press.

Bréard, Andrea (2012). Divination with Hexagrams as Combinatorial Practice. A Paradigmatic Model in Mathematics. *Zhouyi Studies* (English Version) *8*, 157–174.

Bréard, Andrea (2015). What Diagrams Argue in Late Imperial Chinese Combinatorial Texts. *Early Science and Medicine 20*(3), 241–264.

Bréard, Andrea (2019). Inductive Arguments in the Midst of Smoke: 'Proving' Rhetorically and Visually that Algorithms Work. In Ari D. Levine, Joachim Kurtz, and Martin Hofmann (Eds.), *Powerful Arguments: Standards of Validity in Late Imperial China*. Leiden: Brill.

Chen, Houyao 陳厚耀 (late 17th). *Cuozong fayi* 錯綜法義 (The Meaning of Methods for Alternation and Combination). Reprint in (Guo et al. 1993) 4:685–688.

Chen, Song (Mengshi) 陳崧 (夢石) (1898–1904). Duoji bilei houji 垛積比類後記 (Afterword to the *Comparable Categories of Discrete Accumulations*). In *Dongxi congshu* 東溪叢書

---

[65] See Kim (2014) p. 35–52, previously published as Kim (2004).

(Collectanea from the Eastern Creek). Chaojun jingxian tang Dongxi ru gu ge 潮郡敬賢堂東溪茹古閣.

Clavius, Christophorus (1574). *Euclidis Elementorum Libri XV Accessit XVI de Solidorum Regularium Cuiuslibet Intra Quodlibet Comparatione, Omnes Perspicuis Demonstrationibus, Accuratisque Scholiis Illustrati, ac Multarum Rerum Accessione Locupletati*. Rome: V. Accoltum.

Daston, Lorraine J. (1988). *Classical Probability in the Enlightenment*. Princeton, NJ: Princeton University Press.

Diofanto (2011). *De polygonis numeris* (introduzione, testo critico, traduzione italiana e commento di Fabio Acerbi), Volume 1 of *Mathematica graeca antiqua*. Pisa, Roma: Fabrizio Serra.

Edwards, A.W.F. (1987). *Pascal's Arithmetical Triangle*. London & New York: Charles Griffin & Oxford University Press.

Guo, Shuchun 郭書春 et al. (Eds.) (1993). *Zhongguo kexue jishu dianji tonghui: Shuxue juan* 中國科學技術典籍通彙: 數學卷 (Comprehensive Collection of Ancient Classics on Science and Technology in China: Mathematical Books), 5 vols. Zhengzhou: Henan jiaoyu chubanshe 河南教育出版社.

Horng, Wann-Sheng (1991). *Li Shanlan: The Impact of Western Mathematics in China during the Late 19th Century*. Ph. D. thesis, Graduate Center, City University of New York.

Høyrup, Jens (2012). Mathematical Justification as Non-Conceptualized Practice: the Babylonian Example. In Karine Chemla (Ed.), *The History of Mathematical Proof in Ancient Traditions*, 362–383. Cambridge: Cambridge University Press.

Jami, Catherine (1998). Traductions et synthèses : Les mathématiques occidentales en Chine, 1607–1782. In Dominique Tournès (Ed.), *L'Océan Indien au carrefour des mathématiques arabes, chinoises, européennes et indiennes*, 117–126. Saint-Denis: I.U.F.M. de la Réunion.

Kim, Yung Sik (2004). 'Analogical Extension' (*leitui*) in Zhu Xi's Methodology of 'Invesitgation of Things' (*gewu*) and 'Extension of Knowledge' (*zhizhi*). *Journal of Song-Yuan Studies 34*, 41–57.

Kim, Yung Sik (2014). *Questioning Science in East Asian Contexts. Essays on Science, Confucianism, and the Comparative History of Science*. Number 1 in Science and Religion in East Asia. Leiden; Boston: Brill.

Legge, James (Trans.) (1885). *The LîKî, I–X*, Volume IV of *The Sacred Books of China. The Texts of Confucianism*. New York, NY: At the Clarendon Press.

Li, Shanlan 李善蘭 (1867). *Duoji bilei* 垛積比類 (Comparable Categories of Discrete Accumulations), 4 scrolls. In *Zeguxizhai suanxue* 則古昔齋算學 (Mathematics from the Studio Devoted to the Imitation of the Ancient Chinese Tradition) (Jinling 金陵刻本 ed.), Volume 4.

Li, Shanlan 李善蘭 (2019). *Catégories analogues d'accumulations discrètes* (*Duoji bilei* 垛積比類), traduit et commenté par Andrea Bréard. La Bibliothèque Chinoise. Paris: Les Belles Lettres.

Li, Shanlan 李善蘭 and Alexander Wylie 偉烈亞力 (1859). *Dai weiji shiji* 代微積拾級 (Elements of Analytical Geometry and of the Differential and Integral Calculus) 18 scrolls. Shanghai: Mohai shuguan 墨海書館. Original by Elias Loomis 羅密士(Loomis 1851).

Li, Zhizao 李之藻 et al. (1965). *Tianxue chuhan* 天學初函 (First Collectanea of Heavenly Studies), 6 vols. (Reprint ed.), scroll 23, First Series (Chubian 初編卷二十三) of Wu Xiangxiang 吳相湘 (Ed.), *Zhongguo shixue congshu* 中國史學叢書. Taipei: Taiwan xuesheng shuju 臺灣學生書局.

Loomis, Elias (1851). *Elements of Analytical Geometry and of the Differential and Integral Calculus*. New York: Harper & Brothers.

Loomis, Elias (1868). *Elements of Analytical Geometry and of the Differential and Integral Calculus* (19th ed.). New York: Harper & Brothers.

Martzloff, Jean-Claude (1993). Eléments de réflexion sur les réactions chinoises à la géométrie euclidienne à la fin du XVIIe siècle. *Historia Mathematica 20*, 160–179.

Martzloff, Jean-Claude (2006). *A History of Chinese Mathematics*. Berlin, Heidelberg: Springer. Corrected second printing of the first English edition of 1977.

Mueller, Ian (1981). *Philosophy of Mathematics and Deductive Structure in Euclid's Elements*. Mineola, NY: Dover.

Mueller, Ian (2012). Generalizing about Polygonal Numbers in Ancient Greek Mathematics. In Karine Chemla (Ed.), *The History of Mathematical Proof in Ancient Traditions*, 311–326. Cambridge: Cambridge University Press.

Pascal, Blaise (1665). *Traité du triangle arithmétique avec quelques autres petits traitez sur la mesme matière*. Paris: Guillaume Desprez.

Pólya, George (1954). *Induction and Analogy in Mathematics*. Number 1 in Mathematics and Plausible Reasoning. Princeton, NJ: Princeton University Press.

Ricci, Matteo and Guangqi Xu (1607). *Jihe yuanben* 幾何原本 (The Elements). Beijing: [n.p.]. Translation of the first six books of (Clavius 1574), included in (Li et al. 1965).

Schironi, Francesca (2007). Ἀναλογία, *analogia, proportio, ratio*: Loan Words, Calques and Reinterpretations of a Greek Technical Word. In Louis Basset, Frédérique Biville, Bernard Colombat, Pierre Swiggers, and Alfons Wouters (Eds.), *Bilinguisme et terminologie grammaticale greco-latine*, Volume 27 of *Orbis / Supplementa*, 321–338. Leuven, Paris, Dudley, MA: Peeters.

Selin, Helaine (Ed.) (2008). *Encyclopaedia of the History of Science, Technology, and Medicine in Non-Western Cultures* (2nd ed.). Springer.

Tian, Miao (2003). The Westernization of Chinese Mathematics: A Case Study of the *duoji* Method and its Development. *EASTM 20*, 45–72.

Todhunter, Isaac (1865). *A History of the Mathematical Theory of Probability from the Time of Pascal to that of Laplace*. Cambridge and London: Macmillan and Co.

Volkov, Alexeï K. (1992). Analogical Reasoning in Ancient China: Some Examples. *Extrême-Orient, Extrême-Occident 14*, 15–48.

Wang, Robin R. (2012). *Yinyang: The Way Of Heaven And Earth In Chinese Thought And Culture*, Volume 11 of *New Approaches to Asian History*. Cambridge University Press.

Wang, Lai 汪萊 (1854). *Dijian shuli* 遞兼數理 (Mathematical Principles of Sequential Combinations). In *Hengzhai suanxue* 衡齋算學 (Hengzhai's Mathematical Learning), Volume 4, 6B–12B. Chine: Jiashutang 嘉樹堂. Reprint in (Guo et al. 1993) 4:1512–1516.

Wu, Jing 吳敬 (fl. 1450). *Jiu zhang suanfa bilei daquan* 九章算法比類大全 (Great Compendium of Comparable Categories to the Nine Chapters on Mathematical Methods). Reprinted in (Guo et al. 1993) 2:1–333.

Yang, Hui 楊輝 (1261). *Xiangjie jiu zhang suanfa* 詳解九章算法 (Detailed Explanations of the Nine Chapters on Mathematical Methods) (Yijiatang congshu 宜稼堂叢書 1842 ed.). Reprinted in (Guo et al. 1993) 1:949–1043.

Yang, Hui 楊輝 (1842). Chengchu tongbian suanbao 乘除通變算寶 (Mathematical Treasure of Variations on Multiplication and Division). In *Yang Hui suanfa* 楊輝算法 (Yang Hui's Mathematical Methods) (Yijiatang congshu 宜稼堂叢書 ed.).

Yunzhi 允祉 (Ed.) (1723). *Yuzhi shuli jingyun* 御製數理精蘊 (Essence of Numbers and their Principles). [Beijing?]: [n.p.]. Reprint in (Guo et al. 1993) 3.

Zhang, Yong 章用 (1939). Duoji bilei shuzheng 垛積比類疏證 (=Proofing the Formulas in the Duoji Bilei). *Kexue* 科學 *23*, 647–663.

Chapter 6
Fate Calculation 算命: The Mathematics of Divination

Contents

In one scroll of a Ming dynasty edition of a household encyclopaedia, the *Dragon Head Carved in an Ancient Vessel–An Ocean of Learning at a Single Glance without Consulting the Help of Others* (*Dingqie longtou yilan xuehai buqiuren* 鼎鍥龍頭一覽學海不求人),[1] one finds a treatise on physiognomy on the upper half of the pages, and a manual on the calculation of different kinds of geometric surfaces on the lower half. Some of the surfaces from the mathematical "Methods for measuring the surfaces of fields" (*Zhangliang tianmu fa* 丈量田畝法)[2] have striking similarities with the kind of face shapes discussed and illustrated in the upper half of the chapter (see Fig. 6.1).

This combination of geometry and a divination technique is rare in the encyclopaedic genre of the *Complete Books of a Myriad Treasures* (*Wanbao quanshu* 萬寶全書) from the Ming to the Republican era,[3] but "fate calculation" (*suan ming* 算命), the association of divination with what we would consider "mathematics," is a widespread cultural phenomenon in China. Numerical patterns and their

---

[1] Anonymous (1644), scroll 17. This encyclopaedia carries many different book titles at the beginning of the altogether twenty-two scrolls and Cui (2011) p. 140 assumes that it could even be a one-of-a-kind self-made compilation of chapters from different book titles.

[2] The text on "Methods for measuring the surfaces of fields" opens with a poem identical to the one found at the beginning of the geometric section *Suan zhangliang tianfa* 算丈量田法 (Calculation methods for measuring the surfaces of fields) in Xu (1573) scroll 2.

[3] See Bréard (2016).

© Springer International Publishing AG, part of Springer Nature 2019
A. Bréard, *Nine Chapters on Mathematical Modernity*, Transcultural
Research – Heidelberg Studies on Asia and Europe in a Global Context,
https://doi.org/10.1007/978-3-319-93695-6_6

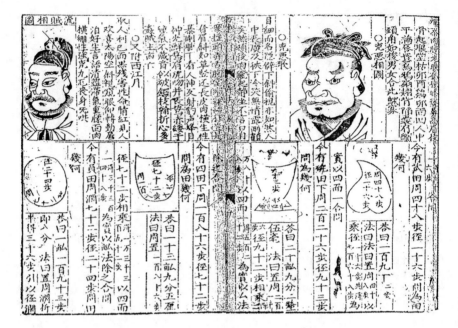

Fig. 6.1 Physiognomy and Geometry on one page in Anonymous (1644) scroll 17

interpretation are at the heart of the construction of philosophically legitimate and successful divination techniques and games with dice and with dominos. The importance of numerical patterns as a rational principle for establishing truth in the late imperial period is evident not only in analogies between physiognomical types and geometric figures but extends to other associations of mathematical objects (such as the arithmetic triangle treated in Sect. 6.2 and polar coordinates described in Sect. 6.3) to numerological and cosmological patterns. This chapter will explore these associations from the late Ming until the early twentieth century, when scholars were particularly eager to prove the real scientific character of traditional divination schemes in the context of new mathematical knowledge and a prohibition of mantic arts as part of a struggle for China's modernity.

6.1 Mathematical Problems Before the Qing

Although there was no explicit mention of a concept of probability in the Chinese mathematical tradition, the presence of divination problems in mathematical texts seems to imply a belief in the possibility to master chance mechanisms through deterministic mathematical algorithms. The following two examples show how divination problems were presented in a mathematical context.

In particular in the first example, the efficiency and reliability of the procedure are underlined, supported by the argument that the chance for failure in prognostication was one of zero per cent. The "Rhyme to calculate if the sick person will die or live" determines numerically the survival of a patient according to his date of birth and duration of disease[4]:

Rhyme to calculate if the sick person will die or live

若還逢九鬼來催 盧醫扁鵲難治　逢三便是病見活 逢六不死淹遲　三因九除見安危 此乃神仙做的　先下年康歲數 次加病月日期　算病死生訣

First, put down [on the calculation board] the age according to the time of birth;
next, add the term of sickness, months and days.
Multiplication by three and division by nine will make apparent if there will be peace or danger;
such is the oeuvre of immortals!
If you come upon [the number] three, then the sick will live,
if you come upon [the number] six, then he will not die soon.
But, on the contrary, if you come upon [the number] nine, the ghosts will come in a haste;
even for Bian Que, the Doctor from Lu, it will be hard to cure [the disease].

Due to the operation of multiplication by three, followed by a division by nine, there are only three possibilities of a remainder: 3, 6 or 9.[5] When the result is three, the patient is cured, when it is six, he survives, but he dies for certain when there is a remainder of nine. Even Bian Que,[6] the legendary physician from the fifth century BCE, could not help, as a line in the versified algorithm emphasizes.

The verse for memorization, composed following the pattern of a well-known tune, *Moon of the Western River* (*Xijiang yue* 西江月), is followed by an example with concrete numerical values. The method and the arithmetic operations are made explicit, and the effectiveness of the prediction method underlined by way of reference to long-term experience:

Let us suppose that there is a man, born in the seventh moon of a *jiazi* year.[7] This man was struck by a disease on the sixth day, twelfth moon in a *guichou* year[8]. He came to ask: "How to know if I will be in peace or not?" I have calculated the number three and consequently, he did indeed not die.

The method says: From the time of birth in a *jiazi* year until a *guichou* year, there are 40 years. By adding the twelve months and six days of sickness, one obtains altogether the number 418.[9] Place it [on the calculation device, here an abacus] and use the "multiplication

---

[4]Translated from the Mathematical Section (*Suanfa men* 算法門) in Yu (1599), scroll 22, p. 13A–13B (Reprint in Sakade et al. (2000) vol. 2, p. 373–374).

[5]Performing the two operations: division by nine after multiplication by three might seem arithmetically redundant, but in contrast to the single operation of division by three, they allow to have precisely the three remainders: 0, 3 or 6, or, as stated here: 3, 6 and 9.

[6]Nickname of Qin Yueren 秦越人, also known as "the Doctor from Lu" (*Lu yi* 盧醫).

[7]The first year in the sexagenary cycle of the Chinese calendar.

[8]The fiftieth year in the sexagenary cycle. There seems to be a mistake here. According to the following calculation, it should be the fortieth year, a *guimao* 癸卯 year.

[9]Addition here is not performed as usually in a decimal place value system. Here: 12 added to 40 makes 412, furthermore adding 6 makes 418.

by 3." 3 times 8, 24. 1 times 3 equals 3. 3 times 4, 12. Altogether one obtains 1254. This number is laid down and furthermore division by 9 is used. [...] One has a remainder of 3. Another [example]: Wang, who held an administrative post and who was born in a *wuwu*[10] year. In a *guichou* year on the seventh day of the twelfth month, my village, named Hengtang Village 橫塘村, caught fire. Calculation of this outburst [of fire] produced the number five[11] and [Mister Wang] did indeed die. First, put down 56 years. Add twelve months and seven days to this. Together one obtains 5727.[12] This number in place, multiplication by three gives 17,181. This number has to be used for division by 9. [...] As a remainder the number nine is left. This method is based on much experience. I have calculated for many persons and there was neither deviation nor error. That is the reason why it is noted here to the attention of scholars.[13]

今有一人甲子七月生人至癸丑年十
二月初六日得病來問吾安否若何吾
算得是三數後果不死
法曰甲子生人至癸丑是四十歲加上
十二月初六日得病共得四一八數在
位用三因　三八一四　一三如
三　三四　十二　共得一二五四
數在位復用九除初位順除至末

[...]

又王掌管戊午生至癸丑年十二月初
七日本村名橫塘村失火被爆算得五
數果然死矣先下五十六歲加上十二
月初七共得五七二七數在位三因得
一七一八一　數得用九除

其法甚驗吾算人多矣並無
差謬故書以知學者

[...]

In the two numerical examples, there is a positional difference in how numbers were laid out for "summation." This "summation" does not follow literally the mathematical rules of addition in a system of decimal positional notation of numbers but rather arbitrarily shifts the units:

Example 1 :   Example 2 :

40   56

12 months   12   months

+   6 days   +   7 days
―――――   ―――――
= 418   = 5727

---

[10]The fifty-fifth year in the sexagenary cycle.

[11]"Five" is certainly a typo and should read "nine," because a remainder of five is not possible, and the following details of the procedure give a division without rest, or as stated here a remainder of 9.

[12]Here, the numbers are placed differently, as in the above example. Adding 12 to 56 gives 572 as above, but contrary to the above example the number of days are not added to the units of the result, but placed to its right.

[13]Translated from the Mathematical Section (*Suanfa men* 算法門) in Yu (1599), reproduced in Sakade et al. (2000) vol. 2, p. 373–375.

This could possibly be an indicator that the diviner manipulated the procedure in order to comply with the intended result. But, had he placed the numbers shown to the right as in the first example shown to the left, he would have obtained 579, multiplied by 3 gives 1737, divided by 9 leaves a remainder of 9, just as by using the other layout on his calculation device. So the final result, a remainder of 9, obtained after multiplication by three and division by nine, is the same in either case. The situation is the same inversely. If we place the numbers to the left as in the second example shown to the right: "summation" would lead to 4126, multiplied by 3 and divided by 9 would equally leave a remainder of 3. It might well be that the layout of numbers has been reconstructed here for a written form and that the diviner's method was only orally transmitted in an unstable textual form. This possibility is underscored because the same mathematical problem is found in an earlier Ming dynasty source, Wu Jing's *Great Compendium of Comparable Categories to the Nine Chapters on Mathematical Methods* (*Jiu zhang suanfa bilei daquan* 九章算法比類大全) from around 1450. Wu gives the same algorithm in versified form and a different numerical example, in which the strict rules of arithmetic are respected[14]:

占病法

先置病人年幾歲　次加月日得病時
三因除九多餘數　三輕六重九難醫
今有病人年四十七歲三月初九日得病問証如何
答曰重也
法曰先置病人年四十七加得病三月九日共得五十九以
三因之得一百七十七卻以九除之先除九十次除八十一餘
六乃難醫
即重也若是初八日得病餘三即輕也若初十日得病餘
九難醫

Method to divine illness

First, put down [on the calculation board] the person's age in number of years;
next, add the months and days of the time she fell ill.
Multiplication by three and division by nine, how much is the remainder;
three: [the illness] is light, six: it is heavy, nine: it will be hard to cure.
Let us suppose that there is a sick person, 47 years of age. He was struck by a disease on the ninth day, third moon. The question is: what is the pattern of the disease?
The answer says: it is heavy.
The method says: First put down the age of the person. 47 Add when he fell ill. third moon ninth day Together one obtains 59. With 3 multiply this, one obtains 177. Then eliminate with 9. First, eliminate 90, then, eliminate 81, the remainder is 6. Thus, it is heavy. If it was on the eighth day, one would obtain as remainder of the illness 3. Thus, it would be light. If it was on the tenth day, one would obtain as remainder of the illness 9. Then it would be difficult to cure.

Another often found example of fate calculation in a mathematical context is the "Method to determine if pregnancy will give a boy or girl." The embryo's gender is

---

[14]Introductory chapter in Wu (1450) 起例, p. 69A–69B.

predicted according to a certain algorithm leading to an odd or even result.[15] The text is quoted here from the most popular mathematical manual of the late imperial era, Cheng Dawei's 程大位 (1533–1606) *Unified Lineage of Mathematical Methods* (*Suanfa tongzong* 算法統宗), originally published in 1592 and reprinted many times during the Qing:

> Method to determine if pregnancy will give a boy or a girl
>> Song
> To the number 49 add the months of pregnancy;
> subtract the age in the current year, it is determined without doubt!
> Eliminate one, up to nine in case of big remainders;
> coming upon an even number, it will be a girl, coming upon an odd number, it will be a boy.
>
> Let us suppose that there is a pregnant woman, who in the current year is 28 years of age and has been pregnant for eight months. The question is will she give birth to a boy or a girl?
> The answer says: She will give birth to a boy.
> The method says : Put down 49, add the months of pregnancy 8. This totals 57.[16] Subtract the number of years 28, this leaves 29. Eliminate [the number for] heaven 1, eliminate [the number for] earth 2, eliminate [the number for] man 3, eliminate [the number for] the four seasons 4, eliminate [the number for] the five elements 5, eliminate [the number for] the six pitches 6, eliminate [the number for] the seven planets 7. If there is a remainder that is odd, it will be a boy, if it is even, it will be a girl. 1, 3, 5, 7, 9 are odd. 2, 4, 6, 8, 10 are even. If the number is big, then furthermore eliminate the eight winds 8.[17]

孕推男女法

歌

四十九數加孕月　減行年歲定無疑

一除至九多餘數　逢雙是女隻生兒

今有孕婦行年二十八歲八月有孕問所生男女

答曰生男

法曰置四十九加孕月共五十七減年二十八餘二十九減天

除一地除二人除三四時除四五行除五六律除六七

星除七不盡奇為男偶為女也一三五七九皆奇二四六八十皆偶

如數多再以八風除八

---

[15]Since there are as many odd as even numbers that lie between one and ten, this reflects approximately the natural relative frequencies of male and female births over a long period of time—something one could easily observe without maintaining birth registers.

[16]The text wrongly gives 56, but continues the calculation with the correct value 57.

[17]Translated from Cheng (1882) scroll 17, p. 25B.

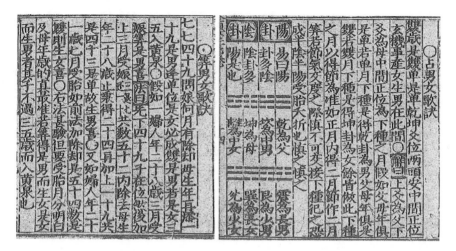

Fig. 6.2 Rhymes for determining if child will be a girl or boy by divination (right) and by calculation (left) in Anonymous (1612) scroll 27 "On Childbirth"

As in the previous example, this verse for memorization is followed by an example with concrete numerical values. The method and the arithmetic operations are made explicit, except for the final steps to the result, which leaves a certain margin for the diviner to obtain a result that he intuitively wants to achieve. The sequential subtraction of the numbers one to seven is indicated and eventually followed by subtraction of eight "if there are many numbers," i.e. if the remainder has several decimal positions. This corresponds to the versified algorithm, which generally prescribes subsequent subtraction of the numbers one up to nine.

The same technique for gender prediction can be found in many other texts, not only of the encyclopaedic genre but also in purely mathematical ones. An identical rhyme with a slightly different formulation of the procedure for the same numerical problem is, for example, included in the introductory chapter in Wu Jing's *Great Compendium of comparable categories to the Nine Chapters on Mathematical Methods*.[18] Here, the additional division by the number 9 is made explicit at the end, in the case of a large remainder. A similar versified problem to "calculate if it will be a boy or a girl" (*Suan nannü gejue* 算男女歌訣) is found in one of the household encyclopaedias, the *Complete book of a million treasures, magnificently embroidered (Miaojin wanbao quanshu* 妙錦萬寶全書).[19] There (see Fig. 6.2), it is not included in the mathematical chapter but in the section on childbirth (*Quanying men* 全嬰門) and preceded by a rhyme and explanations for determining the child's gender "by divination" (*Zhan nannü gejue* 占男女歌訣) using the hexagrams from

---

[18]Wu (1450) 起例, p. 68B–69A.

[19]Anonymous (1612) scroll 27.

the *Classic of Changes* (*Yijing* 易經).[20] Mathematical calculation and calculation by divination are both actions designated with the verb *suan* 算—to calculate—in the two procedural texts.

The above examples show that manipulations of numbers within divinatory contexts are considered mathematically valid arithmetic operations. The insertion of problems for prognostication within mathematical chapters as well as the juxtaposition of arithmetic and divinatory calculations techniques reveals an epistemological proximity that became an entire programme of research for scholars like Jiao Xun 焦循 (1763–1820) who systematically brought mathematical concepts close to the system of *Yijing* numerology.

6.2   Hexagrams as Symbolic Algebra

Jiao Xun was one of the major mathematicians during the Qianlong-Jiaqing period (1736–1820). He was part of the so-called Qian-Jia School, which devoted philological research to textual criticism (*kaozheng xue* 考證學). The leading scholars of the school played major roles in promoting mathematical studies for comprehending more technical passages of the classics.[21] Jiao Xun was particularly interested in applying concepts from the *Nine Chapters* and Song dynasty algebra to reread the system of the divinatory hexagrams in the canonical *Classic of Changes* (*Yijing* 易經):[22]

> If one does not understand the equalization and harmonization [of fractions] or ratios, [which are methods] from the *Nine Chapters on Mathematical Procedures*, it is impossible to comprehend the line movements in the hexagrams.
> 非明九數之齊同，比例，不足以知掛畫之行。[23]

A hexagram is a diagram composed of six superposed lines, where each line is either continuous or broken. Thus, there are $2^6 = 64$ distinct diagrams. Each has its name and a challenging interpretation that has given rise to millennia of exegetic commentaries, including Leibniz's interpretation as a system of binary arithmetic. Although the dual nature of lines was understood in China in terms of Yin and Yang, odd and even, Leibniz was wrong in assuming a mathematical correspondence with a base 2 number system where the two kinds of lines can represent the numbers 0 and 1.[24] Figure 6.3 shows ten of the hexagrams with their respective names on top. Varying divinatory techniques have allowed the production of such a

---

[20] See below for a short explanation of the Chinese hexagrams.

[21] See Horng (1993).

[22] On the cultural significance of this text, its exegetic traditions and its convoluted history, see Smith (2008).

[23] Translated after a quote in Chen (2000) p. 237.

[24] See Bréard (2008).

Fig. 6.3 Some hexagrams as drawn in Jiao Xun's *Précis of Diagrams in the Changes* (1974)

hexagram. They involve numerological procedures and eventual alternations of the lines before arriving at the final mantic figure. Both the techniques as well as the interpretations and judgements of the hexagrams given in the *Classic of Changes* and its commentaries, have been subject to controversies, not least because their analysis allowed for a considerable interpretive latitude. As the case of Jiao Xun will demonstrate, it was even possible to regard them from a purely rational, structural point of view, without even considering the social, psychological or philosophical implications in the relative spatial organization of the six lines.

Besides the concepts used by Liu Hui 劉徽 in his 263 commentary to the *Nine Chapters* for finding the lowest common denominator and the methods for proportional distribution according to certain ratios mentioned in the above quote,[25] Jiao Xun finds more analogies between algebra and the transformations of lines in a hexagram. In his *Explanation of Addition, Subtraction, Multiplication and Division* (*Jiajian chengchu shi* 加減乘除釋, 1797) he links the structure of the arithmetic triangle[26] shown up to the sixth power of a binomial to the hexagrams. The bottom line contains the coefficients 1, 6, 15, 20, 15, 6 and 1 which sum up to sixty-four (see Fig. 6.4). This is precisely the total number of different diagrams in the *Classic of Changes*:

> That it ends with the fifth square-multiplication (*wu cheng fang* 五乘方)[27] is to carry the signification of hexagrams which end with sixty-four.[28]

---

[25]For more details on how Jiao connected the operations of "equalization and harmonization" (*qitong* 齊同) and ratios (*bili* 比例) to the lines in the hexagrams, see Chen (2000) p. 247–252. See also Li (2002).

[26]Literally, "The ancient diagram of the origins of root extraction" (*Gu kaifang benyuan zhi tu* 古開方本原之圖). In China, the Arithmetic Triangle first appears in Yang Hui's 楊輝 *Detailed Explanations of The Nine Chapters on Mathematical Methods* (*Xiangjie jiu zhang suanfa* 詳解九章算法, completed in 1261), but we know that it must have been circulating a century earlier in Jia Xian's 賈憲 work.

[27]In Chinese algebra, the "*n*-th square multiplication" of say x corresponds to x^{n+1}.

[28]Jiao (1799) scroll 2, p. 18B.

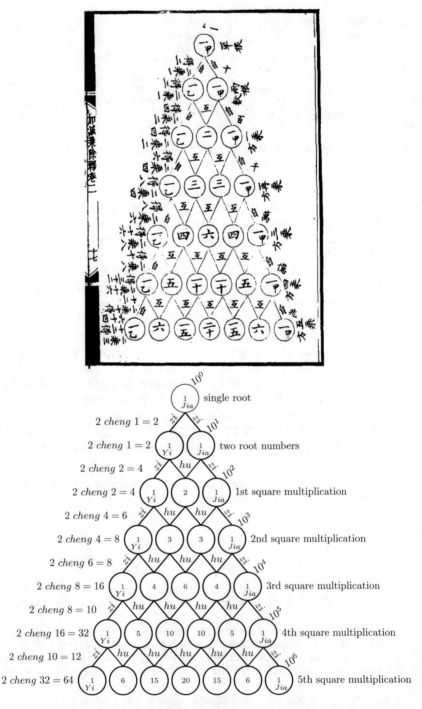

Fig. 6.4 "The ancient diagram of the origins of root extraction" in Jiao Xun's *Explanation of Addition, Subtraction, Multiplication and Division* (1799)

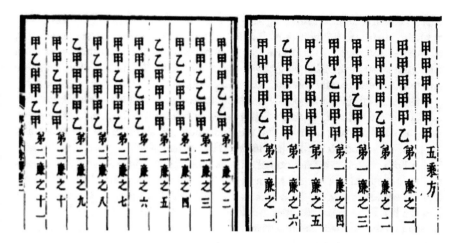

Fig. 6.5 Symbolic interpretation of "The ancient diagram of the origins of root extraction" in Jiao Xun's *Explanation of Addition, Subtraction, Multiplication and Division* (1799)

The numbers in the cells of the last line shown correspond to the coefficients of the expansion of $(a + b)^6$:

$$(a + b)^6 = 1 \cdot a^6 + 6 \cdot a^5 b + 15 \cdot a^4 b^2 + 20 \cdot a^3 b^3 + 15 \cdot a^2 b^4 + 6 \cdot ab^5 + 1 \cdot b^6$$

In addition to this algebraic interpretation of the triangle, for which Jiao gives extensive verbal and diagrammatic explanations, he reads the last line in an abstract manner as the possible mutations of the two types of lines of a hexagram. But he dares to do so only in a manuscript version of his *Précis of Diagrams in the Changes* (*[Yi] tu lüe* [易]圖略), where he explicitly uses the arithmetic triangle as a generator of a new order for the hexagrams.[29] In his *Explanation of Addition, Subtraction, Multiplication and Division*, he argues merely through the intermediary of the two symbols *Jia* 甲 and *Yi* 乙.[30] This allowed him to represent each cell of the triangle as a combination of *Jia*s and *Yi*s (see Fig. 6.5). The last line, for example, would be read as a representation of the hexagrams where lines were replaced by symbols.[31] The term $6a^5b$, for example, were the six possibilities to combine five *Jia* with one *Yi*. When *Jia* and *Yi* are interpreted as the continuous and interrupted line respectively, the coefficient 6 quantifies the six possibilities to mute one line out of six in a hexagram. When we mute two lines out of six, there are fifteen possibilities (cf.

---

[29] See Chen (2000) p. 242. Jiao actually only does so for forty-five hexagrams, the remaining nineteen are reconstructed analogically. See *idem* p. 243–244.

[30] These were used in the Jesuit translations, since the *Elements* Ricci and Xu (1607), to represent for example α and β in geometric diagrams.

[31] Had Jiao read the symbols as the numbers 0 and 1, he would have found the same interpretation of the hexagrams as Leibniz with his binary arithmetic. See Bréard (2008).

$15a^4b^2$ and $15a^2b^4$). Again, using Jiao Xun's symbols *Jia* 甲 and *Yi* 乙, this can be abstracted to fifteen possible combinations of four *Jia* and two *Yi*, or, symmetrically, of two *Jia* and four *Yi*. Jiao illustrates systematically all the sixty-four cases from the bottom line of the triangle with the two symbols *Jia* 甲 and *Yi* 乙 following the order of the coefficients in the last line of the arithmetic triangle from right to left.

Besides building a conceptual framework extending beyond the *Nine Chapters*,[32] one of the main themes in Jiao's *Explanation of Addition, Subtraction, Multiplication and Division* was to show that multiplication and addition are fundamentally the same operations. The idea is reinforced by his use of the very same verb *cheng* 乘 for both operations. Traditionally in mathematics, this was the term to express multiplication only. When Jia explains the formation of the cells of the arithmetic triangle shown in Fig. 6.4 he expresses the fact that, for example, the coefficients 6 in the last line are the result of "mutually multiplying" (*hu cheng* 互[33]乘), or of adding together in mathematical terms, the coefficients 1 and 5 above the respective cells containing the number 6:

> The [coefficient of] squared *Jia* of the fourth square-multiplication and the first edge [coefficient] of the *Yi* are mutually "multiplied," the degree of their numbers is used to form the coefficient 6 of the first edge coefficient of the fifth square-multiplication. The [coefficient of] squared *Yi* [of the fourth square-multiplication] and the fourth edge [coefficient] of the *Jia* are mutually "multiplied," the degree of their numbers is used to form the coefficient 6 of the fifth edge coefficient of the fifth square-multiplication.[34]

The legend to the left of the triangle confirms that Jiao did not understand "multiplication" *sensu stricto*. He noticed that the cells in the triangle were connected through specific additive and multiplicative relations:

1. the sum of each line is a constant multiple, i.e. a multiple of 2, of its previous sum, as expressed in:

   "2 *cheng* 32 makes 64,"

2. the number of connections of cells from one line to the other is in an additive relation, augmenting by the number 2, as expressed in:

   "2 *cheng* 10 makes 12."

Jiao's arithmetic triangle therefore is not to be understood as an illustration of "pure" mathematics. The significance of the operation *cheng* goes beyond multiplication as one of the four arithmetic operations. Its polyvalent mathematical meaning reflects Jiao's interest in stable numerical patterns of transformation. This

---

[32] See Jiao's letter to another Qian-Jia scholar quoted in Horng (2001) p. 390.

[33] The expression *hu* is reproduced in the arithmetic triangle in Fig. 6.4 to indicate the "mutual" additive concatenation of cells from one line to the next. The coefficients along the outer diagonals of the triangle have a constant value of 1 since they are associated with the terms multiplied "by themselves" (*zi* 自). Mathematically they are the coefficients of the pure terms a^i and b^i.

[34] Translated from Jiao (1799) scroll 2, p. 16B, Chen (2000) p. 239 quotes this passage falsely as from scroll 3 and omits the former half of the first sentence.

is also at the heart of the interpretations of the divinatory diagrams of the *Classic of Changes* and allowed authors like Jiao Xun to connect mathematical concepts found in the newly rediscovered literature to numerological divinatory practices.[35] As Jiao remembers, "it was in the fifty-second year of Qianlong reign [ca. 1787] that I began to practice the Nine-Nine procedures [for multiplication] in order to understand the *Nine Chapters*. Also, I obtained Qin Jiushao's and Li Ye's books, [. . .]."[36] Having studied earlier mathematical writings, Jiao Xun spent much of his research on uncovering the analogies between arithmetic structures and the hexagrams:

> Reading the words of King Wen and the Duke of Zhou is like reading the detailed calculation sketches of the "Nine Inscriptions from the Dongyuan School" (*Dongyuan jiu rong* 洞淵九容).[37] The detailed calculation sketches serve to elucidate the method of the "celestial element"; the interpretations of the divinatory trigrams and their lines serve to elucidate the transformations of the hexagrams. They [calculations and hexagrams] can be mutually observed and compared to each other.[38]

With this novel interdisciplinary approach, Jiao Xun made a revolutionary discovery. Too revolutionary probably, since he did not publish his new order of the hexagrams, which he generated following the mathematical patterns of the coefficients in the sixth line of the arithmetic triangle. As pointed out earlier, this remained a partial sketch in his manuscript version of the *Précis of Diagrams in the Changes*.[39]

6.3 Proving the Scientificity of Correlative Cosmology

Although some authors ascribe to Jiao Xun's era the birth of what they call "the Science of Yijing Learning" (*Yijing kexue* 易經科學) due to the rapprochement of mathematics and numerology,[40] more explicit efforts to prove the scientific character of the cosmological basis of traditional divination techniques were undertaken in the early Republican period. These efforts developed in defence of fortune-telling, when popular and ritual practices including their theoretical

---

[35] Jiao Xun's ideas on mathematics may also have been reshaped by his reading of the *Elements* and of Xu Guangqi's other mathematical works. See Horng (2001).

[36] Jiao (1829) scroll 5, p. 12B–13A.

[37] Reference to a method to solve mathematical problems where a figure is inscribed in another figure. Li Ye was one of the representatives of this methodology ascribed to a Daoist group. For more details, see Li (2002) chap. 2.

[38] Jiao (1829) scroll 6, p. 11B.

[39] For a reproduction of his manuscript, *Tu lüe* 圖略 6, see Jiao (1974) vol. 1, p. 328–329. I will discuss in more detail Jiao Xun's epistemological approach to numbers in an upcoming publication.

[40] See, for example Dong (1993).

underpinnings came under harsh attack from the anti-superstition campaigns of the Republican government.[41] One possible strategy to cope with the state's modernist ideology was to show that the theories on which divination techniques were based were scientific and hence modern.

One actor who devised one such innovative approach was Yuan Shushan 袁樹珊 (1881–1952?), a famous fortune-teller from Shanghai, who had studied sociology in Japan and made his living from medical and divinatory practice.[42] In 1929 he compiled an *Account of the Theories of Divination and Astrology* (*Shu bushi xingxiang xue* 述卜筮星相學) in reaction to the hostile political climate that led to the "Procedure for the Abolition of the Occupations of Divination, Astrology, Physiognomy, Shamanism and Geomancy" (*Bushi xingxiang wuxi kanyu feichu banfa* 卜筮星相巫覡堪輿廢除辦法) promulgated by Nanjing's Ministry of Interior in 1928. Drawing on Chinese classical literature and Western sources from various natural and human sciences, he sets out to demonstrate the scientific character of fortune-telling:

> The present compilation in total counts more than a myriad words, packed together to make eight scrolls. The *Zhou Changes*, Taiyi, Dunjia, Liuren,[43] spirit chess, character divination, choosing auspicious [days] all belong to divination. Fate calculation, physiognomy, house shapes, grave locations, all belong to astrology. They will be explained purely with scientific methods. Copiously quoting from the classics, tracing their origins, summarizing their profound principles, no words are left unclear. Even Chinese, Eastern and Western books that have rarely been seen at present times are entirely recorded one by one. This will not only cater for those who carefully study these topics, but should also be brought to the knowledge of scholars who do research in astronomy, geology, psychology, logic, law, politics, economy, biology, chemistry, mineralogy, history, mathematics, medicine and philosophy.
> 是編計十餘萬言，釐因為八卷。以周易、太乙、遁甲、六壬、棋卜、字卜、選吉，屬卜筮；以推命[44]、相人、相宅、相墓，屬星相。純粹以科學方法說明之。且引經據典，尋流溯源，提要鉤玄，語無泛設。至我國及東西各國，卜筮星相學之書目，其世所罕見者，本書均一一備錄。非唯足供留心斯學者之參考，即研究天文、地質、生理、心理、論理、法律、政治、經濟、生物、化學、礦物、歷史、算術、醫學、哲學等學者，亦所當知也。[45]

Proving that the various techniques of "prognostication and the sciences are closely linked to each other and [the former] not [the result of] daydreaming drawn on one's own imagination" is the theme of the third scroll in Yuan Shushan's

---

[41]Nedostup (2009) gives a detailed picture of the anti-superstition campaign in Jiangsu province during the Nanjing decade (1927–1937), for Guangzhou, see Poon (2011) esp. chap. 3, "Politicizing Superstition and Remaking Urban Space."

[42]See Lang (2013) p. 266.

[43]Taiyi, Dunjia and Liuren are the three models (*san shi* 三式, lit. the three cosmic boards) of Chinese predictive arts. See chap. 3, 4, and 5 in Ho (2003).

[44]In scroll 4, p. 5B where all the listed techniques are treated the expression *xingming* (lit. [determining] fate through star constellations) is used.

[45]Translated from the abstract (*tiyao* 提要) at the beginning of the book (Yuan 2014) p. 2.

Account.[46] In the following, I will limit myself to showing how Yuan instrumentalized, in particular, mathematical theories to demonstrate the scientific foundations of fortune-telling techniques. One section entitled "The Mathematics of the *Ganzhi*-System[47] and the *Five Elements*" is, except for a final commentary, entirely quoted from the *Research on the Mathematics of the Ganzhi-System and the Five Elements* (*Ganzhi wuxing zhi shuxue yanjiu* 幹枝五行之數學研究) by a certain Jin Wenqi 金雯琦 from Wujin 武進.[48] Jin's idea was to apply complex numbers expressed in polar coordinates to the various relations between the Five Elements, water, wood, fire, earth and metal, that played a crucial role in prognostication.

In Jin Wenqi's own words, one goal was to prove that astrology is a legitimate scientific practice. But what additionally shines through is the impression that Jin's enterprise was probably also an aesthetic one, showing the perfect and beautiful match of a mathematical theory and an ancient Chinese system of ordering the world:

> Jin Wenqi from Wujin Shengrui 聖瑞[49] in his *Research on the Mathematics of the Ganzhi-System and the Five Elements* says: "When I, Rui, read all sort of books on astrology and fortune-telling, it appears that from the principles/doctrines that the Ancients excelled in, they could generally know numbers. Only there are all those books that relate to arithmetic and those works on astrology that are truly transmitted forever. I started to suspect that astrology and mathematics form an integral part. Recently, I looked at the theory of complex numbers, and had the feeling that a mutual correspondence in its principle of periodicity and the cyclic repetitions of the *ganzhi*-system is hidden. Thus, I positioned the Heavenly Stems and the Five Elements as divisions in angles of equal parts. Calculating with complex numbers, indeed there was an obvious order. When using this to explain the production, conquering, combinations and conflicts, there even was no error and none that was not exhausted. As for the *ganzhi*-system, it has its origins earlier than the Emperor [Fu] Xi, but complex numbers were created only 200 years ago. Long ago and recent are more than one thousand dynasties apart. I investigated their principles, and found that they match perfectly together. This reveals that astrology may be traced to a foundation, hence it has never been worn out. What one might wonder is, if the [book-] burning of the Qin dynasty did not reach books on divination, how is it that none of the famous writings of the ancient sages was preserved? Now, when we want to pursue the search for the definite essence of production, conquering, combinations and conflicts, it is so remote that it is still impossible to obtain. Isn't this a very strange affair? Recently I heard that there are scholars who study and know the principles, they often elucidate astrology through science and have quite a few results. I am particularly thankful for this. Later on, things will continue to grow, many a little make a lot. I know for a fact that astrology will create an autonomous science, and this can be asserted. For what follows I am not ashamed of being ignorant, and I am humbly offering what I know, hoping only that those within the seas who fully understand will rectify it. First, I

---

[46]命學與科學。有息息相通之妙。非嚮壁虛造者. (Yuan 2014) p. 156.

[47]The so-called *ganzhi*-system combines the ten Heavenly Stems (*tiangan* 天干) with the twelve Earthly Branches (*dizhi* 地支) to form a cycle of sixty elements. This sexagenary cycle was used to denote days and years in the Chinese lunar calendar.

[48]I have found no further details on Jin, except for a jotting in Xu (1996), section 2 on "Recipes and Techniques" 方伎類二, which relates an anecdote about a prognostication based upon Jin's dream involving a white chicken.

[49]Probably Jin Wenqi's style name according to a footnote in Yuan (2010) p. 88.

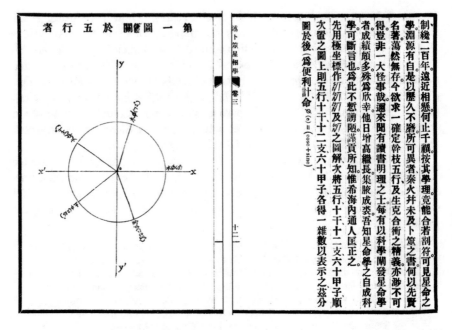

Fig. 6.6 Yuan Shushan's *Account of the Theories of Divination and Astrology* (*Shu bushi xingxiang xue* 述卜筮星相學), 1929 in Yuan (2014)

use polar coordinates to make $\sqrt[5]{1}$, $\sqrt[10]{1}$, $\sqrt[12]{1}$ and $\sqrt[60]{1}$ for the sections of the diagrams. Then I place the Five Elements, the ten Heavenly Stems and the twelve Earthly Branches and the sixty *Jiazi* on the diagram, thus each of the Five Elements, the ten Heavenly Stems and the twelve Earthly Branches and the sexagesimal [cycle] *Jiazi* obtains a miscellaneous number to represent it. The partitioned figures are below." [For convenience of calculation, we denote $\phi(e) = \cos e + [i] \cdot \sin e$.][50]

Jin Wenqi's model is the representation of suitable numbers on the unit circle in the complex plane. By positioning the five elements, ten Heavenly Stems, the twelve Earthly Branches, and the sixty binomials for the cyclical stem-branch combinations as points in a certain way on the unit circle, one can—through mathematical operations—map certain, but not all the philosophical correlations between these points.

As for the mathematical theory of complex numbers (*xushu* 虛數), Yuan's or Jin's sources are unclear. Trigonometry was already introduced much earlier in China by the Jesuits and complex numbers were certainly known through nineteenth-century translations on algebra where imaginary numbers are introduced in the context of

---

[50]Translated from Yuan (2014) p. 156–157. Formulas in the text are originally expressed in modern notation. See Fig. 6.6. The final formula is erroneous and lacks of the imaginary unit i, I added it here in square brackets.

Fig. 6.7 John Fryer's and Hua Hengfang's translation of the *Algebra* article from the *Encyclopaedia Britannica* (Fu and Hua 1873)

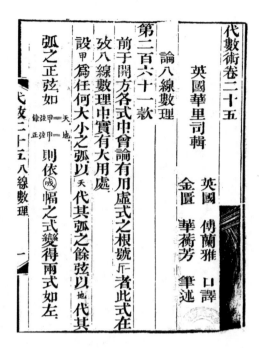

trigonometry, as in the translation of William Wallace's entry on algebra in the *Encyclopaedia Britannica* (see Fig. 6.7):

> The introduction of the imaginary symbol $\sqrt{-1}$ into analysis, has given great assistance in all investigations connected with the calculus of sines. Let x denote the cosine of any arc a, and y its sine, then, by formulæ (B) [...]"[51]

Before describing how Jin used imaginary numbers in the complex plain in his work, I will provide some mathematical preliminaries necessary for understanding.

Points in a plane can generally be described in polar coordinates (*jizuobiao* 極坐標) with a radial coordinate r and an angular coordinate θ. For conversion to Cartesian coordinates x and y (x and y are real numbers) the trigonometric functions are used:

$$x = r \cos \theta$$

$$y = r \sin \theta$$

Every complex number z can be represented as a point in the complex plane, and can therefore be represented by specifying either the point's Cartesian coordinates:

$$z = x + iy$$

---

[51] Article 261 at the beginning of the chapter on trigonometry in Wallace (1842) p. 494.

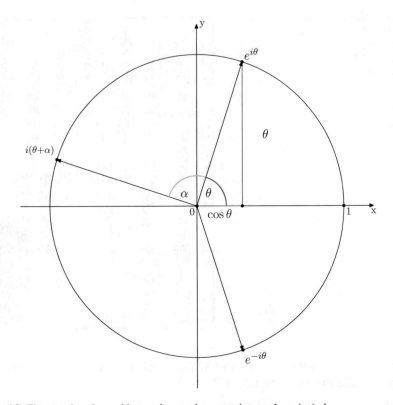

Fig. 6.8 The complex plane with complex numbers as points on the unit circle

or the point's polar coordinates (Fig. 6.8):

$$z = r \cdot (\cos\theta + i\sin\theta) = re^{i\theta}$$

where i is the imaginary unit $i = \sqrt{-1}$ and e is Euler's number.[52]

Thus, all the points on a unit circle (with radius $r = 1$) can be represented by:

$$e^{i\theta} = \cos\theta + i\sin\theta.$$

$\bar{z} = x - iy$ is defined as the conjugate complex number of $z = x + iy$. Their sum is a real number: $\bar{z} + z = 2x$. In polar form, the conjugate of $re^{i\theta}$ is $re^{-i\theta}$. Geometrically speaking, $\bar{z}$ and z lie symmetrically with respect to the x-axis.

---

[52]Euler's formula states that for any real number x:

$$e^{ix} = \cos x + i\sin x.$$

The advantage of working with polar coordinates is that it makes it much easier to multiply, divide or potentiate complex numbers. For $r = 1$ all these arithmetic operations correspond geometrically to rotations of points along the circumference of the unit circle:

$$e^{i\theta} \cdot e^{i\alpha} = e^{i(\theta+\alpha)} = \cos(\theta + \alpha) + i\sin(\theta + \alpha)$$

$$e^{i\theta} \div e^{i\alpha} = e^{i(\theta-\alpha)} = \cos(\theta - \alpha) + i\sin(\theta - \alpha)$$

$$(e^{i\theta})^n = e^{in\theta} = \cos(n\theta) + i\sin(n\theta)$$

What Jin Wenqi says in the above-quoted passage about the roots of unity, relates to the fact that the equation $z^n = 1$ has *exactly* n solutions. We thus obtain the division of the unit circle, for example, into five sectors by considering $\theta = 2\pi (= 360°)$:

$$e^{i(2\pi)} = \cos(2\pi) + i \cdot \sin(2\pi) = 1 + 0 \cdot i = 1$$

and finding, for example, the fifth-roots of unity. There are exactly five of them ($k = 0, \ldots, 4$):

$$e^{i \cdot (k \cdot \frac{2\pi}{5})} = \sqrt[5]{1}.$$

The five, ten, twelve and sixty resulting points when extracting the fifth, tenth, twelfth and sixtieth roots of unity then can be associated respectively with the Five Elements, the ten Heavenly Stems, the twelve Earthly Branches and the sixty binomials of the sexagenary cycle. For all four sets of points, Yuan provides diagrams which show the location of the respective elements and, through formulation of a general procedure and one or two examples, he explains how polar coordinates perfectly model the traditionally defined mutual relationships between the points. That this is the case, Jin concludes, "does not seem to be an accidental affair!" (*si fei ouran zhi shi* 似非偶然之事)[53] In the case of the first diagram concerning the Five Elements (see Figs. 6.6 and 6.9), for example, the following mutual relationships can be "calculated" using their polar coordinates as follows:

1. The Five Elements' mutual ratio (*xiangbi* 相比): relating one Element to itself, or geometrically speaking mapping it onto itself, corresponds to a rotation by $0°$. Thus the quotient of the "miscellaneous number" corresponding to the Element under consideration equals unity:

$$e^{i\theta} \div e^{i\theta} = 1$$

[53] Yuan (2014) scroll 3, p. 17B.

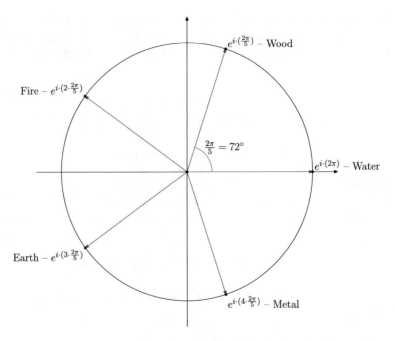

Fig. 6.9 Arrangement of the Five Elements by division of the unit circle into five equal segments

2. The Five Elements' mutual productions (*xiangsheng* 相生): rotation by 72° maps one Element upon the one it produces, rotation by −72° maps an Element upon the one it is produced by. For example, Fire produces Earth, thus:

$$\frac{\phi(216°)}{\phi(144°)} = \frac{e^{i \cdot (3 \cdot \frac{2\pi}{5})}}{e^{i \cdot (2 \cdot \frac{2\pi}{5})}} = e^{i \cdot \frac{2\pi}{5}} = \phi(72°)$$

3. The Five Elements' mutual conquests (*xiangke* 相剋): rotation by 144° maps one Element upon the one it conquers, rotation by −144° maps an Element upon the one it is conquered by. For example, Wood conquers Earth, thus:

$$\frac{\phi(216°)}{\phi(72°)} = \frac{e^{i \cdot (3 \cdot \frac{2\pi}{5})}}{e^{i \cdot (\frac{2\pi}{5})}} = e^{i \cdot (2 \cdot \frac{2\pi}{5})} = \phi(144°)$$

The calculations and examples accompanying the following three diagrams which map the Heavenly Stems (see Fig. 6.10), the Earthly Branches and the sexagenary cycle proceed in a similar way. The ordering of the elements on the unit circle is always such that the mathematical operations maintain the traditionally valid transformations between the elements within one set and in relation to another. All diagrams follow—in counter-clockwise sequence—the established order for

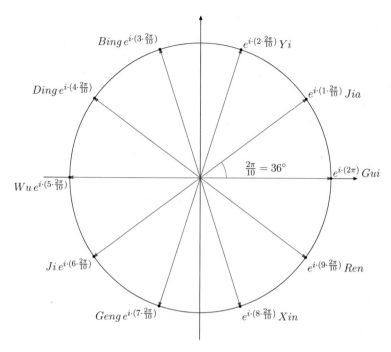

Fig. 6.10 Arrangement of the Ten Heavenly Stems by division of the unit circle into ten equal segments

enumerating the five elements (see Fig. 6.9), the ten Heavenly Stems, the twelve Earthly Branches and the sixty stem-branch combinations.

The question of where the starting point of each sequence is located is discussed in the legends to the figures. The reasons given for choosing a certain arrangement are mainly mathematical:

> The reason why we begin with Wood, and continue with Fire, Earth, Metal and Water is that we grasp the order of mutual production and because it is more convenient to seek conformity with the second diagram.[54]
> [...] The reason why the locus of *wu* 戊 and *ji* 己 is not in the centre, is because it is more convenient for the mathematics.[55]
> [...] The reason why one begins with *jiayin* 甲寅 and not with *jiazi* 甲子 [i.e. placing it at 6°], is because it is more convenient for the mathematics to seek conformity with the second and third diagrams.[56]

---

[54]Legend to the diagram for the arrangement of the Five Elements, *idem* scroll 3, p. 13A.

[55]Legend to the diagram for the arrangement of the ten Heavenly Stems, *idem* scroll 3, p. 14A.

[56]Legend to the diagram for the arrangement of the sixty stem-branch combinations, *idem* scroll 3, p. 16B.

Only the arrangement of the diagram for the twelve Earthly Branches seems to
be justified by cosmological considerations:

> On the four axis are placed the Four Graveyards (*si mu* 四墓).[57] It [the diagram] is arranged
> this way, because Earth is the most special among the Five Elements. Furthermore, it
> corresponds to the meaning of "Heaven begins with *zi* 子, Earth starts at *chou* 丑, Man
> is produced from *yin* 寅."[58]

In his final comment Yuan Shushan praises Jin's "formulas, which appear both
natural and unaffected" (*gongshi ziran bu jia zaozuo* 公式自然不假造作). They are
also revolutionary in his eyes because they establish a more scientifically advanced
relationship between the theoretical basis of traditional divination techniques and
mathematics, which hitherto had simply been framed in shared concerns with
"numbers" (*shu* 數). Yuan firmly believed that "this would open a new era in the
history of fortune-telling!"[59]

Scientific modernity and beliefs in prognostication which are only apparently
irrational, have indeed proven to be sustainably compatible in the Chinese realm.
The professionalization of fortune-telling did not retreat in spite of the repeated
governmental campaigns during the entire twentieth century. Neither did the
efforts that search for matches between scientific theories and *Yijing* numerology
disappear.[60] They were even considered more legitimate through the authority of
Fritjof Capra's *Tao of Physics*, translated in 1985 into Chinese. Indeed, some years
ago I received a paper on the Genetic Code and the ancient Chinese *Book of Changes*
for review from the American Mathematical Society.[61] The article was devoted to
the genetic coding of biochemical systems and its parallels with the symbolic system
of the hexagrams. The author associated with each of the sixty-four genetic triplets,
classified into eight groups, a 6-digit binary number, 0 and 1 being assumed to be
equivalent with the broken and unbroken lines in the Chinese diagrams. He showed
the numerological and structural analogies between the Chinese classic and modern
biomechanics. For the readers of the *Mathematical Reviews*, I could not help but
dismiss the article as anachronistic,[62] but this chapter, I hope, has shown that there
is a complex and entangled history behind such approaches.

---

[57] These are *chen* 辰, *xu* 戌, *chou* 丑 and *wei* 未, placed here on the axis, in the four cardinal
directions. All four belong to the Element Earth 土.

[58] Legend to the diagram for the arrangement of the twelve Earthly Branches, *idem* scroll 3, p. 15A.
The quoted phrase "天開於子, 地闢於丑, 人生於寅" appears repeatedly in the writings of the
Neo-Confucian scholar Zhu Xi 朱熹(1130–1200). It can for example be found in the *Classified
Dialogues of Master Zhu* (*Zhuzi yulei* 朱子語類, compiled in 1270), 論語二十七, 衛靈公篇, 顏
淵問為邦章.

[59] Yuan (2014) scroll 3, p. 17B.

[60] For an overview of the discipline of "the Science of Yijing Learning" in China and particularly
the *Yijing*-fever (*Yijing re* 易經熱) that started in the 1980s, see Dong (1993) and Amelung (2003)
p. 261–262.

[61] Petoukhov (1999).

[62] See my review MR1943948 in the AMS MathSciNet database http://www.ams.org/mathscinet/.

References

Amelung, Iwo (2003). Die "vier grossen Erfindungen": Selbstzweifel und Selbstbestätigung in der chinesischen Wissenschafts- und Technikgeschichtsschreibung. In Iwo Amelung, Matthias Koch, Joachim Kurtz, Eun-Jeung Lee, and Sven Saaler (Eds.), *Selbstbehauptungsdiskurse in Asien: Japan —China —Korea*, 243–274. München: Iudicium.

Anonymous (1612). *[Xinban quanbu tianxia bianyong wenlin] Miaojin wanbao quanshu* [新板全補天下便用文林] 妙錦萬寶全書 (Complete Book of a Million Treasures, Magnificently Embroidered [and Entirely Supplemented for the Convenient Use of all the People in the World, Newly Printed]). Jianyang: [n.p.]. Housed in the Institute of Oriental Culture, University of Tokyo, Niida Collection (Tōkyō Daigaku Tōyō Bunka Kenkyūjo Niida Bunko 東京大學東洋文化研究所仁井田文庫). Reproduced in (Sakade et al. 2004) vols. 12–14.

Anonymous (Ming 1368–1644). *Dingqie longtou yilan xuehai buqiuren* 鼎鍥龍頭一覽學海不求人 (A Dragon Head Carved in an Ancient Vessel – An Ocean of Learning at a Single Glance without Consulting the Help of Others). Reprint in (Cui 2011) 14:139–318.

Bréard, Andrea (2008). Leibniz und Tschina—ein Beitrag zur Geschichte der Kombinatorik? In Hartmut Hecht, Regina Mikosch, Ingo Schwarz, Harald Siebert, and Romy Werther (Eds.), *Kosmos und Zahl. Beiträge zur Mathematik- und Astronomiegeschichte, zu Alexander von Humboldt und Leibniz*, Volume 58 of *Boethius*, 59–70. Stuttgart: Franz Steiner Verlag.

Bréard, Andrea (2016). Eclectic or Essential Knowledge? Mathematical Methods in the *Wanbao quanshu* 萬寶全書. Paper presented at the International Workshop *Late Ming through early Republican Wanbao Quanshu* 萬寶全書: *Texts and Readers*, organized by Joan Judge, Joachim Kurtz and Barbara Mittler at the Karl-Jaspers Centre, Universität Heidelberg.

Chen, Juyuan 陳居淵 (2000). *Jiao Xun ruxue sixiang yu yixue yanjiu* 焦循儒學思想與易學研究 (Jiao Xun's Confucian Thought and Research into Yi-Learning). Jinan: Qi Lu Shushe 齊魯書社.

Cheng, Dawei 程大位 (1533–1606) (1882). *Zengbu Suanfa tongzong daquan* 增補算法統宗大全 (Complemented Collection of the Unified Lineage of Mathematical Methods) (Taishantang 泰山堂 ed.). Jingdu.

Clavius, Christophorus (1574). *Euclidis Elementorum Libri XV Accessit XVI de Solidorum Regularium Cuiuslibet Intra Quodlibet Comparatione, Omnes Perspicuis Demonstrationibus, Accuratisque Scholiis Illustrati, ac Multarum Rerum Accessione Locupletati*. Rome: V. Accoltum.

Cui, Jinming 翠金明 (Ed.) (2011). *Mingdai tongsu riyong leishu jikan* 明代通俗日用類書集刊 (Compendium of Popular Everyday Encyclopaedias from Ming Dynasty). Chongqing: Xinan shifan daxue chubanshe 西南師範大學出版社.

Dong, Guangbi 董光壁 (1993). *Yixue kexue shigang* 易學科學史綱 (=A Brief History of the Yijing Learning Science). Wuhan chubanshe 武漢出版社.

Fu, Lanya 傅蘭雅 (Fryer, John) and Hua Hengfang 華蘅芳 (1873). *Daishu shu* 代數術 (Algebra). Shanghai: Jiangnan jiqi zhizao zongju 江南機器製造總局. Original: (Wallace 1853), or probably the earlier edition (Wallace 1842) which was identical.

Guo, Shuchun 郭書春 et al. (Eds.) (1993). *Zhongguo kexue jishu dianji tonghui: Shuxue juan* 中國科學技術典籍通彙: 數學卷 (Comprehensive Collection of Ancient Classics on Science and Technology in China: Mathematical Books), 5 vols. Zhengzhou: Henan jiaoyu chubanshe 河南教育出版社.

Ho, Peng Yoke (2003). *Chinese Mathematical Astrology. Reaching out to the Stars*. London: Routledge.

Horng, Wann-Sheng (1993). Chinese Mathematics at the Turn of the 19th Century: Jiao Xun, Wang Lai and Li Rui. In Cheng-Hung Lin and Daiwie Fu (Eds.), *Philosophy and Conceptual History of Science in Taiwan*, Volume 141 of *Boston Studies in the Philosophy of Science*, 167–208. Springer Netherlands.

Horng, Wann-Sheng (2001). The Influence of Euclid's Elements on Xu Guangqi and his Successors. In Catherine Jami, Peter Engelfriet, and Gregory Blue (Eds.), *Statecraft and Intellectual Renewal in Late Ming China: The Cross-Cultural Synthesis of Xu Guangqi (1562–1633)*, 380–397. Leiden: Brill.

Jiao, Xun 焦循 (1799). *Litang xue suan* ji 里堂學算記 (Collection on Mathematical Learning from the Hall of Li). Jiangdu 江都: Jiaoshi 焦氏.

Jiao, Xun 焦循 (1829). *Yi tu* lüe 易圖略 (*Précis of Diagrams in the Changes*), 8 vols. Guangzhou: Xue hai tang 學海堂.

Jiao, Xun 焦循 (1974). *Diaogulou jingxue congshu si shi er juan* 雕菰樓經學叢書:四十二卷 (Collectanea from the Diaogu Hall of Classical Studies), 42 scrolls (Jiangdu 江都 ed.), Volume 21 of *Qingdai gaoben bai zhong huikan. Jing bu* 清代稿本百種彙刊. 經部. Taipei: Wenhai chubanshe 文海出版社.

Lang, Mixie 朗宓榭 (Lackner, Michael) (2013). Bielei de kexue: Minguo shiqi de Zhongguo chuantong xiangshu yu xixue 别类的科学: 民国时期的中国传统相术与西学 (Another Kind of Science: Traditional Physiognomy and Western Learning in Republican China). In Xu Yan 徐艳 (Ed.), *Lang Mixie Hanxue wenji* 朗宓榭汉学文集 (Collected Writings in Chinese Studies by Michael Lackner), 263–271. Shanghai: Fudan daxue chubanshe 复旦大学出版社.

Li, Yaqing 李雅清 (2002). Jiao Xun Yi xue zhi shuli siwei 焦循《易》學之數理思維 (Jiao Xun's Mathematical Thinking of Yi Learning). Master's thesis, Guoli zhengzhi daxue Zhongguo wenxuexi 國立政治大學中國文學系, Taipei. Advisor: Dong Jinyu 董金裕.

Li, Zhizao 李之藻 et al. (1965). *Tianxue chuhan* 天學初函 (First Collectanea of Heavenly Studies), 6 vols. (Reprint ed.), scroll 23, First Series (Chubian 初編卷二十三) of Wu Xiangxiang 吳相湘 (Ed.), *Zhongguo shixue congshu* 中國史學叢書. Taipei: Taiwan xuesheng shuju 臺灣學生書局.

Nedostup, Rebecca (2009). *Superstitious Regimes. Religion and the Politics of Chinese Modernity*. Harvard East Asian Monographs. Cambridge, Mass.: Harvard University Press.

Petoukhov, V., Sergei (1999). Genetic Code and the Ancient Chinese *Book of Changes*. *Symmetry: Culture and Science 10*(3–4), 211–226.

Poon, Shuk-wah (2011). *Negotiating Religion in Modern China: State and Common People in Guangzhou, 1900–1937*. Hong Kong: Chinese University Press.

Ricci, Matteo and Guangqi Xu (1607). *Jihe yuanben* 幾何原本 (The Elements). Beijing: [n.p.]. Translation of the first six books of (Clavius 1574), included in (Li et al. 1965).

Sakade, Yoshinobu 坂出祥伸, Yōichi Ogawa, 小川陽一, and Tadao Sakai, 酒井忠夫 (監修) (Eds.) (1999–2004). *Chūgoku nichiyō ruisho shūsei* 中國日用類書集成 (Compendium of Chinese Daily Life Encyclopedias). Tokyo: Kyūko shoin 汲古書院.

Sakade, Yoshinobu 坂出祥伸, Yōichi Ogawa, 小川陽一, and Tadao Sakai, 酒井忠夫 (監修) (Eds.) (2000). *Chūgoku nichiyō ruisho shūsei* 中國日用類書集成 (Compendium of Chinese Daily Life Encyclopedias). Tokyo: Kyūko shoin 汲古書院.

Smith, J., Richard (2008). *Fathoming the Cosmos and Ordering the World. The* Yijing (I-Ching, or Classic of Changes) *and its Evolution in China*. University of Virginia Press.

Wallace, William (1842). Algebra. In *The Encyclopædia Britannica or Dictionary of Arts, Sciences, and General Literature* (7th ed.), Volume II, 420–502. Edinburgh: Adam and Charles Black.

Wallace, William (1853). Algebra. In *The Encyclopædia Britannica or Dictionary of Arts, Sciences, and General Literature* (8th ed.), Volume II, 482–564. Edinburgh: Adam and Charles Black.

Wu, Jing 吳敬 (fl. 1450). *Jiu zhang suanfa bilei daquan* 九章算法比類大全 (Great Compendium of Comparable Categories to the Nine Chapters on Mathematical Methods). Reprinted in (Guo et al. 1993) 2:1–333.

Xu, Ke 徐珂 (1996). *Qingbai leichao* 清稗類鈔 (Classified Anecdotes of the Qing Dynasty) (2nd ed.). Beijing: Zhonghua Shuju 中華書局. Originally published Shanghai: Shangwu yinshuguan 商務印書館, 1917.

Xu, Xinlu 徐心魯 (1573). *Panzhu suanfa* 盤珠算法 (Methods for Calculating on an Abacus) (Xiong Tainan 熊台南 ed.). Complete title: 新刻訂正家傳秘訣盤珠算法士民利用 (*Newly engraved and corrected edition of the* Method of Calculating on an Abacus, *with verses for memorization transmitted secretly through family, for the use of scholars and commoners.* Preserved in Japan, Naikau bunko,子 56 函 5 號. Reprinted in (Guo et al. 1993) 2:1143–1164.

Yu, Xiangdou 余象斗 (Ed.) (1599). *[Xinke tianxia simin bianlan] Santai wanyong zhengzong* [新刻天下四民便覽] 三台萬用正宗 (Santai's Orthodox Instructions for Myriad Uses [for the Convenient Perusal of all the People in the World, Newly Engraved]). Housed in Tōkyō Daigaku Tōyō Bunka Kenkyūjo Niida Bunko 東京大學東洋文化研究所仁井田文庫 (Institute of Oriental Culture, University of Tokyo, Niida Collection). Reproduced in (Sakade et al. 2000) vols. 3–5.

Yuan, Shushan 袁樹珊 (2010). *Shu bushi xingxiang xue* 述卜筮星相學 (Account of the Theories of Divination and Astrology), 8 vols. Beijing: Yanshan chubanshe 燕山出版社.

Yuan, Shushan 袁樹珊 (2014). *Shu bushi xingxiang xue* 述卜筮星相學 (Account of the Theories of Divination and Astrology). Xinyitang shushu guji zhenben congkan 心一堂數術古籍珍本 叢刊. Hong Kong: Sunyata 心一堂. Reprint of Runde tang congshu 潤德堂叢書 ed.

Chapter 7
Data Management and Knowledge Production in Late Qing Institutions

Contents

In 1907, the first central Statistical Bureau in China was founded within the *Commission to Draw up Regulations for Constitutional Government* (*Xianzheng biancha guan* 憲政編查館), and was backed up by a nationwide network of Statistical Offices at central government ministries and Investigation Bureaus in the provinces. The issues involved in the modernization process of China's long administrative and mathematical practices at the turn of the twentieth century not only concerned the collection of data and the statistical treatment that allowed the vast amounts of numerical data stored in the imperial archives to be turned into knowledge urgently needed for the late Qing educational and political reforms, modernization also concerned a "mental revolution." This mental revolution was related to the conception of quantitative data, its collection, production, representation and dissemination. Providing accurate and timely statistical data for the state and sharing it with a larger public revealed the social and technical aspects of communication that are reflected in the archival management of quantitative material within the imperial bureaucracy and administration.

© Springer International Publishing AG, part of Springer Nature 2019
A. Bréard, *Nine Chapters on Mathematical Modernity*, Transcultural Research – Heidelberg Studies on Asia and Europe in a Global Context,
https://doi.org/10.1007/978-3-319-93695-6_7

In this chapter, I shall analyse the profound changes that occurred (or were wished for) in the conceptualization of numerical data that the Chinese state had collected routinely since early times, and which during late Qing reforms were considered to be central for political decision making. These conceptual changes played out on several levels and were paralleled by the formation of a new communication network allowing for an accommodation of information in the new statistical format. First, reformers insisted on the objectivity of the reported numbers, underscored by the slogan "to seek the truth through the facts" (*shi shi qiu shi* 是實求事). Second, numerical data was no longer simply stored in the archives but proved useful for the newly created knowledge-producing statistical institutions. And finally, access to statistical information was no longer restricted to high-ranking clerks with access to the imperial archives, but became public knowledge through the first available government gazettes and printed statistical tables.

Based on official communications, I shall in the following first analyse the historical context within which the late Qing government developed a certain enthusiasm for social counting, what kind of information the reformers were particularly interested in and what purposes the produced numbers were meant to serve in the specific setting of late Qing educational and constitutional reforms (Sect. 7.1). Given the long administrative tradition of collecting numbers in China, lobbying both high- and low-ranking officials to recognize the novelty of the statistical approach imported from abroad was an important task. In the second part of this chapter (Sect. 7.2), I will look more precisely at how numerical data and statistical science was conceptualized more generally by the bureaucrats and by those who authored the first Chinese language theoretical manuals on social statistics. Statistics on civil affairs will serve as a case study, allowing us to assess the modernization of and the difficulties in the actual production and dissemination of statistical data during the last years of the Qing dynasty and how statistics then was linked to the modernization of China, which I will address in the conclusion (Sect. 7.3).

7.1   Reform as Context

In China, the introduction of statistics as a branch of learning and as an effective way to communicate in aggregate format a large amount of information, took place in the context of institutional, political and legal reforms after the Chinese defeat in the first Sino-Japanese War (1894–1895). These reforms provided fertile territory for the placement of statistical institutions in the capital and the provinces and for the introduction of statistical education in the curriculum of newly created modern-style schools that mainly followed Meiji Japan's institutional model.[1] Although the

---

[1]On how, shortly after the war, China and Japan "moved from mortal enmity to apparent friendship," see Reynolds (1993) p. 17–23.

Imperial Maritime Customs Service had its own specialized Statistical Department since 1873, with a vast amount of statistical tables and reports being published in both English and Chinese, it seems that its statistical expertise had little or no influence on late Qing institutional reforms.[2]

The creation of a central Statistical Bureau (*Tongji ju* 統計局) as part of the Grand Secretariat (*Nei ge* 內閣) was first suggested in November 1906.[3] Its task was to coordinate and normalize statistical work from the Boards and compile further statistics on topics outside their domains. It is unclear, however, whether this proposal was ever put into practice before August 1907 when the Commission to Draw up Regulations for Constitutional Government memorialized to the emperor on the establishment of a Statistical Bureau within the Commission. Proposing by-laws concerning its missions, the memorial outlines the cycle of statistical production, an information framework allowing the gathering, reporting, processing, circulation, analysis and use of quantitative knowledge about the Chinese state:

> As to statistics, they are a method with which to examine the state of affairs of our fiscal administration and fathom our national strength so that comparisons can be made and appropriate policies developed. Therefore, the state of the empire has to be analysed internally, and the competition of the world has to be observed externally. From now on, the Boards and Ministries at the Capital and the provinces shall establish detailed tables on their domain of activity and submit these regularly for consultation to our office. Our office will synthesize all the tables for the purpose of estimation of the actual state of the empire. Examination of all the states, which publish statistical annals, showed that statistics are compiled in each domain in order to be informed about its population at a single stroke.[4]

The memorial received by the Grand Council (*Junjichu* 軍機處) on 24 August 1907, and examined the same day by the emperor, was endorsed with the simple comment : "Let it be done as recommended" (*ru suo yi xing* 如所議行). The emperor obviously had no objection to the commissioners' suggestions. Later that year, another memorial from the Constitutional Commission, referring to the German and Japanese model, requested that each province be commanded to establish an Investigation Bureau (*Diaocha ju* 調查局) and each Board and Ministry in the Capital to establish its own Statistical Service (*Tongji chu* 統計處). The creation of this first nationwide network of specialized statistical institutions was just one, yet one of the most important, of a number of reforms that took place during the modernization processes of the New [governmental or administrative] Systems movement (*Xinzheng* 新政) (1898–1911).[5] I will limit myself here to addressing the constitutional and educational reforms within which statistics played a crucial role.

---

[2]See Bréard (2006).

[3]Memorial proposing administrative reforms at the capital, dated GX 32.9.20 (6/11/1906) 釐定京內官制摺 in *GXXF* (1909) no. 20, p. 97A–118B.

[4]Translated from National Palace Museum Ming-Qing Archives (1979) 第一編二 預備立憲的宣布和策劃 1: vol. 1, p. 47–51.

[5]On the meaning of the term *Xinzheng*, see Reynolds (1993) p. 207n1.

7.1.1   Educational Reform

The main target of China's education reform at the turn of the twentieth century was the replacement of a 1200-year-old civil service examination system (*keju* 科舉) with overseas studies and higher education. One of the most influential essays of reformist literature in late Qing China was *Exhortation to Study* (*Quanxue pian* 勸學篇), written in 1898 by Zhang Zhidong 張之洞 (1837–1909), then Governor General of Hunan Province. Zhang advocated the adoption of Western-style industries and the modification of the traditional educational and examination system, recommending study in and about Japan. But he vehemently opposed the adoption of Western democratic institutions or concepts of social egalitarianism that would not allow the survival and strengthening of the Chinese tradition.[6] His basic philosophy was summed up in the conservative slogan "Chinese studies for fundamental principles—Western studies for practical applications" (*Zhong ti xi yong* 中體西用). Zhang presented this treatise to the throne during the One Hundred Days Reform of mid-1898. The Emperor Guangxu was so pleased that he ordered it to be copied and distributed to provincial leaders and directors of education throughout the empire.

Consequently, substantial numbers of Chinese students flowed into Japan after 1898. Either self-financed or with a government scholarship, they studied in Japanese universities or specialized schools which accommodated their particular needs.[7] On 2 September 1905, China's venerable civil service examination system was finally abolished and gradually replaced by a standardized educational system at all levels. Up to that point, the civil service examinations had consisted of compositions of essays on topics selected from the Confucian *Four Books* and the *Five Classics*, and policy questions on mainly philological issues.[8] But institutions created within the New Systems institutional reforms required more "men of talent" (*rencai* 人材), who were specialists trained in the new domains of government administration, such as law, political economy or statistics. The introduction of statistics to the curricula of New Schools of Law and Administration (*Fazheng Xuetang* 法政學堂), was one of the outcomes of the educational reforms.

In Japan, a specialized school for statistics was first founded in 1881, where the director of the Statistical Bureau, Torio Koyata 鳥尾小弥太 (1848–1905), and the statistician Sugi Kōji 杉亨二 (1828–1917)[9] were the two professors teaching the first fifty-three students, of whom twenty-seven graduated in 1886. In China, the situation was quite different. Statistics as a field of knowledge was certainly not considered part of what then was known as "Western studies," i.e. mathematics,

---

[6]Bays (1978).

[7]There is a vast amount of literature on the Chinese study in Japan programme. Reynolds (1993) p. 41–64 is a good introduction. For further details, see Huang (1975) and Tan and Lin (1983). Harrell (1992) analyses how the Chinese students' experience in Japan politicized them.

[8]See Miyazaki and Schirokauer (1994) and for many more details, see Elman (2001).

[9]A collection of his early teachings is reprinted in Sugi (1902).

chemistry, physics or mechanics.[10] At the turn of the twentieth century, when the first higher education institutions for law, political or economic sciences were founded, where statistics should have been taught exclusively, neither qualified teaching personnel, nor Chinese language manuals were available. In the early phase, Japanese manuals served as a basis for teaching statistics, as did the Imperial Maritime Customs' statistics in the Economy Department of the Metropolitan University (*Jingshi Daxuetang* 京師大學堂)[11] and in the Ministry of Finance's School of Political Economy.[12]

The Constitution of the Metropolitan University,[13] Peking University's forerunner established in 1898, mentions statistics, next to banking and insurance, as a subject to be taught to future government officials in the third year of a course on political economy (*licaixue* 理財學)[14] in the School for Administration (*Shixue Guan* 仕學館). Japanese doctors of law taught this course,[15] assisted by Chinese translators. These were Chinese students returning from Japan, who were subsequently hired to teach the course by themselves. Generally speaking, the Metropolitan University had difficulty finding qualified teachers for the newly introduced courses in modern fields of knowledge. Archival documents from 1905 refer to two teachers of "difficult courses," Qian Chengzhi 錢承誌 and Lin Qi 林棨, both of whom returned as students from Japan and later became members of the central Statistical Bureau.[16] When the Grand Minister of Educational Affairs (*Xuewu dachen* 學務大臣) requested the extension of Qian's and Lin's services, he stressed that "teachers of sciences are few."[17]

In the specialized New Schools of Law and Administration (*Fazheng Xuetang* 法政學堂), the regulations also provided for a course on statistics. Zhang Zhidong's memorial of 1904 on the Regulations for New Schools (*Xuetang zhangcheng* 學

---

[10]See, for example, the curriculum in those fields at the Shanghai Polytechnic Institute in Fu (1895).

[11]See the communications between the Ministry of Foreign Affairs and the Metropolitan University on May, 13th and 16th, 1903 on sending the trade volumes of the previous years' customs for editing textbooks. Reprinted in Beijing University and N° 1 Historical Archives (2001) p. 194–195.

[12]Statistical Secretary Despatch n° 940, 1910. N° 2 Historical Archives, Nanjing.

[13]Constitution (*Jingshi Daxuetang zhangcheng* 京師大學堂章程) sent on 15 August 1902 as an attachment to a Palace Memorial by the Chancellor of the Metropolitan University, Zhang Baixi 張百熙, dated 4 August 1902. Reprint in Beijing University and N° 1 Historical Archives (2001) p. 146–173.

[14]In Song dynasty the term *licai* referred to the techniques of economic management promoted by Wang Anshi 王安石 (1021–1086). The term was reintroduced into China via Japan at the end of the nineteenth century, referring to political economy.

[15]Ye (1974) p. 37 mentions Sugi Eizaburō 杉栄三郎, and Iwaya Magozō 厳容孫蔵. See Reynolds (1993) p. 80–81 for more details on their academic background and activities.

[16]See the documents in Beijing University and N° 1 Historical Archives (2001) n° 265 (15 August 1905), n° 273 and n° 274 (12 September 1905).

[17]See the Palace Memorial, dated September 12, 1905 in Beijing University and N° 1 Historical Archives (2001) n° 273.

Fig. 7.1 Zhang Zhidong et
al., *Memorial to Fix the
Regulations for New Schools*
(1904)

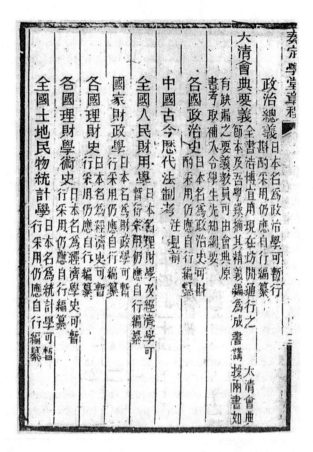

堂章程) referred in particular to a weekly 1-h course on "Statistics on Civil Affairs
and Land of the Whole Empire" during the third and fourth year of study of political
science (see Fig. 7.1). He suggests that "temporarily Japanese language manuals
may be used, but subsequently Chinese teaching material should be compiled."[18]

When the Metropolitan University's School for Administration[19] became the
independent Metropolitan School of Law and Administration (*Jingshi Fazheng
Xuetang* 京師法政學堂) in 1907, statistics was taught for two weekly hours in the

---

[18]Memorial, received on 16 January 1904 (GX 29/11/29) by the Grand Secretariat 內閣奉. Reprint
in Zhang et al. (1904) p. 109–304, see here in particular section 2 p. 133–141 (*di er jie zhengfa ke
daxue* 第二節 政法科大學)

[19]The School for Administration (*Shixue Guan* 仕學館) was integrated into the School for
Metropolitan Graduates (*Jinshi Guan* 進士館) in March–April 1904 (GX 30/2). See Ye (1974)
p. 36.

third year of the regular courses (*zhengke* 正科) of the politics department.[20] Later in 1910, when the general curriculum of New Schools of Law and Administration was revised in the course of the constitutional reforms, statistics became a weekly 2-h course in the first year of study in the political and the economic section.[21] This underscores the growing importance attached to statistical education for expectant state officials. For the preparation of constitutional government, for example, one pressing issue was the task of counting the population for an adequate organization of elections of local councils.

The abolition of the traditional examination system, along with the establishment of a nation-wide education system, was related to general change in early twentieth-century China. Changes in education were only one part of a broad programme promoted by a group of reformers in Chinese society. Among the New Systems reforms, education reform was the earliest and probably the most far-reaching. It produced the leaders of the future, who had not followed the purely traditional Confucian path of learning, but were exposed to modern political, legal and economic thought. Followed in chronological order by military reforms, legal and judicial reforms, the final effort to save the dynasty through constitutionalism, brought about the first specialized statistical institutions in China.

7.1.2 Constitutional Reform

The reform movement in late Qing China, which called for institutional changes, was sanctioned by an imperial edict from the Empress Dowager issued on 19 January 1901. The transformation of state institutions, however, could only be envisaged if their political power was left intact.

Reform-minded officials, such as Huang Zunxian 黃尊憲 (1848–1905) dreamed of seeing the establishment of a constitutional monarchy in China following Meiji Japan's example.[22] Huang was appointed Counselor to the Imperial Chinese Legation in Tokyo in 1877. During his stay in Japan he composed *Poems on Miscellaneous Subjects from Japan* (*Riben zashi shi* 日本雜事詩, 1879) and *Treatises on Japan* (*Riben guozhi* 日本國志, published 1898). The Emperor Guangxu 光緒 (r. 1875–1908) subsequently requested copies of these poems because of "Huang's progressive attitudes and eye-witness accounts of Japanese modernization."[23] Both a poem on statistical tables and a short account of the Japanese Grand Council of State's Bureau of Statistics (*Dajōkan Tōkeiin* 太政官統計院) are included in

---

[20]See the Memorial submitted February 2, 1907 (GX 32/12/20) to fix the regulations for the Metropolitan School of Law and Administration (*Zouding Jingshi Fazheng Xuetang zhangcheng zhe* 奏定京師法政學堂章程摺) in Board of Education (1909) p. 188.

[21]Ye (1974) p. 238, 242 and *XTXF* (1911), n° 27, p. 4B.

[22]Noriko (1981).

[23]Lynn (1997) p. 113.

Huang's writings.[24] Huang stated that statistical tables revealed the power of the nation (*guoshi* 國勢) and extensively cited statistical tables in the *Treatises*. The poem compares the statistical yearbooks which Huang had seen in Japan with the work of ancient China's annalists, implying that the beginnings of modern statistical tables can be found in the compilations of the first official dynastic histories:

> *On Statistical Tables*
> Grasping the essential point of a matter, calculations do not vary.
> Collecting the bits and pieces, how the quantities compare!
> Horizontal writing of occidental languages equals the methods of the Zhou.
> Who would have thought that the regularization of register books is based upon [ancient China's] historiographers?[25]

Huang Zunxian's official career ended with the crushing of the 1898 reform movement, which he had joined enthusiastically; but his writings, the first to refer to a Statistical Bureau in Japan, might have had an influence beyond the Hundred Days Reform during the constitutional movement. Book registers in the Beijing archives show that Huang's *Treatises on Japan* circulated within the central government agencies in Beijing concerned with constitutional reform.[26]

In the summer of 1904, reform-minded officials forwarded a printed translation of the Meiji constitution to the Empress Cixi. In the same year, Zhang Zhidong printed copies of his *Exposition of the [Japanese] Constitution* ([*Riben*] *xianfa yijie* [日本]憲法義解) and *History of the Japanese Diet* (*Riben yihui shi* 日本議會史), which he personally discussed with the high Manchu official Tieliang 鐵良 (1863–1939), who later became "a key proponent for constitutionalism."[27] I will not rewrite the entire history of constitutionalism in late Qing China here, but instead concentrate on the events related to statistical institutions.[28]

In November 1905, a temporary institution, the Office for the Investigation of the Principles of Modern Politics and Government (*Kaocha Zhengzhi Guan* 考察政治館), was established to supervise everything connected with the preparation of constitutional government. It was renamed by imperial edict on 13 August 1907 as the Commission to Draw up Regulations for Constitutional Government.

---

[24] Both reprinted in Huang (1985) p. 639–640. I am particularly grateful to Prof. Iwo Amelung for pointing this reference out to me.

[25] The editor points out that the version in the first edition of the poems is different for the last four sentences:

> Horizontal writing of occidental languages equals the record books of the Zhou.
> Who would have thought that the origin of their excellent methods is based upon [ancient China's] historiographers?

[26] See the *Register Book of Written Material* (*Shuji ce* 書籍冊), March 1911 (XT 2/2). Beijing N° 1 Historical Archives, *Xianzheng choubeichu* 憲政籌備處 *zaji* 雜記 174.

[27] Reynolds (1993) p. 187.

[28] Two entire monographic studies are dedicated to this subject; Meienberger (1980) is complemented by Chen (1978), who focuses on the Japanese influence during constitutional reforms in China.

Fig. 7.2 Anonymous, *Explanation of Statistics in Japan* (ca. 1907)

Despite being traditionally translated as "Constitutional Commission," a convention that I shall adopt in this chapter, a vast amount of the work conducted by this commission concerned larger questions of public administration. Official Chinese documents referred to the commissioners as "high officials sent to diverse countries to investigate their governments" (*chu shi geguo kaocha zhengzhi dachen* 出使各國考察政治大臣).[29] One such document communicated by an anonymous "high official investigating governments" introduces Japan's statistical annals and gives an outline for each of the 558 tables. It is published in a series of books explaining the Japanese constitutional and administrative system and was meant to prepare the Qing government's subjects for the coming constitution (Fig. 7.2).[30]

When constitutional commissioners suggested the creation of a first central Statistical Bureau within the Constitutional Commission, we do find allusions, similar to Huang Zunxian's, to the most ancient Chinese administrative traditions of compiling numerical data.[31] Referencing the old in a proposal to reform justified the adoption of a foreign concept or institution and underlined its compatibility with Chinese traditional administrative practices.[32] As mentioned earlier, the emperor

---

[29] See Horowitz (2003).

[30] Anonymous (high official investigating governments) (1907).

[31] For a full translation of the memorial, see Bréard (2008) Appendix A1 p. 152–160.

[32] For a detailed discussion of early Chinese administrative statistics, see Bréard (2008) p. 97–98.

indeed had no objection to the commissioners' suggestions and endorsed the memorial with a simple "Let it be done as recommended."

A detailed 9-year plan was worked out for the preparation of constitutional government, and the role of prefectural Statistical Offices and provincial Information Bureaus would be to follow, report on, and summarize all of the events scheduled to occur between August 1908 and 1916, thus measuring the advancement of the reforms.[33] To name just a few, their tasks were to prepare Provincial Assemblies, proclaim regulations for local self-government,[34] for a population census and for financial administration; request the establishment of a Commission for the Revision of the Banner Organization; compile a manual for a simplified way to recognize characters; compile a compulsory reading list for the people; revise the new penal code; compile a civil code, a commercial code, and laws relating to civil suits, the code of criminal procedure and other codes; coordinate provincial assembly elections (scheduled for summer 1909), election of local councils, followed by county councils; spread literacy with ultimately twenty per cent of the population being able to read; and determine a budget and accounting system. The work of the Constitutional Commission thus covered a wide range of public administration issues, with universal education forming one very ambitious goal in a continuation of the earlier educational reforms. In the following, I will refer in particular to statistics on civil affairs, to illustrate the issues involved in the ongoing modernization of statistics as a mirror of societal phenomena and as a tool for the late Qing government's civil administration.

7.2   Modernizing Statistical Practices

A tighter administrative structure should not only allow better control of society and of finances in the provinces, it should also guarantee better monitoring of the reform processes through the statistical methodologies applied to specifically investigated questions of social, political or financial importance. Soon after their establishment in 1907, the new statistical institutions began to generate a profusion of tables and data. Although regular reporting to the Centre was nothing new in the imperial communication system, officials pressured the young bureaus to respond more promptly and not fall so seriously behind schedule, as was observed through an inventory of unreported data for the Guangxu and earlier reigns.[35]

---

[33]List with a Chronology of Preparing all Matters appended to a joint Memorial by the Constitutional Commission and the National Assembly (*Zizheng yuan* 資政院), dated 27 August 1908 (GX 34/8/1) published in National Palace Museum Ming-Qing Archives (1979) vol. 1, p. 61–67.

[34]On the idea of local self-government in late Qing China, see Thompson (1995).

[35]See the long "List enumerating the documents not reported from the provinces during cumulative years" (*Ge zhi sheng jinian wei bao ge an kai lie qingdan* 各直省積年未報各案開列清單),

Population numbers, for example, were no longer simply important for fiscal reasons, but urgently needed to determine the distribution of seats in local assemblies. The government's systematic campaign to attack, under foreign pressure, the opium problem—not only as an economic but also as a social problem—after 1906 relied upon a system of registering opium smokers and statistical techniques to obtain reliable information on opium production, distribution and consumption. The first set of data appearing in the *Statistical Tables of Shuntian Prefecture for the year 1908* gives information on the field sizes used for plantation and the numbers of opium stores and smokers that had reduced or given up smoking, according to their social status and the county where they were located.[36]

But modernizing statistics was not limited to institutional aspects of data communication and the making of new contents, but also pertained to the format and the accuracy of the provided data. In order to suit bureaucracy, the production of "good" statistics requires a complete set of numbers, from all the provinces, based on common standards of measurement and tabular layout. That such ideal statistical knowledge was not easily produced in a situation where a government in crisis required, within a very short lapse of time, precise numbers that were central to the financial, political, educational and social landscape of the empire, can easily be observed in the available archival sources. A lot of effort went into designing tabular layouts, and into the management of delays, regionalism[37] and errors and the extrapolation of missing figures. The first statistical tables published by the Board of Education for the year 1907, for example, were based upon an incomplete set of numbers. Since three provinces—Zhili, Guangdong and Sichuan—had not reported on time, the corresponding figures, the legend says, "were not calculated but approximated, in order to preserve the correctness of the comparison."[38]

7.2.1 New Modes of Presentation

Although we do not know what mathematical methods were used to estimate the students in Zhili, Guangdong and Sichuan, the three provinces assumed to have the most students in the empire according to a pie chart in the tables for the year 1907, we certainly notice in Fig. 7.3 that the exploding number of students in schools from 10,000 in 1902 to 51,000 in 1907 (right to left) is visualized in a dramatic way.[39] By choosing a graphic representation of the resulting number of students

---

attached to a memorial received February 9, 1907. Archives of the Office of Military Archives 方略館, Printed Materials n° 41. Beijing, N° 1 Historical Archives.

[36] See Shuntian Prefecture (1908) p. 97A–101A.

[37] The provincial investigation bureaus did not always respect the imperially edited table patterns, but used self-made questionnaires or different tabular forms in their investigations, which they believed to be more suitable to local conditions. See Zhang and Mao (2005) p. 87 in case of the "Customs Survey" completed towards the end of the Qing dynasty.

[38] Legend to illustration shown in Fig. 7.3 from Board of Education, Department of General Affairs (1973) p. 9.

[39] Reprinted in Board of Education, Department of General Affairs (1973) p. 9.

Fig. 7.3 Growing number of
students in schools (Board of
Education, *First Statistical
Tables on Education*, 1907)
Board of Education,
Department of General
Affairs (1973)

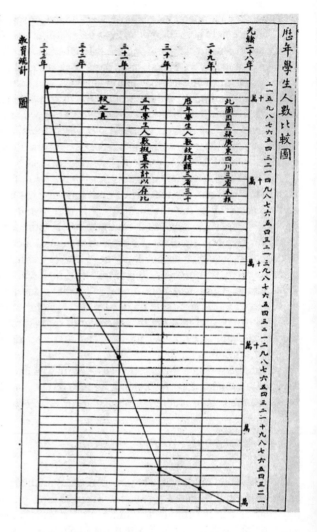

in public schools all over the empire, the success of educational reforms could
easily be grasped. Such would not have been the case had the new statisticians
adopted the traditional list format, which requires a linear reading and a certain
level of numeracy to compare the totals of different years. Other graphic layouts
of statistical data, such as bar diagrams or pie charts for distributions, were placed
along with Fig. 7.3 at the beginning of the first compiled educational statistics for
the year 1907. Descriptive and numerical tables follow these diagrams, but their
history still needs to be written. Chinese authors insisted on the origins of tabular
layout going back as early as the Xia or Zhou dynasty. Yet it seems that in the early
twentieth century, Chinese consumers of statistical material still needed to get used

to tabular statements, which was not the prevailing format for communicating data to the central administration during the Ming and Qing.[40]

As mentioned above, I do not know how accurate the published educational numbers were, but it is not the purpose of this chapter to evaluate the available data from a quantitative point of view.[41] I am instead interested in what standardization efforts were made to allow the production of accurate and reliable numbers, how the authors of official documents tried to communicate a new concept of numbers to its first generation of statistical experts and what strategies they used to convey an impression of accuracy and truth.

7.2.2 Publicizing 'True' Numbers

Through statistical publications the Chinese state became not only an administrative but also a cognitive space, one that was observed and described through coherent matrices. A fundamental prerogative then was that the reported numbers had to correspond to the actual social, economic, moral or any other reality concerned. The in China well-known slogan "seeking the truth through the facts" (*shi shi qiu shi* 事 實求是) defined what counted as objective knowledge and was officially linked to statistical figures:

> Your Majesty's servant, looking upward he implores the glance of your sacred Majesty upon a memorial respectfully reporting upon the obedient establishment of a Statistical Office at our Ministry and proposing the regulations. [...] In the beginning, when our Ministry started to establish [a Statistical Office], it was the Correspondence Section that both managed statistical and other affairs. Afterwards we received an imperial mandate commanding to depute a special employee to manage the full manifestation of the Imperial Court's policy of "seeking the truth through the facts," which emphasizes the utmost significance of statistics. It was last year during the eleventh moon, that this Ministry obediently established a Statistical Office and temporarily selected officials to share its responsibilities.[42]

Reform-minded high officials regularly reminded their provincial investigators about the importance of accurate investigations.[43] But accuracy was also desired for

---

[40]See, for example, Hart to Morse, 13 September 1905. Robert Hart Letters. Houghton Library, Harvard University, Cambridge, MA (MS Chinese 4):

> Chinese are clamouring for quicker information and the Ministry of Commerce and Industry wants statistics: they are also growing accustomed to tabular statements.

[41]So far, except for population numbers, there are no quantitative studies available on the reliability of late Qing statistics.

[42]Memorial from the *Ministry of Civil Affairs* concerning the establishment of a Statistical Office (August 7, 1908). Translated from *ZZGB* (1911), n° 280 (GX 34/7/11) p. 4–5.

[43]For example, Duan Fang 端方, Governor of Liangjiang (Zhejiang, Jiangsu and Jiangxi Provinces) in a memorial on the establishment of an Investigation Bureau, received by the Grand Council (*Junji chu* 軍機處) October 16, 1908 (GX 34/9/22). Beijing N° 1 Historical Archives, Grand Council Archives n° 166675:

> Your servant ordered the Bureau's personnel to make the search of the truth through facts their first principal.

Fig. 7.4 Statistical table showing the total number of students in Shuntian Prefecture for the year 1910

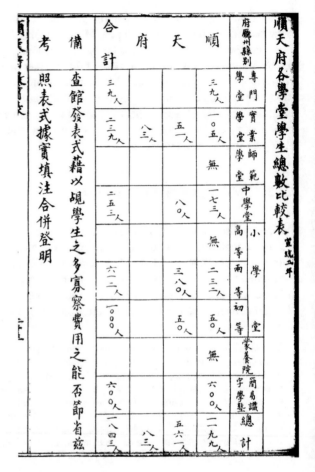

the final output, and was certainly more objective when it came to calculating the totals of several items in a table. Certain layouts even had an inherent mechanism to double-check the numbers, as can be seen in a table giving the total number of students in Shuntian Prefecture (now Bejiing), i.e. 1843, in Fig. 7.4. The same sum should be obtained when adding up horizontally the last row of cells, giving the total number of students in each column, and when adding up the cells in the leftmost column, indicating the total of students per type of school.

Furthermore, only common standards according to which data was to be collected in the provinces and submitted to the central agencies allowed uniformity and comparability within China's large territory. As pointed out in the outline of the *Statistical Tables on Education for the Year 1907*, monetary units were not yet standardized among China's provinces: six provinces reported in silver dollars (*yinyuan* 銀元 or 銀圓), eleven in silver taels (*yinliang* 銀兩), one used both currencies and four provinces did not even specify the underlying monetary units.[44]

The same problems of normalization occurred for measures and weights and the emperor generally showed concern about a lack of coordination among provincial officials and between the central and local statistical institutions, which might limit the success of the modernization programmes and centralization efforts:

> In all provinces, the popular sentiments of customs and all the honoured practices of conservatism and reform are unequal and not uniform. Now, the said Commission [for Drawing up Regulations for Constitutional Government] opened two Bureaus: one for compilation and one for statistics, but there is no cooperative work between the centre and the periphery, there is no means to carry into operation with utmost benefit what has already been decided. [...] According to the by-laws fixed in an earlier memorial, [all matters] shall be managed with great care and honesty. Each document of investigation shall be submitted officially, according to a schedule, to the said Commission. Concerning statistical affairs, it is suitable that each Ministry and Court will first collect and then compile.[45]

In a later communication, the same concerns in knowledge management are addressed by the memorialist:

> Chinese frontier walls are broad and long, and our people and products are abundant. Government administration is so vast that coordination is difficult. Communication is so difficult that centre and periphery become alienated. Viceroys and Governor-Generals cannot exhaustively know the affairs of their provinces; the Boards and Departments at the capital cannot exhaustively understand the affairs of the Viceroys and Governor-Generals. [...] In civil administration the most difficult task is to examine accurately the population; in financial administration the most important task is to manage expenses effectively. Today long standing practices remain unchanged. Even if audits provide against confusion, our accounting practices are the result of flexible accommodations. When we investigate households, we find that many have persons omitted from our registers. When we investigate the Treasury's accounts, we find that in many places there are irregular numbers. They would not be accepted in court. With so many omissions and irregularities, we simply cannot have a good understanding of the complex multiplicity of the whole system.[46]

In the course of drafting a constitution and a legal code, formal legislation of statistical procedures enacted by the central state authority defined a uniform and just basis for diverging local practices, thus theoretically enabling numerical comparisons between different branches of government activity and comparisons over time. Following the institutional establishment of central and local statistical agencies, the precise layouts and contents of a multitude of statistical tables concerning civil and financial administration were proposed and imperially approved by the Constitutional Commission in March 1909. A month earlier, the

---

[44] See Board of Education, Department of General Affairs (1973) Preface 例言, p. 1a.

[45] Imperial Edict, dated 22 October, 1907.

[46] Memorial of the Commission for Constitutional Government for settling table patterns for civil and financial administration (received XT 1/2/20, March 11, 1909). Reprint in *ZZGB* (1911) n° 493 (March 15, 1909) p. 3–5.

Commission had sent a short communication to all the provincial Investigation Bureaus urging them to start classifying their findings without waiting for the promulgation of statistical table patterns.[47] The Commission stressed the difference between documents resulting from investigations and the ultimate compilation of statistical tables, which were subject to specific imperially prescribed standards. These included such details as width and breadth of paper, where it was "not allowed to make unauthorized changes."

The first statistical tables for the year 1907 published by Statistical Offices established at central government ministries, were on education,[48] law,[49] agriculture industry and commerce,[50] and posts and communications.[51] Bureaucratic efficiency is mirrored in the amazing speed with which some of the respective ministries could produce the required tables so quickly after the establishment of a Statistical Office: the Ministry of Postal Administration alone produced six volumes of about fifty tables each, the Ministry of Education 334 tables complemented by ten diagrams for their first tables for the thirty-third year of Guangxu reign (ca. 1907) and the Ministry of Law 830 tables for thirty-fourth year (ca. 1908). The Ministry of Civil Administration could only provide eighty tables due to missing information from certain provinces.[52]

Following the new rules for designing statistical tables, the format shifted from traditionally compiled *Clear Lists* (*qingdan* 清單 or *qingce* 清冊) to a new layout. Modern style statistics for the monthly reporting of grain prices for example now followed a tabular layout, where administrative units are represented in columns and crossed with the different types of grain. A bottom line indicates the excess or deficit as compared with the previous month's prices (see Fig. 7.5 for the prices in Yunnan in 1911). Numerical data continued to be expressed with Chinese characters, which could take up more space than provided for by the table patterns. But the regulations pertaining to the layout and rules for the preparation of a total of 326 tables on financial and civil administration strictly prohibited adapting the width or breadth of each cell.[53]

---

[47] See *XTXF* (1911) n° 1, p. 50B–51A.

[48] Board of Education, Department of General Affairs (1973).

[49] See the memorial accompanying the submitted tables, endorsed on 16 July 1908 (GX 34/6/18) in *XTXF* (1911) n° 18, p. 10A–11A.

[50] Statistical Office of the Board of Agriculture Industry and Commerce (1908).

[51] See *ZZGB* (1911) n° 948 (XT 2/5/14) p. 5–7.

[52] See *ZZGB* (1911) n° 867 (XT 2/2/21) p. 3–4.

[53] See articles 13 and 14 of the Memorial from the Constitutional Commission concerning general rules for the design of statistical tables (received March 15, 1909) in *XTXF* (1911) n° 2, p. 2A–3B. For a full translation of the memorial, see Bréard (2008) Appendix A.5. p. 179–186.

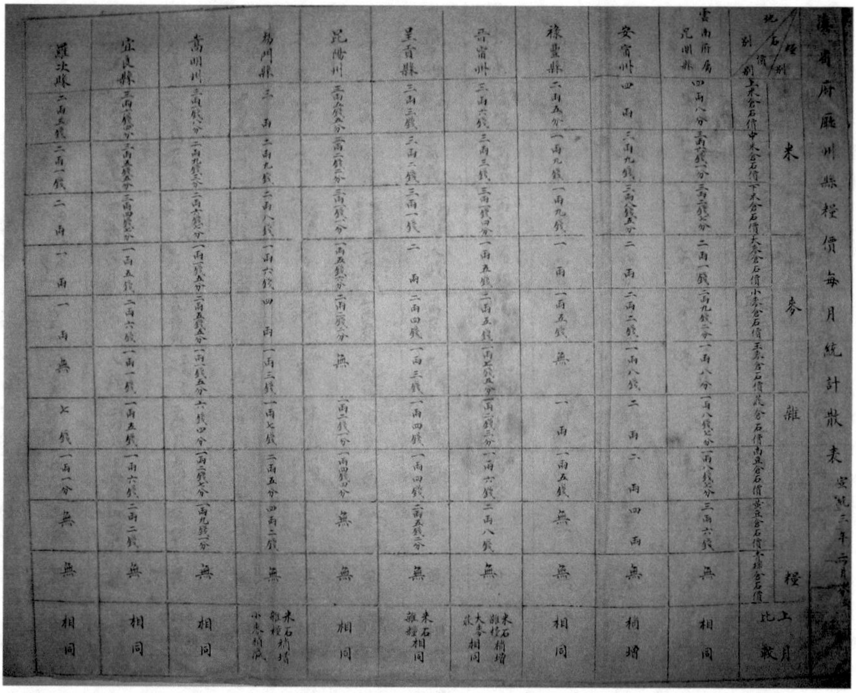

Fig. 7.5 Statistical table showing grain prices for Yunnan Province (1911)

7.2.3 Conceptualizations

In the year following the promulgation of statistical legislation, numerous communications between Ministries and Governors or Governor-Generals of the Provinces refer to delays in reporting and problems in the compilation of national statistics. But their difficulties were not only due to an un-unified monetary and metrical system, problems of personnel and a lack of incentives for local officials to share information. I would argue that there was a conceptual problem for the provincial and metropolitan bureaucrats in grasping the novelty of the statistical approach. Although confusion was partly due to a terminological problem,[54] it was mainly conceptual: apparent similarities between the traditional reporting system of numerical lists and the compilation of statistical tables were many, but analysing the reported statistical data with specific methods and using the results in the process of political decision-making was something new:

---

[54]The term *tongji* 統計 used for "statistics" was not a neologism, but existed in earlier times. It traditionally was used to designate a total or the sum of a series of numbers.

It is our humble opinion that the reason why statistics in all the Eastern and Western States are published is not only that they allow to revise accounts, or just sketch exercises for mathematical formulas. What they want is to research the strength of the nation, the condition of the populace, find out the reasons for its waning and waxing, its advancing and retiring; they regard them [i.e. statistics] as the basis of comparison for devising programmes. It is the importance of this connection which made [statistics] one of the branches of science.[55]

This also required a new understanding of the very object of statistics, which was concerned with social dynamics and not with static entities. Meng Sen 孟森 (1868–1937), an activist in the constitutional movement and translator of the most influential Japanese manual on social statistics in China , was among the first in 1909 to underline this "mental revolution." Confucian learning, he said, is too much plastered onto politics and tied down by conventions, but "what counts for statisticians are the myriad transformations of human affairs."[56] Although Meng did not pursue any research in statistics, he certainly did strongly support the introduction of statistics in the curriculum and in government administration. His manual, a Chinese translation of Yokoyama Masao's 横山雅男 (1861–1943) *General Discussion of Statistics*[57] circulated widely in China, not only in government circles but also in institutions of higher education where statistics were taught in the law, politics and economics curriculum (Fig. 7.6).[58]

The manual is divided into nine chapters, starting with a historical outlook on the evolution of statistics in the West and in Japan. Yokoyama defines statistics as a science that "uses quantities to observe social phenomena. Or to state it in more detail: the science of statistics is the research of the causality and regularities of the phenomena of a society's and a state's stability and movements, based on mass observations with combined methods."[59] The following chapters—all equally descriptive in nature—are dedicated to more specific topics dealing with statistical theory and methods, statistical institutions, population statistics, economic statistics, administrative statistics, social statistics, moral statistics, educational and religious statistics. The chapter on population statistics, presented as the most important type of statistics, takes up the major part of the book.

---

[55] Memorial of the Constitutional Commission dated 11/3/1909, 擬定民政財政統計表式酌舉例要摺 reprinted in: *ZZGB* (1911) 摺奏類 n° 493, 1909 (宣統元年二月二十四日).

[56] Meng (1909) p. 4. Meng had studied law and politics in Japan and translated several other works on constitutional government and law. Upon his return, he was in charge of the courses on Politics and Law organized by the Jiangsu Education Association. See Bastid (1988) p. 168.

[57] Yokoyama (1906), Meng (1909). Both the original and the translation, were reprinted and edited many times up to the 1930s.

[58] The Constitutional Commission lists two copies of the translation in its 1911 book inventory. See the *Register Book of Written Material* (*Shuji ce* 書籍冊), XT 2/2 (March 1911). Beijing N° 1 Historical Archives, *Xianzheng choubeichu* 憲政籌備處 *zaji* 雜記 174.

[59] Yokoyama (1921) p. 47–48 or Meng (1909) p. 53.

Fig. 7.6 Title page of Meng
Sen's translation of
Yokoyama Masao's *Tōkei
tsuron* (1909)

Meng's translation was not a translation in a strict sense. Interspersed with his personal commentaries, it reflected his critical attitude towards China's conservative forces and praised the use of statistical methods as an indispensable tool in governance[60]:

> The translator comments: In the twenty-eighth year of Meiji [1895], the fourth month of *Yimo* in the cyclic calendar, which is about in our second or third month, at the time when China was defeated by the Japanese in Port-Arthur [Lüshun], if the author's [Yokoyama's] writings and additional commentaries had been transmitted abroad, their science would have earned highest praise. Now, not even mentioning Korea, the people of my nation go to Japan for study. This opened the path for statistics. This branch of learning is a universal instrument open to all. Those who first become aware of this certainly will have glory, but even those who awake later, must not be ashamed. In my [China's] antiquity, those who transmit statistics have done so for more than 4000 years. Who knows that in my country the arts and sciences have declined and not progressed? This also applies to statistical theory. Nowadays there are few who have even heard its name, and those who investigate its content are as if they were both in bright and dark territory. Seeing that in the West, care and diligence was given to this for 300 years, and that Japan has hastily started to catch up with it for 40 years, we cannot miss another day in studying it. The gentry of my [country] is still dreaming. Alas, what a pity!

---

[60]Meng (1909) p. 43.

Such innovation in imperial bureaucracy was, of course, a hard sell and several reform-minded thinkers tried to subtly introduce the establishment of statistics as an accepted part of governmental functions by pointing out the parallels with ancient Chinese administrative tradition. In Japan, the contemporary Prussian conception of descriptive statistics as a "Staatswissenschaft," a science serving the state, had been introduced in the 1870s primarily through Sugi Kōji and his student Yokoyama Masao. Both used or translated German works on social statistics in their teachings and applied methods of investigation in their statistical work. In a historical chapter in one of Yokoyama's writings, precisely the book which was translated into Chinese by Meng Sen, the author follows the Jesuit missionaries in tracing the history of statistics in East Asia back to the times when statistics had an actuality well before they were a concept:

> During the eighteenth century, the French missionary Gaubil translated the *Tributes of Yu* chapter of the *Book of Documents*,[61] thus transmitting it to Europe. When the people of these countries saw it they were astonished and took it as some kind of East Asian ancient statistics. [...] When we evaluate it from the viewpoint of contemporary statistical learning, it does not have the least value, but in terms of the descriptive school it has nothing that does not provide material for study. Indeed, several thousand years ago there was already that kind of investigations ![62]

Equalizing statistics with geographical descriptions in tabular form as they can be found in the *Tributes of Yu*, obviously all depends on the interpretation of the term "statistics" itself. Seen merely as an activity of collecting and registering data for administrative purposes, the history of statistics in China certainly has its origins in early antiquity, and German authors of the nineteenth century shared that view.[63] But statistics as a discipline, with systematic methods of investigation and an institutional framework does not begin to develop in China until the introduction of Western-style administrative statistical methods via Japan during the constitutional reforms. This introduction was officially justified by the claim that it was not introducing anything foreign but rather restoring and extending the most authentic Chinese tradition in compiling statistical data. Innovations designed to support and improve the status quo, with the hope of strengthening the dynasty, fit well into the concept of constitutional reform within tradition sanctioned by the court. In a memorial of March 1909, where the Constitutional Commission proposed the guidelines for designing statistical tables for civil and financial administration, the idea that statistics could find its origins in Chinese antiquity was propagated to convince the emperor diplomatically of the necessary change in administrative procedures:

---

[61] The *Yu gong* 禹貢 chapter of the *Book of Documents* (*Shang shu* 尚書) is the most authoritative ancient Chinese terrestrial description of the Chinese state. Its authorship is legendarily attributed to Yu the Great (5th–3rd cent. BCE).

[62] Translated from Meng (1909) p. 7–8.

[63] See for example Richthofen (1877) p. 278: "Im Ganzen kann man diese erste Abtheilung als eine politische und statistische Geographie betrachten, welche mit besonderer Beziehung auf öffentliche Arbeiten und Verwaltung verfasst ist."

The origins of the methods of statistics stem from ancient times. The yearly accounts of the officials under the Zhou Dynasty were called *kuai* 會, monthly accounts were called *yao* 要, and daily accounts were called *cheng* 成. [...] Although history does not record how they were compiled and transmitted, it is clear that antiquity valued statistics. Han emperors ordered the compilation of registers in the prefectures and principalities. Tang emperors instructed provinces and districts to report to the centre. Up to today, at each end of a term of the year, offices inside and outside the capital must present memorials on population registers, city temples, revenues from taxes, and granaries. These all involve the great principles of civil and financial administration. It was always agreed upon that ancient and new regulations were specially to be handed down through the course of time. Account books and astronomical records were handled with customary familiarity, and merely regarded as pure formalities reported to the government. Gradually their original meaning became lost. Having received Imperial sanction, we have established a Bureau of Statistics.[64]

In a similar tone, the Ministry of Civil Affairs constructs a historical narrative where statistical practices in China can be traced back to the Xia dynasty, but only a well-defined statistical methodology allows for the sought after "investigation of facts":

I venture to comment upon the methods of statistics: although China does not have a particular theory, when we examined the ancient *Documents from Xia Dynasty* (*Xia shu* 夏 書), we found that the classified series of recorded facts in the *Tributes of Yu* chapter attain the essential meaning of statistics. After this, all the books of historical records also used tables and diagrams extensively to facilitate their consultation and comparison. Certainly, if matters are confusing and mysterious and if there are no methods to bring out the essential points, this can hardly receive the merit of checking and investigating the actual facts. Our Ministry is officially in charge of civil affairs. These generally concern: the successes and failures of interior policies, the width of borders and frontiers, the growth of population and the conditions of society, which equally rely on a detailed reporting of statistics.[65]

7.3 What's New in a Number?

During the late Qing political and educational reforms, traditional approaches to the collection, use and preservation of numerical and descriptive data were particularly challenged. When visiting the N° 1 Historical Archives in Beijing with the intention of studying the institutionalization of modern style statistical institutions in late Qing China during the Constitutional Reforms (1906–1911), I first noticed the following: specific archives of the first statistical offices at the central government bureaucracy were only preserved (or available) for the Imperial

---

[64]Memorial of the Commission for Constitutional Government for settling table patterns for civil and financial administration. Received XT 2/2/20 (March 3, 1909). Reprint in *ZZGB* (1911) n° 493 (March 15, 1909), p. 3b–5a. For a full translation of the memorial, see Bréard (2008) Appendix A.4, p. 173–179.

[65]Memorial from the Ministry of Civil Affairs concerning the establishment of a Statistical Office (August 7, 1908). Translated from *ZZGB* (1911), n° 280 (GX 34/7/11) p. 4–5.

Household Department (*Nei wu fu* 內務府) and the Imperial Clan Court (*Zongren fu* 宗人府), exactly those two agencies closely related to the Imperial Kindred. My first guess was that promoting administrative modernization was best possible in circles close to the emperor, and that statistical institutions there were thus run and supervised more efficiently. But when I looked at the documents in these two specific archives (dating from 1907 to 1911), I realized that the prevailing format was the traditional record book, keeping track of all incoming and outgoing funds for regular and unusual expenses, there were also registers of documents, records of lateral communications, account books and numerical lists. Their high standard of organization allows us to believe that archival record-keeping in these statistical offices was done according to conventions which were already well established. The administrative procedures to keep such records seemed unaltered by the reforms, and the documents produced after 1907 simply changed their names: they now carried the neologism "statistical tables" (*tongji biao* 統計表) in their title, which was then a recently imported loan word from Japanese. On the other hand, scattered communications and numerical tables involving personnel of statistical institutions in newly reformed central ministries and in the provinces after 1906 show that these were more receptive to the modernization of numerical data collection and knowledge production for the State. But only a small portion of documents from statistical institutions that were created in newly established ministries or commissions at the end of the Qing are retrievable today. Their archives were probably never constituted because administrative routines were not entirely defined before the Nationalist Revolution of 1911 and the end of the imperial bureaucracy. It might also be the case that, because the final product, the statistical tables, were soon published in printed form by the Qing government, the underlying manuscript tables were no longer regarded as worthy of conservation.

It thus seems that by the end of the Qing dynasty several traditions of managing statistical information coexisted. They were institutionally subsumed under a common name but represented by an heterogeneous first generation of "statisticians." Some were traditionally trained, but others were Chinese students of law, politics and economy hired after their return from Japan where they became acquainted with statistical theories. These arrived in China, where an old administrative tradition of collecting numerical data had prevailed in the imperial bureaucracy, during a period of major reforms in the central government. Statistics attracted the attention of reformers who were in search of guidance in political decisions to save the dynasty in crisis and to centralize control over the provinces. It was then that numbers started to enjoy a prominent position. Compiling annual tables and publishing them allowed for comparisons with other nations and for measuring success in modernizing Chinese society through educational, political and legal reforms. Statistical aggregates became a fundamental mode of representing China's financial, economic, geographic, demographic, educational and legal features. But all this could not be achieved without taking a radically new approach to numbers: submitting numerical lists to the throne was no longer viewed as a ritual act of fiscal relevance performed by provincial officials in order to comply with their statutory obligations. What was urgently needed were numbers that corresponded to an actual situation inquired about. Even if these numbers remained a social construction, they were supposed to quantify reality and a changing society.

References

Anonymous (high official investigating governments 考察政治大臣) (around 1907). *Riben tongji shili* 日本統計釋例 (Explanation of Statistics in Japan). Beijing: Government Gazette Bureau (*Guanbao ju* 官報局).

Bastid, Marianne (1971). *Aspects de la réforme de l'enseignement en Chine au début du 20ᵉ siècle, d'après les écrits de Zhang Jian*, Volume 64 of *Faculté des lettres et sciences humaines de Paris-Sorbonne, publications, série 'recherche'*. The Hague: Mouton.

Bastid, Marianne (1988). *Educational Reform in Early Twentieth-Century China*, Volume 53 of *Michigan Monographs in Chinese Studies*. Ann Arbor: University of Michigan, Center for Chinese Studies. Trans. by Paul J. Bailey from (Bastid 1971).

Bays, Daniel H. (1978). *China Enters the Twentieth Century: Chang Chih-tung and the Issues of a New Age, 1895–1909*. Ann Arbor: The University of Michigan Press.

Beijing University and Nº 1 Historical Archives (Ed.) (2001). Beijing Daxue 北京大學 and Zhongguo di yi lishi dang'an guan 中國第一歷史檔案館. *Jingshi daxuetang dang'an xuanbian* 京師大學堂檔案選編 (=Selected Archives of Metropolitan University). Beijing: Peking University Press 北京大學出版社.

Board of Education (Ed.) (1909). Xuebu 學部. *Xuebu zouzi jiyao* 學部奏咨輯要 (Important Memorials and Communications from the Board of Education). Beijing: Xuebu Zongwusi 學部總務司 (Board of Education, Department of General Affairs). Reprint in *Jindai Zhongguo shiliao congkan san bian di shi ji* 近代中國史料叢刊三編第十輯, vol. 96. Taipei: Wenhai chubanshe 文海出版社, 1986.

Board of Education, Department of General Affairs (Ed.) (1973). Xuebu Zongwusi 學部總務司. *Di yi ci jiaoyu tongji tubiao: Guangxu sanshisan nian fen* 第一次教育統計圖表光緒三十三年分 (First Statistical Tables on Education for the 33ʳᵈ year of Guangxu reign [1907]) (Reprint ed.). Taipei: Zhongguo chubanshe 中國出版社 (China Press).

Bréard, Andrea (2006). Robert Hart and China's Statistical Revolution. *Modern Asian Studies 40*(3), 605–629.

Bréard, Andrea (2008). *Reform, Bureaucratic Expansion and Production of Numbers. Statistics in Early 20th Century China*. Habilitation (History of science), Technische Universität Berlin.

Chen, Fengxiang 陳豐祥 (1978). *Riben dui Qing ting qinding xianfa zhi yingxiang* 日本對清廷定憲法之影響 (The influence of Japan on the Qing court's imperially prescripted constitutional law). Master's thesis, Guoli Taiwan Shifan Daxue, Lishi Yanjiusuo 國立臺灣師範大學歷史研究所 (Taiwan Normal University, Dept. of History), directed by Prof. Lin Mingde 林明德.

Elman, Benjamin A. (2001). *A Cultural History of Civil Examinations in Late Imperial China* (Reprint ed.). Taipei: SMC Publishing Co. Originally published Berkeley: University of California Press, 2000.

Fu, Lanya 傅蘭雅 (Fryer, John) (Ed.) (1895). *Gezhi shuyuan xixue kecheng* 格致書院西學課程 (Curriculum of Western Studies in the Shanghai Polytechnic Institute). Shanghai: [n.p.].

Harrell, Paula (1992). *Sowing the Seeds of Change. Chinese Students, Japanese Teachers, 1895–1905*. Stanford, Calif.: Stanford University Press.

Horowitz, Richard (2003). Breaking the Bonds of Precedent. *Modern Asian Studies 37*(4), 775–797.

Huang, Fuqing 黃福慶 (1975). *Qingmo liu Ri xuesheng* 清末留日學生 (=Chinese Students in Japan in Late Ch'ing Period), Volume 34 of *Monograph Series* 中央研究院近代史研究所專刊. Taipei: Academia Sinica Institute of Modern History 中央研究院近代史研究所.

Huang, Zunxian 黃遵憲 (1985). *Riben zashi shi guangzhu* 日本雜事詩廣注 (Poems on
    Miscellaneous Subjects from Japan, with expanded commentaries) (Zhong Shuhe 鍾叔河 ed.).
    Zou xiang shijie congshu 走向世界叢書, English series title: From East to West: Chinese
    Travelers Before 1911. Changsha: Yuelu shushe.

Index (1910). *Da Qing Guangxu Xuantong xin faling fenlei zongmu* 大清光緒宣統新法令分類
    總目 (Classified Index to New Administrative Methods and Regulations of the Guangxu and
    Xuantong Reign, Qing Dynasty). Shanghai: *Shangwu yinshuguan* 商務印書館.

Lynn, Richard John (1997, Dec.). "This Culture of Ours" and Huang Zunxian's Literary Ex-
    periences in Japan (1877–82). *Chinese Literature: Essays, Articles, Reviews (CLEAR)* 19,
    113–138.

Meienberger, Norbert (1980). *The Emergence of Constitutional Government in China
    (1905–1908). The Concept Sanctioned by the Empress Dowager Tz'u-Hsi*, Volume 1 of
    Schweizer Asiatische Studien. Bern: Peter Lang.

Meng, Sen 孟森 (1909). *Gaiding zengbu tongji tonglun* 改訂增補統計通論 (A General
    Discussion of Statistics) (3rd ed.). Shanghai: Shangwu yinshuguan 商務印書館. Original:
    (Yokoyama 1906).

Miyazaki, Ichisada and Conrad Schirokauer, (Trans.) (1994). *China's Examination Hell*. New
    Haven: Yale University Press.

National Palace Museum Ming-Qing Archives (Ed.) (1979). Gugong bowuyuan Ming-Qing
    dang'an bu 故宮博物院明清檔案部. *Qingmo choubei lixian dang'an shiliao* 清末籌備立憲
    檔案史料 (Historical Material from the Archives for Preparation of Constitutional Government
    at the End of Qing Dynasty), Volume 1–2 of *Zou xiang shijie congshu* 走向世界叢書, English
    series title: *From East to West: Chinese Travelers Before 1911*. Beijing: Zhonghua shuju 中華
    書局. Reprint Taipei: Wenhai chubanshe 文海出版社, 1981 (*Jindai Zhongguo shiliao congkan
    xubian* 近代中國史料叢刊續編; 81).

Noriko, Kamachi (1981). *Reform in China. Huang Tsun-hsien and the Japanese Model*, Volume 95
    of *Harvard East Asian Monographs*. Cambridge, Mass.; London: Harvard University Press.

Reynolds, Douglas Robertson (1993). *China, 1898–1912: The Xinzheng Revolution and Japan*,
    Volume 160 of *Harvard East Asian Monographs*. Cambridge, Mass.: Council on East Asian
    Studies, Harvard University.

Richthofen, Freiherr von, Ferdinand (1877). *China. Ergebnisse eigener Reisen und darauf
    gegründeter Studien*, Volume 1 (Erster Band): Einleitender Theil. Berlin: Verlag von Dietrich
    Reimer.

Shuntian Prefecture (*Shuntianfu* 順天府) (Ed.) (1908). *Shuntianfu Guangxu sanshisi nian tongji
    biao* 順天府光緒三十四年統計表 (Statistical Tables of Shuntian Prefecture for the 34th year
    of Guangxu Reign). [Beijing].

Statistical Office of the Board of Agriculture Industry and Commerce (*Nong gong shang bu tongji
    chu* 農工商部統計處) (Ed.) (1908). *Nong gong shang bu tongji biao di yi ci* 農工商部統計表
    第一次 (First Statistical Tables of the Board of Agriculture, Industry and Commerce), 6 vols.
    六冊. Beijing.

Sugi, Kōji 杉亨二 (1902). *Sugi Sensei kōenshū* 杉先生講演集 (Collected Lectures of Sir Sugi),
    Sera, Taichi 世良太一 (Ed.). Tokyo: Yokoyama Masao 橫山雅男.

Tan, Ruqian 潭汝謙 and Lin Qiyan 林啓彦 (Trans.) (1983). *Zhongguoren liuxue riben shi* 中國人
    留學日本史 (History of Chinese Students in Japan). Beijing: Shenghuo dushu xinzhi sanlian
    shudian 生活．讀書．新和．三聯書店. Original: Sanetō Keishū 實藤惠秀, Chūgokujin
    Nihon ryūgakushi 中国人日本留学史.

GXXF (1909). *Da Qing Guangxu xin faling* 大清光緒新法令 (New Administrative Methods and
    Regulations of the Guangxu Reign, Qing Dynasty). Shanghai: Shangwu yinshuguan 商務印書
    館. 20 vols., indexed in (Index 1910).

XTXF (1911). *Da Qing Xuantong xin faling* 大清宣統新法令 (New Administrative Methods and
    Regulations of the Xuantong reign, Qing Dynasty). Shanghai: Shangwu yinshuguan 商務印書
    館. 35 vols. Indexed in (Index 1910).

ZZGB (from 1908 to 1911). *Zhengzhi guanbao* 政治官報 (Government Gazette). Beijing:
    Xianzheng biancha guan guanbaoju 憲政編查館官報局. Published daily.

Thompson, Roger R. (1995). *China's Local Councils in the Age of Constitutional Reform 1898–1911*, Volume 161 of *Harvard East Asian Monographs*. Cambridge, Mass. and London: Harvard University Press.

Ye, Longyan 葉龍彥 (1974). *Qingmo Minchu zhi fazheng xuetang* 清末民初之法政學堂 (1905–1919) (*New Schools for Law and Politics at the End of Qing Dynasty and during the Early Republican Period*, 1905–1919). Ph. D. thesis, Taipei: Sili Zhongguo wenhua xueyuan 私立中國文化學院 (Shixue yanjiusuo 史學研究所).

Yokoyama, Masao 横山雅男 (1906). *Tōkei tsuron* 統計通論 (A General Discussion of Statistics) (8th ed.). Tokyo: Senshūgakkō 專修學校. Other editions: 1904[1], 1921[41].

Yokoyama, Masao 横山雅男 (1921). *Tōkei tsuron* 統計通論 (A General Discussion of Statistics) (21st ed.). Tokyo: Yūhikaku Shobō 有斐閣書房.

Zhang, Qing 张勤 and Mao Lei 毛蕾 (2005). Qingmo gesheng diaochaju he xiuding falüguan de xiguan diaocha 清末各省調查局和修訂法律館的習慣調查 (Customs Survey Towards the End of the Qing Dynasty). *Xiamen daxue xuebao* (*Zhexue shehui kexue ban*) 廈門大學學報 (哲學社會科學版) (Journal of Xiamen University (Arts & Social Sciences) 6, 84–91.

Zhang, Zhidong 張之洞 et al. (1904). *Zouding xuetang zhangcheng* 奏定學堂章程 (Memorial to Fix the Rules and Regulations for New Schools). Reprint Taipei: Tailian guofeng chubanshe 台聯國風出版社 (Shixue yanjiusuo 史學研究所), 1970.

Chapter 8
Applied Versus Pure Mathematics

Contents

Fan Huiguo 范會國 (1899–1983), who participated in the 1936 foundation of the Chinese Mathematical Society and studied function theory with Charles Émile Picard at the Faculté des Sciences in Paris,[1] began his article on "Mathematics and its Applications" with the following caricature of the quixotic mathematician:

> The general view of a mathematician is one of a crazy freak who has absolutely no access to the real world (*yu xianshi shijie juelu zhi fengkuang guaiwu* 與現實世界絕路之瘋狂怪物). Assiduous day and night, year in and year out, he is ever in to deducing abstract principles; he spends his life in the middle of bizarre formulas. Twisting his brain, searching for happiness, provoking annoyance (*re fannao* 惹煩惱), when hungry he cannot eat; in the winter he cannot dress; dangers at home or national crisis, there is nothing he can ease, and much more besides. But the origins of mathematics are in fact of experimental nature. Geometry began from being a branch of physics. In high antiquity, people in Babylon already knew how to inscribe a regular hexagon in the middle of a circle, with each side of the hexagon being half of the circle's diameter. In Egypt, [...][2]

The main body of Fan's article then is a collection of historical examples from Leibniz to Henri Poincaré, which illustrate the experimental origins of mathematical

---

[1] He obtained his doctorate there in 1929 with his thesis *Recherches sur les fonctions entières quasi exceptionnelles et les fonctions méromorphes quasi exceptionnelles*. Lyon: Bosc Frères & Riou, 1929. See Li and Martzloff (1998) p. 189–190.

[2] Fan (1937) p. 1.

© Springer International Publishing AG, part of Springer Nature 2019
A. Bréard, *Nine Chapters on Mathematical Modernity*, Transcultural
Research – Heidelberg Studies on Asia and Europe in a Global Context,
https://doi.org/10.1007/978-3-319-93695-6_8

laws and the applications of mathematical theories in physics and mechanics.[3] In his conclusion, Fan argues for the study of theoretical mathematics in parallel with its possible fields of applications. He reveals a critical perspective on Chinese past science policies, which he accuses of focusing too much on the applied sciences. Like many of his compatriots, Fan believed that this focus was responsible for China's strongly felt backwardness in the sciences during the early twentieth century:

> Altogether, from the above, one can see that theory and application indeed have an intimate relation, and that they can mutually use each other. They are of one piece, of a chain-like structure; they share the work and collaborate at the same time because they co-exist and they flourish together.
> At the unprecedented occasion of [China's] national danger and the rising menaces, my countrymen very energetically promoted "applied sciences." A necessary and inevitable matter of course also has its whys and wherefores. Only "theoretical sciences" still require simultaneous and similar attention. Doing two things at a time, with two approaches advancing together, only then we can "try hard to catch up," without resulting in eternally yielding behind others.[4]

This chapter is about the conflictual relations among pure and applied mathematical disciplines in the century from circa 1850 to 1950. It shows the historical variations of the ways in which boundaries were drawn in this period, and how, often under the influence of the political orientations of China in a global context, the focus upon one or the other aspect of mathematics was shifted. A first section will describe the official debates about the nature of mathematics as it evolved into a discipline by the early twentieth century. However, it would be anachronistic to attempt a historicization of the relationship between what we now consider "pure" or "applied" mathematics in China at the turn of the twentieth century. Adopting a China-centred approach, which is in general more fruitful when dealing with non-European science and traditions, I will instead discuss a distinction between what in Chinese terms are the "fundamental principles" (*ti* 體) and "applications" (*yong* 用).

In the second part of this chapter, I will use the example of statistics to illustrate how the pure–applied paradigm was instrumentalized in scientific and political discourses from the 1930s to the early 1950s. Statistics, as one mathematical discipline, is a good case study with which to analyse the global and political entanglements between different kinds of theories and practices involved in the collection, analysis and use of numbers. Roughly speaking, there were three approaches to statistics in circulation:

1. administrative statistics: which had a very long tradition in China itself as a tool of fiscal management;
2. social statistics: were imported, translated and naturalized from Germany via Japan in the early twentieth century;

---

[3]Fan later wrote a treatise on *Rational Mechanics* (Fan 1944).
[4]Fan (1937) p. 9.

3. mathematical statistics: this theoretical branch of statistical learning involves the application of probability theory to the analysis of statistical data. This approach originated in the Anglo-Saxon tradition and was brought to China by students returning from their studies abroad, but it was severely criticized by the Communist regime after 1949.

The ideal of the Republican statisticians engaging social and mathematical statistical work and research in parallel fizzled when Mao declared formalist approaches to be bourgeois. Socialist statistics became the official style, Russia the model, and the state statisticians either adjusted or became *persona non grata*. In a third part of this chapter, I will discuss some of the strategies of important statistical actors who had the versatility and talent needed to find consensus with the hostile state, oscillating between complementary and at times conflictual approaches to statistics.

8.1 Mathematics Before it Becomes a Discipline

In late imperial China, the notions of pure and applied mathematics were not actors' categories, a fact that remained true until shortly before the 1930s. Mathematics was not even a well-established discipline until the late nineteenth century, when it began to be taught in government schools.[5] Applications of mathematics, such as mechanics, physics, or optics, were not an independent branch of learning in ancient China. Thinking and knowledge concerning mechanics or physics only existed in scattered philosophical, technological, astronomical and mathematical texts.

A look at the bibliography of the official history of the Qing (1644–1911) dynasty, compiled after its fall, reflects the intellectual landscape of mathematical learning in late imperial times. There were two categories related to numbers, "astronomy & mathematics" (*tianwen suanfa lei* 天文算法類)[6] followed by "magical computations" (*shushu lei* 術數類).[7] The latter contains titles from the numerological genre, which is often linked to impostors and to heterodoxy, whereas the former lists mathematical and astronomical texts and compilations of Chinese and foreign origins.

In order to understand the place held by Qing mathematics in the Chinese intellectual landscape we need to go back in time to when the Jesuits introduced Western science to China. In the late Ming (first half of the seventeenth century), the sciences were part of "concrete learning" (*shixue* 實學). Characterized by an empiricist approach and an emphasis on statecraft to address actual problems,

---

[5]The only exception was during the Sui and Tang dynasty when a School for Computation was founded and functioned (with interruptions) for approximately 100 years. See Keller and Volkov (2014) p. 59–63 and Bréard and Horiuchi (2014) p. 159–166.

[6]Zhao et al. (1981) 志一百二十二, 藝文三, 子部, 天文算法類, 算書, esp. p. 4343–4347.

[7]Ibid. p. 4348.

"concrete learning" was opposed to speculative approaches and dismissed intro-
spective reflections as empty words. According to Ruan Yuan's 阮元 (1764–1849)
biographic note on Xing Yunlu 邢雲路 (*jinshi* 1580), who composed the *Researches
on pitchpipes and the calendar old and new* (*Gujin lüli kao* 古今律曆考 in 72 *juan*)
some time in the Wanli period in view of an astronomical calendar reform:

> the complexity and the literary richness of the redaction is fundamentally not adapted to
> concrete learning (*gai wenzhang fanfu ben wu dang yu shixue* 蓋文章繁富本無當於實
> 學).[8]

Ruan Yuan accused Xing of wanting to magnify his writings, of adding even more
volumes by citing from the classics and dynastic histories, which he condemned as
a method to deceive the public. Jesuit mathematical science in late Ming and early
Qing China, too, was significantly shaped by a demand for concrete studies among
scholars. The example of reactions to the introduction of Euclidean geometry by the
Jesuits, who translated the first six books of Clavius's version of Euclid's *Elements*
into Chinese in 1607, illustrates the implications of the concept of "concrete
learning" for mathematics quite well. As Martzloff has shown, "Chinese scholars
had the idea to abridge the text of the *Elements* by discarding its hypothetico-
deductive rhetoric so as to offer to novice mathematicians a set of simple and
efficient geometrical techniques, in conformity with the ideals of the movement of
promotion of 'concrete sciences', *shixue*, so important in China in the 17th and 18th
centuries."[9] If geometry had to be truly useful, in the context of "concrete learning,"
it was more appropriate to propose a set of complete geometric directives to the
Chinese readers so that they could go right to the goal.

This practical outreach towards foreign mathematics was further amplified by the
slogan "Chinese studies for fundamental principles—Western studies for practical
applications" (*Zhong ti xi yong* 中體西用). During the late Qing, it served
increasingly as a framework for rationalizing the import of foreign knowledge.[10]
"Chinese studies for fundamental principles" referred to Chinese culture and
its Confucian underpinnings. It primarily implied the maintenance of traditional
Chinese political and economic systems and their corresponding ideologies while
adopting "Western studies for practical applications." After the Opium Wars, the
Qing turned in particular towards mechanics, physics, chemistry, mathematics and
military technology among the Western sciences and technologies, but Western
institutions and thought were subject to rejection. This state-sponsored movement,
the so-called Self-Strengthening Movement, was geared toward embracing modern
technologies and foreign industries as a means of empowering the crumbling empire
after 1861.[11] But some scholars' vision of the Self-Strengthening Movement went
beyond the mere goal of producing good weapons. They called for adjustments in

---

[8] Ruan and Luo (1882) scroll 31, p. 16A.

[9] Quoted from Martzloff (1993) p. 160.

[10] See Ding and Chen (1995) and Levenson (1968) vol. 1, p. 59–78.

[11] See Fairbank (1978) chap. 10 "Self-Strengthening: The pursuit of Western Technology."

Fig. 8.1 Fryer and Hua (trans.), *A Treatise on Probability* (1898 ed.)

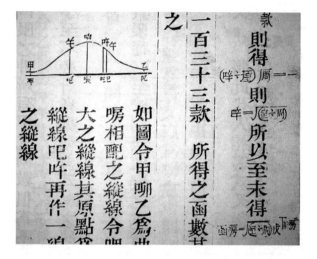

China's existing institutional system of government, education and recruitment for government service. Working within the "Chinese principles–Western applications" paradigm, translators and reformist statesmen engaged in a range of projects, including the building of technical schools, arsenals, shipyards and translation bureaus. It was in such institutional settings that mathematics were taught for the first time in China in a governmental institution whose sole purpose was not to prepare for the state examination based on the classical preference for moral textual knowledge. More concretely, the emphasis in these schools was laid on calculus as the basis of a strong empire. Calculus was considered crucial for learning the casting of cannons, the construction of steamships, and other modern mechanical equipment. But mathematics and astronomy more generally were subject to controversy over the question of whether or not they were mere matters of "technique" (*jiqiao* 技巧), which had little esteem in a Confucian setting and might thus have an adverse effect on the spirit and moral principles of the literati. In addition to such anxieties about the decline of virtue, science teaching was at first left to foreign instructors. Many of whom were ex-missionaries and thus suspected of corrupting the students with Christian doctrines.[12]

As already shown in Chap. 5, mathematical language was not undermined by the presence of a foreign symbolic system. In a certain sense, one could even say that in late Qing translations of English language books on calculus, algebra or probability theory (see Fig. 8.1) the preservation of a Chinese essence was to be respected at any cost. This meant the rejection of mathematical formulas expressed with Latin letters and Arab numerals, and the creation of a proto-grammatical symbolism based on characters and syntactic features from the Chinese written language.[13] This kind of mathematical discourse remained prevalent during the entire latter half of the

---

[12]See *idem* p. 528–530.

[13]See Bréard (2001).

Table 8.1 Li Shanlan's 李善蘭 mathematical examination problems in Xi and Gui (1880)

Juan	Content	No. of problems	Percentage
1	Astronomy	20	10.1
	Mechanics	30	15.2
2	Plane geometry	23	11.6
	Spherical geometry	3	1.5
	Finite series	9	4.5
	Indeterminate equations	5	2.5
	Logarithms	1	0.5
	Binomial expansion	1	0.5
3	Circle measurement	42	21.2
4	Right-angled triangles	25	12.6
	Combinations & Permutations	1	0.5
2 & 4	Problems for daily use	38	19.2
Total		198	100

nineteenth century—even if it produced paper-wasting layouts of pages containing sometimes only a single horizontal formula and one line of vertical text on a page.

Looking precisely at the way that mathematics was taught by the first professional Chinese instructor, Li Shanlan, we can investigate further what the Chinese principles–Western applications dichotomy meant in terms of mathematical theory and applications. Li was teaching at the Interpreter's College (*Tongwen guan* 同文館), an institution that was the outpost of the Self-Strengthening Movement and the *ti-yong*-paradigm,[14] when he—who, according to his preface to the translation of William Whewell's *Elementary Treatise on Mechanics* (*Zhongxue* 重學) "translated the geometry in the morning and mechanics in the afternoon"[15]—complained that

> nowadays, mathematician-astronomers and the later generations can all thoroughly understand geometry. But they know nothing yet about mechanics.[16]

Table 8.1 gives an idea of the distribution of exam problems among the applications of mathematics to astronomy, mechanics, and Chinese traditional mathematics that co-existed in Li Shanlan's teachings.[17] Although integral and differential calculus was on the mathematics curriculum in the Interpreter's College, only a single problem on the binomial expansion of $(a + x)^{1/2}$ can be found among the exam problems,[18] and some problems on conics figure among the geometry problems. Other topics with a partly modern, and from a terminological

---

[14]Li was affiliated with the Interpreter's College from 1868 to his death in 1882.

[15]Whewell (1859) "Preface", p. 1A.

[16]*Idem.*

[17]Table adapted from Guo (2003). For an analysis of the exam problems of other late Qing schools where mathematics were taught, see Li (2012).

[18]See Chap. 4, p. 89.

perspective, apparently foreign label, such as "finite series," "logarithms," "right-angled triangles," "indeterminate equations," "circle measurement," "combinations and permutations," are in fact all related to the Chinese mathematical tradition, and the problems are borrowed from classical Chinese texts dating from the Han to the Yuan dynasty. The problem on logarithms for example asks the following:

A gourd and a bean start sprouting and growing on the same day. On the first day, the gourd creeps up a length of 1 *chi* 6 *cun*.[19] Afterwards every day it steadily grows less, half every day. The bean, on the first day, creeps up a length of 1 *cun*. Afterwards every day it steadily grows more, half every day. On which day are the two trailing plants equal [in length]?

瓜豆同日發芽生蔓瓜蔓初日長一尺六寸以後每日所長遞半豆蔓初日長一寸以後每日所長遞加半二蔓第幾日相等[20]

This problem of pursuit is modelled upon problem 7.11 from the *Nine Chapters* asking the same question with different initial values for two different kinds of plants found in wetlands, the calamus (*pu* 蒲) and *Schoenoplectus validus* (*wan* 莞).[21] Yet the solution suggested by Cai Xiyong 蔡錫勇 (1850–1896)[22] in the exam problem is not the traditional one; rather, it is a solution using algebra and logarithms:

The explanation says: This is with continued proportions. What the gourd creeps up on the first day is taken as the final *lü*.[23] What the bean creeps up on the first day is taken as the initial *lü*. The number of *lü* one obtains is the number of days the two trailing plants are equal [in length]. I will use algebra to make this clear:

$$B = \text{initial } l\ddot{u}$$

$$H = \text{multiplier}$$

$$A = \text{final } l\ddot{u}$$

$$B = 1$$

$$BH = 2$$

$$BH^2 = 4$$

$$BH^3 = 8$$

$$BH^4 = 16$$

$$BH^5 = 32$$

---

[19]Measures of length, where 1 *chi* = 10 *cun*.

[20]Translated from Xi and Gui (1880) scroll 2, p. 59A–59B.

[21]On problems of pursuit in the Chinese mathematical tradition and their possible astronomical significance, see Bréard (2002).

[22]Cai, after graduation from the Interpreter's College, worked as a translator for the China-based news agency of the Associated Press. From 1893 to 1897 he headed the New School for Self-Strengthening (*Ziqiang xuetang* 自强學堂), Wuhan University's precursor.

[23]On the concept of *lü*, see p. 43.

From this one can see that the power of H must be one less than the number of terms. Let us take the number of terms as the heavenly [unknown, here: x]. Then, the final *lü* is always:

$$A = BH^{x-1}$$

Therefore:

$$H = \sqrt[x-1]{\frac{A}{B}}$$

In accordance with the principles of algebra, one can transform the above formula into:

$$x - 1 = \frac{\log_{10} A - \log_{10} B}{\log_{10} H}$$

$\log_{10} B$ is the logarithm of the initial *lü* 1, it is equal to 0. Thus with the logarithm of 2 (301) divide the logarithm of what the gourd creeps up on the first day 1 *chi* 6 *cun* (1204).[24] One obtains 4 and by adding 1, one obtains 5. This makes the number of days when [the two plants] are equal.[25]

Probably not all students solved the problem in this way, but instead followed the model of the *Nine Chapters*. That the editors chose to publish Cai Xiyong's solution as exemplary can be read as an ideological manifesto: the Chinese essence is preserved by copying an ancient problem from the *Nine Chapters*, while Western applications, in this case algebra and the logarithmic function, are used to solve it.

8.2   Mathematical and Other Approaches to Statistics

It is within this *zhongti-xiyong* (Chinese principles—Western applications) framework that the history of science, and more specifically the history of mathematics, evolved in late Qing China. The paradigm only gradually faded away at the turn of the twentieth century and shifted towards discourses about a divide between social, natural and applied sciences, a modern classification borrowed from the Japanese academic system.[26] The Self-Strengthening Movement was severely criticized and drastic reforms of fundamental Chinese political and educational institutions were advocated. Reformers recognized the importance of Western studies and learning from Western political institutions. They called for the large-scale establishment of schools for universal education, which taught science and Western languages and finally in 1904 abolished the traditional state examination system based on the sole study of the Confucian Classics.

---

[24]The values of the common logarithms of 2 and 16 given here are indicated qualitatively, their values are $\log_2 = 0.301\ldots$ and $\log_{10} 16 = 1.204\ldots$ respectively.

[25]Translated from Xi and Gui (1880) scroll 2, p. 59A-59B, original text shown in Fig. 8.2.

[26]See Elman (2005) p. 409–410 and Li (1995) p. 105–107.

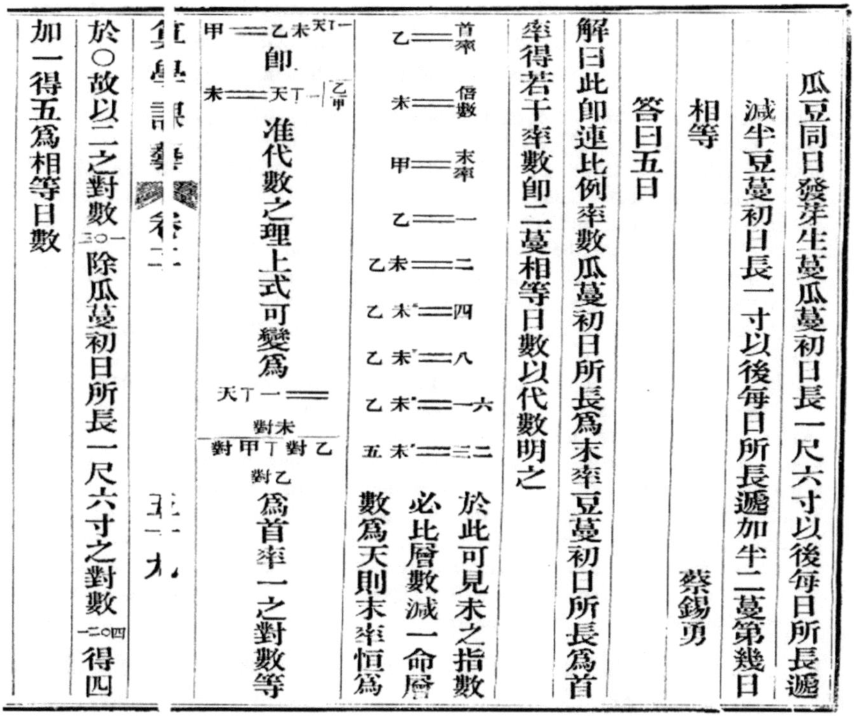

Fig. 8.2 A problem from the *Nine Chapters* solved algebraically with logarithms in Xi and Gui (1880)

Statistics was at the heart of the agenda of institutional and constitutional reforms, and a first nationwide network of statistical institutions was established in China in 1907. Among the publications of the first director of the Central Statistical Bureau is a critique of the Self-Strengthening Movement[27] in response to the Empress Dowager Cixi's Reform Edict issued on 19 January 1901, to which all high officials were ordered to express their opinions. Shen Linyi 沈林一 (1866–1911?), in conservative style and with abundant references to China's antiquity and ancient philosophers, proposed altogether twenty-six items to be reformed or newly introduced. These included the clarification of population and territorial data, for which reporting had fallen far behind schedule during the nineteenth century. Statistical practices, in dire need of reform, had fit particularly well into the neo-Confucian concept of "concrete learning," but this was no longer the case with the arrival of social sciences. They challenged the very notions of "truth" and "fact," and administrators were compelled to produce numbers that would correspond accurately to social or economic actualities. This requirement was often stressed

---

[27]Shen (1901).

in official communication with the court, as a memorial from the *Commission to Draw up Regulations for Constitutional Government* (*Xianzheng biancha guan* 憲政編查館) proposing the guidelines for designing statistical tables shows:

> Although statistics rely on theoretical principles they certainly have to be in accordance with the actual facts (*shishi* 事實). At the time of preparation for constitutional government, among the old habits followed as before, there are also those which have merged with new methods. [. . .] As, for example, the principles of longitude and latitude of borders and frontiers, the growth and mortality of the population, and other items. Although they were measured at a certain time, the investigations could not explore these items in detail. What statistics must certainly not be short of, the table patterns certainly need to contain. We naturally should hurry to pursue partial investigations, first filling out a rough outline, it is then appropriate to proceed from the general to the particular and from the coarse to the fine.
>
> The key intention of statistics is the examination of facts (*heshi* 核實). As for example, in calculating distance one has to distinguish clearly between *li* and *bu*.[28] In calculating receipts and expenditures the amounts have to correspond to each other. In general, when one is able to make all facts accurate in detail, then words will not tread on emptiness. Until now, in each province regulations, reports, registers and legal records have long been looked at as being equal to legends, or stationary that is merely to be preserved. Nowadays, statistical management is the most important task of the New Systems movement. [Statistics] are not equal to ordinary official documents. One does not have to gloss over faults or procrastinate. As before, rather long-standing practices and old habits do bring about trouble and raise difficulties.[29]

I will use statistics in the second part of this chapter as a case study to illustrate how the division between the theoretical and the applied was instrumentalized in scientific and political discourse. When statistical theories for application to social phenomena were imported in China at the turn of the twentieth century, China already had a long administrative tradition of its own, a tradition of collecting numbers and reporting them to the throne. Statistics, although not defined as a discipline, then were sets of descriptive tables produced for the state. Reformers defended the import of foreign statistical theories as something that was merely grafted onto old administrative institutional practices. This legitimizing strategy of defending the new (application) by referencing the old (essence) is made clear in the very first of the general rules for statistical tables, accompanying the same memorial as quoted above. The memorialist nevertheless has to admit that systematic surveys are an innovation imported from the West:

> In ancient times there was no such expression as "statistics," but there were statistical methods. The *Tributes of Yu*[30] is the oldest tradition of the Zhou dynasty officials. Zhang

---

[28] Two units of length.

[29] Memorial dated 15 March 1909. Translated from *XTXF* (1911) n° 2 (1910), p. 2B. For an integral translation of the Memorial, see Bréard (2008) Appendix A.5, p. 179–186.

[30] The *Yu gong* 禹貢, a chapter in the *Book of Documents* (*Shang shu* 尚書), which is the most authoritative ancient Chinese terrestrial description. Its authorship is attributed to Yu the Great (5th–3rd cent. BCE), the legendary founder of the Xia dynasty. The claim that he was the first Chinese statistician can be traced back to at least as early as to the French Jesuit missionary Antoine Gaubil (1689–1759). See Simon (1970) and Richthofen (1877).

Cang 張蒼 from Han dynasty presided over the compilation of registers in the prefectures and principalities. This is an even clearer proof that historical tables started to be made from [Si]ma Qian[31] on, and that the essence of the tables is an imitation of the [style prevalent in the] Zhou dynasty. That statistics had to be set out under the form of tables, was possibly adopted with the intention to make their examination simple. Now, taking statistics and combining them with investigations, is certainly not entirely inherited from Western culture, it is also an extension of ancient methods.[32]

Such historiographic construction, similar to the well-known notion of the "Chinese origins of Western science" (*Xixue zhongyuan* 西學中源) since the Jesuits transmitted elements of Western science to China, helped to justify innovation and import from the West by appealing to the authority of a native Chinese past. Official and unofficial discourse during the reforms at the end of the Qing dynasty abound in justifications of the new by referencing the old. By underlining the existence of a statistical tradition in the Chinese imperial administration since remote antiquity, the memorialists subtly convinced the emperor and other high officials to conduct reforms based on traditional patterns, or in their words, to "reform within tradition." Adopting anything foreign was thus reduced to a simple reintroduction of Chinese methods that were partly lost over the centuries.

Mathematics, too, had enjoyed a long tradition since Chinese antiquity, and its place with respect to statistics was a debate that in twentieth-century China became increasingly framed in terms of geopolitical references. Such was not the case in the earliest writings on statistical theory, which I will demonstrate through the example of two important statistical figures from the end of the Qing: Shen Linyi, the first director of the Central Statistical Bureau, and Meng Sen, translator of the most influential statistical manual. Both men were representative of two complementary and at times conflicting views on statistics. As can be seen in Meng Sen's preface to a book on prison statistics, he entirely subscribes to the view of Japanese social statisticians that statistics is about the quantification of social phenomena, where large-scale surveys result in a condensed numerical description. Meng's commentary to the idea of aggregate numbers is not only an elegant testimony of his talent for classical literary composition; his discourse is also much embedded in Confucian moral principles:

> Heaven and Earth are possessed of the most entire benevolence. At other instances, Heaven and Earth perpetually through floods and draughts bring about suffering from illnesses and diseases to this people. Man's sentiments are possessed of the most entire forebearance. At other instances Man's sentiments perpetually through punishments by cutting off the nose or the ears, by castration or tattooing and other kinds of chastisement confuse one's ears and eyes and dazzle one's willpower. All of these instances cause damage. As for statistics, it is about cumulating a number of these instances, as innumerable as the sand of the Ganges,

---

[31] Ca. 145–ca. 86 BCE. Author of the first Chinese history, the *Records of the Grand Historian* (*Shiji* 史記).

[32] See Article (*tiao* 條) 1: *Xianzheng bianchaguan you zou ding tongji biaoshi zongli* 憲政編查館又奏定統計表式總例 (The Commission for Drawing up Regulations for Constitutional Government further memorializes to fix general rules for the design of statistical tables) in *XTXF* (1911) n° 2 (1910) p. 2A.

giving an extreme that is not made of instances (*ji bu xuyu* 極不須臾).[33] From within the extreme that is not made of instances, one can arrange the ears and eyes, and stabilize one's willpower.[34]

Among the mathematical methods in Meng Sen's influential manual translated from Japanese,[35] one finds only three short sections on summation, proportions and mean values—precisely the mathematical content that Chinese administrators would have mastered entirely had they studied the *Ten Books of Mathematical Classics* (*Suanjing shi shu* 算經十書) from the Tang dynasty.[36] In a chapter on the relation of statistics to other sciences, the author even calls to mind the fact that until recently statistics and mathematics were conceptually incompatible:

> Mathematics originate in deductive methods, statistics originate in inductive methods. These two learnings truly have no relation whatsoever. During the last 10 years though, among statisticians were some who rely on mathematics and in their statistical research moreover on political arithmetic.[37]

As for statistics as a mathematical science involving probability theory, most actors in China were unaware of this rather novel approach at the turn of the twentieth century, although in 1896, the *Jueyi shuxue* 決疑數學, a Chinese translation of the article on probability in the *Encyclopaedia Britannica*, was published.[38] John Fryer (chin. Fu Lanya 傅蘭雅) and Hua Hengfang 華蘅芳 had prepared jointly for several years a Chinese version of Thomas Galloway's book-length article on *Probability*, which was published for the first time in the seventh edition of the *Encyclopaedia Britannica* (Fig. 8.3).[39] Galloway's essay had also been published as a separate treatise[40] and was reprinted in the eighth edition of the *Encyclopaedia Britannica* (1853). It was probably the latter edition that served as a textual basis for Fryer and Hua's translation, since we know that in 1868 Fryer had

---

[33] Later in the text, on p. 105, Meng Sen explains this expression in relation to mass observations (*dashu guancha* 大數觀察):

> The expression "large numbers" has the meaning of what I call an extreme that is not made of instances (*Dashu yu zhe ji wu suo wei ji bu xuyu zhi yi* 大數云者即吾所謂極不須臾之義).

[34] Meng (1907) p. 103.

[35] See Chap. 7, p. 186.

[36] See Bréard (2006).

[37] Meng (1909) p. 70.

[38] See Fu and Hua (1896). The expression *jueyi* chosen by the translators for "probability" existed earlier in classical Chinese, literally meaning "dispelling doubt" or "to resolve questions that are difficult to decide" (*jiejue yihuo* 解決疑惑).

[39] Galloway (1796–1851) was a Scottish mathematician working as an actuary at the Amicable Life Assurance Office. His most famous paper "On the Proper Motion of the Solar System" won a Royal Medal in 1848 for his application of probability theory to determining the proper motion of the solar system. See Galloway (1843) and Obituary (1852).

[40] Galloway (1839).

CONTENTS

I. General principles of the theory of probability
II. Of the probability of events depending on a repetition of trials, or compounded of any number of simple events, the chances on respect of which are known a priori and constant
III. Of the probability of events depending on a repetition of trials, or compounded of any number of simple events, the chances in respect of which are known a priori and vary under different trials
IV. Of mathematical and moral expectation
V. Of the probability of future events deduced from experience
VI. Of benefits depending on the probable duration of human life
VII. Of the application of the theory of probability to testimony, and to the decisions of juries and tribunals
VIII. Of the solution of questions involving large numbers
IX. Of the probable mean results of numerous discordant observations, and the limits of probable error
X. Of the method of least squares

Fig. 8.3 Content of Thomas Galloway's *A Treatise on Probability* (Galloway 1839)

ordered "The Encyclopedia Britannica, complete to present date" for the officials of the Jiangnan Arsenal.[41]

The Chinese translation was published without crediting either Galloway or the *Encyclopaedia Britannica* for the original text. Because in its preface Galloway praises his friend Augustus De Morgan's (1806–1871) article on the "Theory of Probability" in the *Encyclopedia Metropolitana* (1837) as "by far the most valuable work in the [English] language" on the general theory of probability, many historians have erroneously associated de Morgan's work with the Chinese translation.[42] Faithful to the original text, the translation is a complete Chinese version of the introduction and all the chapters contained in Galloway's more than 200-page long essay on classical Laplacian probability theory. It also includes its applications to data furnished by mortality tables and the determination of the most probable mean values of astronomical and physical observations or experiments, in which Galloway followed Poisson's "very general and elegant analysis."[43] Although the book contained both theoretical and applied aspects of mathematics, it was largely ignored by the Chinese public.

Such was probably also the case for Meng Sen. He did have some basic notions of probability theory applied to vital statistics, but when he comments extensively upon the calculation of mortality tables his references are Japanese authors:

---

[41] Bennett (1967) p. 80.

[42] Yan (1990) for example states wrongly that the original of John Fryer's and Hua Hengfang's translation was Augustus De Morgan's work.

[43] Galloway (1839) Preface p. xi. For a detailed discussion of the mathematical content of the book, see Guo (1989) and Wang (2007).

This kind of method has not yet been explained. It is unavoidable that the readers' confusion will be increased. The calculation methods referred to are twofold: [...] See Kameichi Nakaba's book *The most recent statistical learning*.[44]

One can also assume that Shen Linyi ignored the existence of the above-mentioned Chinese treatise on probability theory and the mathematical law of large numbers translated by John Fryer and Hua Hengfang. When in 1910 several high officials, including Shen Linyi, sent a petition to the Ministry of Education suggesting the use of Hua Hengfang's translations or books in the mathematics curriculum, they included a list of twelve titles, but they did not refer to the one on probability theory.[45] More generally, I doubt that Shen Linyi was interested in statistics from a theoretical point of view. His publications do not refer to any theoretical works he might have encountered through contacts with his contemporaries[46] or in the Constitutional Commission's library, which owned two copies of Meng Sen's Chinese language statistical manual but no other books on statistics except for a report on statistics in Japan.[47]

Shen Linyi was more familiar with statistical work than with theory. He was actively involved in the practices that related to the collection and recording of fiscal data or to the standardization of measures and weights with the basic arithmetic operations involved, such as the calculation of mean values or the determination of a surplus or a deficit when comparing numerical data. All this existed in the Chinese administrative tradition and, before becoming director of the newly founded Statistical Bureau, he did some practical work in this direction.[48] In his *Brief Study of Chinese and Western Measures and Weights* (*Zhongxi duliang heng kaolüe* 中西度量衡考略), Shen traced the roots of measurement back to the legendary Xia dynasty, the early Han and Tang astronomical and mathematical classics,[49] and the temperament of pitch pipes as described in late Ming dynasty essays. In order to supervise the prohibition of home-made measures and weights at local markets, Shen very pragmatically insisted upon the establishment of a municipal official at the markets to "establish the lengths, determine the graduations, promulgate

---

[44]Reference to Kameichi (1899). Translated from Meng (1909) p. 270–273.

[45]In Yang (1910) p. 6A–6B.

[46]For example, Shen was in close contact with Xue Fucheng 薛福成 (1838–1894), who, like Shen, was a native of Wuxi, Jiangsu province, and for whose diary, written during his diplomatic mission to Great Britain, France, Italy and Belgium from GX 17/3/1 to GX 20/5/20, Shen wrote a postface. See Lü (1995) p. 57.

[47]See the *Book Register* (*Shuji ce* 書籍冊) from 1911, Archives of the Commission for Constitutional Reform, N° 1 Historical Archives, Beijing, showing two copies each of Anonymous (high official investigating governments) (1907) and Meng (1909).

[48]For a more detailed account of Shen's CV, see his Record of Conduct reprinted in Qin et al. (1996) 8:581 and translated in Bréard (2008) Appendix A.6, p. 186–188.

[49]The *Gnomon of Zhou* (*Zhou bi suan jing* 周髀算經) and *Master Sun's Mathematical Classic* (*Sunzi suan jing* 孫子算經) respectively.

the design, make the patterns." He did "not believe that this cannot rule out the calamities and disorders!"[50]

Among the actors involved in statistics before the 1930s, Gu Cheng 顧澄 (1882–ca. 1947) was an exception. He was the first to envision the mathematization of statistical research and moved institutionally from the social sciences to mathematics. Professor for Statistics at the *Beijing Special School for Law and Politics* (北京法政專門學校統計教授), director of the Statistics Department of the *Ministry of Education* (教育部統計科科長), director general of the *Central Statistical Association* (中央統計協會總幹事), professor for Mathematics at *Beiping University* and *Jiaotong University* and later, in 1935, one of the founders of the *Chinese Mathematical Society*, he translated two important works in this direction: Mansfield Merriman's *Method of Least Squares* and George Udny Yule's *Introduction to the Theory of Statistics*.[51] In the preface of the latter, he lamented that the social statistical approach imported from Japan lacked a strong theoretical, mathematical basis, and he dismissed the existing manuals as useless:

> Among earlier translations on statistics there was only Yokoyama Masao's GENERAL DISCUSSION OF STATISTICS. [...] It only narrates the history of statistics, how to establish tables, draw charts and other things. This does not have any merit for theoretical study. Investigating into other Japanese books on statistics, I found that mostly all of them are like that. By the time I wrote the translation of this book, I had nothing to compare or amend the terminology used in theoretical learning, so I decided upon all the terms according to my personal opinion. Within the seas [China], the great men of refinement may hopefully correct what is not up to the standards.[52]

More generally, the notions of "pure" and of "applied" mathematics appeared in the Chinese language and scientific institutions around the 1930s, when many overseas students returned home, many with PhDs in Mathematics from Europe and America.)[53] It is then that a shift in framing the dichotomy between essence and application in terms of China and the West took place. Diverse textual sources in the 1930s (see Table 8.2) mirror the debates around a dichotomy between pure (or theoretical) and applied mathematics, sometimes framed as a methodological

---

[50]Shen and Li (1908) scroll 5, p. 2B.

[51]Gu (1910) based on Merriman (1884) and Gu (1913) based on Yule (1911).

[52]Gu (1913) Preface p. 2.

[53]After the Boxer Rebellion, 15 million US dollars were available for educational purposes from the remissions of the Boxer Indemnity Fund. This allowed Chinese students to study in the United States after 1908. Although by then many more students went to Japanese universities with government or personal funding, the United States, followed by Europe, was the preferred destination for pursuing advanced degrees in pure and applied mathematics, or science and engineering, subjects in which many Chinese students obtained PhDs from the 1920s on. Jiang Lifu 姜立夫 (1890–1978) for example, left China in 1910 to study mathematics at the University of California, Berkeley and obtained his PhD from Harvard in 1919 on *The Geometry of Non-Euclidean Line Sphere Transformation*. Upon his return to China he was the only mathematics professor in Nankai University in Tianjin, where he founded the Mathematics Department in 1920. Among his many students was the later famous geometer S. S. (Shiing-Shen) Chern 陳省身 (1911–2004). See Xu (2002).

Table 8.2 Chinese translations and compilations of writings on statistics from ca. 1900 to 1940

Year	Translation	Original(s)
1896	Fu and Hua (1896)	T. Galloway, *A Treatise on Probability: Forming the Article under that Head in the 7th Edition of the Encyclopaedia Britannica*, 1839
1903	Niu and Lin (1903)	Yokoyama Masao 橫山雅男, *Tōkeigaku kōgiroku* 統計学講義録 (*Lecture Notes on Statistics*), 1903
1909	Meng (1909)	Yokoyama Masao 橫山雅男, *Tōkei tsuron* 統計通論 (*A General Discussion of Statistics*), 1906[8]
1910	Gu (1910)	M. Merriman, *A Text-Book on the Method of Least Squares*, 1884
1913	Gu (1913)	Partial translation of G. U. Yule, *An Introduction to the Theory of Statistics*, 1911
1924	Zhao (1924)	E. M. Elderton and W. P. Elderton, *Primer of Statistics*
1925	Chen (1925)	compiled on the basis of W. I. King, *The Elements of Statistical Method*, A. L. Bowley, *Elements of Statistics*, M. T. Copeland, *Business Statistics*, and G. R. Davies, *Introduction to Economic Statistics* for use in Schools of Commerce
1927	Zhu (1927)	L. L. Thurstone, *The Fundamentals of Statistics*, 1925
1934	Chen (1934)	Collective work, contains translations of articles published for example in *Biometrika*
1934	Zhu (1934)	H. E. Garrett, *Statistics in Psychology and Education*, 1926
1935	Ning (1935)	W. I. King, *The Elements of Statistical Method*, 1912. Translated earlier into Japanese in Tamura (1927)
1936	Chen (1936)	Partially based on G. U. Yule, *An Introduction to the Theory of Statistics*; R. E. Chaddock, *Principles and Methods of Statistics*; L. L. Thurstone, *The Fundamentals of Statistics*; H. O. Rugg, *Statistical Methods Applied to Education*; Irving Fisher *The Making of Index Numbers*; F. C. Mills and D. H. Davenport, *A Manual of Problems and Tables in Statistics*
1937	Li (1937)	A. L. Bowley, *Elements of Statistics*, 1920[4]
1941	Li and Lu (1941)	F. C. Mills, *Statistical Methods, Applied to Economics and Business*, 1924

distinction in terms of precision or logical rigor. Shen Xuan 沈璿 (1899–1983), for example, in 1933, when he was the director of the Mathematics & Physics Department of the *Shanghai Academy of Scientific Research* (*Shanghai kexue yanjiuyuan* 上海科學研究院), explained the methodological divide in his talk on *The Spirit of Pure and Applied Mathematics*[54]:

> Pure mathematicians rely on theoretical methods of deduction when calculating their results. Throughout their calculations, they emphasize the mathematical rigour and the

---

[54]Shen (1933). Talk given in 1933 at the Mathematics Department of *The Great China University* (*Daxia daxue* 大夏大學) in Shanghai. Later, in 1940, Shen obtained a PhD in Japan from the Tohoku Imperial University 東北帝國大學.

uniformity of the symbols in mathematical formulas. Applied mathematicians, on the other hand, have different interests. They are deeply concerned with the outcome of actual quantities (*shiji zhi shuliang* 實際之數量). When they calculate, they pay attention to how they can simplify their numerical calculations, how they can know the error produced after having economized a certain mathematical term and how, by performing formula checks, they can simplify the identification of the committed error, etc. Therefore, they frequently consider decreasing the logical rigour in order to obtain the goals mentioned above.

For the specific case of statistics, the economist Tang Qingzeng 唐慶增 (1902–1972), who studied economics at the University of Michigan and obtained a master's degree from Harvard University,[55] relativized the notion of accuracy in statistics. According to Tang, who often referred to Western authors, precision has a different epistemic value when approached from the social and the natural sciences:

> Statistics is not just a single method, it is also a kind of science. [...] Stated otherwise: Statistics specifically record the condition of mankind in a particular moment, representing the social life of a certain era. Schulzer calls statistics "the statics of history." This meaning is close to my point.
>
> Perfect statistics should possess the characteristic of accuracy (*zhunque* 準確). What is called accuracy, basically has two meanings. One is absolute, the other is approximate. Approximate accuracy designates a time or something temporary. In human lives things constantly change, so they are based on temporary phenomena. If statistics has an approximate accuracy, it is enough for human usage. If it is said that in next year's budget of the government of a certain state, there will be an income of 2,450,000,000 *Yuan*, it is already clear enough. Therefore, one does not need to determine in advance how many *fen* or how many *li* there are. These are the distinguishing features of the social sciences. As concerns the natural sciences or the pure sciences (*chuncui kexue* 純粹科學), when constructing a new law one needs definite accuracy in the beginning, in statistics such is not the case. This is why in statistics we have assessments and conjectures (*guding yice* 估定 臆測) and other tasks to accomplish what in the natural sciences does not exist.[56]

In spite of the differing values accorded to accuracy in the pure and applied sciences, Tang defended the idea that the two approaches are complementary and that research should be pursued in both directions in parallel.[57] This vision was shared by most Chinese Republican statisticians. Among other factors—political,

---

[55] On Tang Qingzeng's life and thought, see Borokh (2013).

[56] Tang (1931) p. 136.

[57] See Tang (1931) p. 135:

> As for economic statistics, they particularly record the facts (*shishi* 事實) of wealth and industry and commerce, based on the accuracy of mathematical methods, they arrange and put in order with the purpose of deciding upon an accurate plan, they make comparisons in order to gain all sorts of knowledge, they use inductive methods in order to obtain all sorts of experimental laws. These laws, which should rely upon the statistics which have determined them, can only be applied to a specific time and place. [...] The Italian statistician Gabaglio says: "Statistics do choose and analyse economic realities, they represent the material of these realities, arranged and in order in a critical and comparative mode. Accomplished statistics provide for the study of economy a kind of material that is based on experience, and thus complements what cannot be achieved from abstract theories and proofs." etc. These words are true indeed.

economic, intellectual, literary, and historical—the one-sided emphasis of applications in China's past was made responsible for the underdevelopment of
statistical science, in spite of early statistical ideas that authors had identified in
China's heritage many centuries before the 1930s.[58] To overcome the feeling of
backwardness with respect to the level of statistical science that Chinese actors
encountered abroad, combining theory and applications was considered crucial for
China's progress of scientific knowledge:

> No matter what science, they all can be researched from both sides: the theoretical and
> the applied. In the first case, they are called pure (*dunzheng* 純正) sciences, the latter are
> called applied (*yingyong* 應用) sciences. Expounding the fundamental principles (*yuanli
> yuanze* 原理原則) of statistics, is what is "statistical theory" (*tongji yuanli* 統計原理), it
> belongs to the pure sciences. Discussing the applications of statistics to concrete realities, is
> what is called "applied statistics," it belongs to the applied sciences. It is generally assumed
> that applied statistics do not necessarily need profound theories, that, in turn, advanced
> statistical theories are also not necessarily close to applications and that therefore one
> cannot but consider them separately. But, if one does not depend on real circumstances, one
> cannot discover valuable theories and by discarding theories, it is also difficult to expect
> improvement and progress. Therefore, the two [application & theory] have mutual causes
> and effects, and do really not have clear boundaries, yet, each of them has its own focus.[59]

The algorithmic entanglement of the different theoretical and practical aspects
of ideal statistical research, combining the administrative procedures, the design
of a statistic and the mathematical analysis of data, was beautifully illustrated in
1942 by Liu Hongwan 劉鴻萬 in his manual *The Elements of Statistics* (*Tongjixue
gangyao* 統計學綱要). Referring to his illustration shown in Fig. 8.4 and translated
in Fig. 8.5,[60] he states:

> The essentials of statistical methods and the procedures of statistical analysis, since these
> are all things concerned with mathematics, therefore the learning of the mathematical
> foundations and principles contained in research of statistical method is called "math-
> ematical statistics" (*shuli tongjixue* 數理統計學). Also, these "mathematical statistics"
> are one branch of applied mathematics (*yingyong shuxue* 應用數學). The diagrammatic
> presentation that emerges from the entire body of statistics is as in the above figure.[61]

In China, statistical methods were, of course, not confined to mathematical theory
for statistics *per se*, nor were they applied only to social phenomena as Liu's
diagram might suggest. They were also applied theoretically to a variety of human
sciences, in particular demography, psychology, sociology, education and history.
Here, too, the West served as a model for propagating new fields of research, and

---

[58] See, for example, the article "Reasons why China has not developed statistical science and Zheng
Qiao's theory of illustrated registers" (*Wo guo tongjixue bu fada zhi yuanyin yu Zheng Qiao zhi
tupu xueshuo* 我國統計學不發達之原因與鄭樵之圖譜學說) by Wei Jun 盛俊, one of the co-
founders of the China Statistical Society (*Zhongguo tongji xueshe* 中國統計學社), in Chen (1934)
p. 95–99.

[59] Qin (1934) p. 2.

[60] Liu adds translations of the Chinese terms into German or/and English for most of the technical
statistical vocabulary used in Liu (1942). I have reflected these in Fig. 8.5, adding quotation marks
to the German terms.

[61] Liu (1942) p. 14.

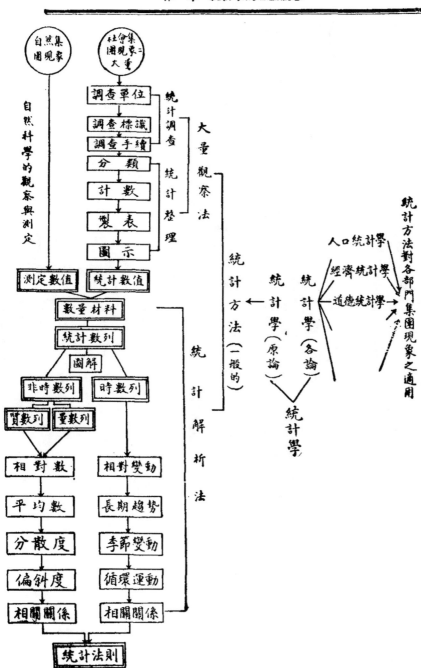

Fig. 8.4 Statistical landscape as portrayed in Liu (1942) p. 13

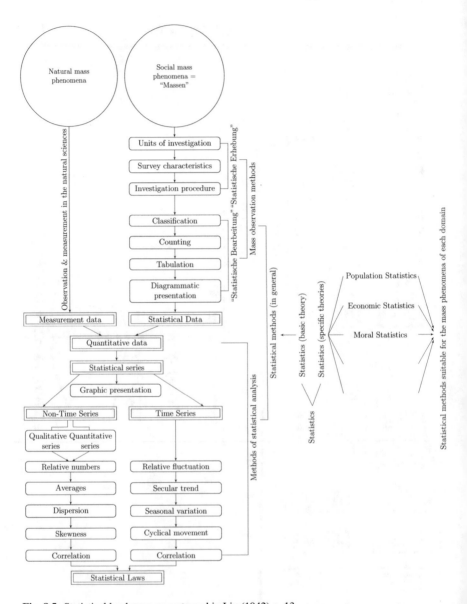

Fig. 8.5 Statistical landscape as portrayed in Liu (1942) p. 13

many manuals published around the 1930s were translations or compilations based on foreign writings (see Table 8.2 for a selective list of translations and manuals up to the 1940s).

Only the case of history is an exception to this. In 1922, the reformer and statesman Liang Qichao 梁啓超 (1873–1929) claimed in his lecture on *Historical Statistics*, given in Nanjing at the South–East University 東南大學, that the

Fig. 8.6 The *Diagram of the Supreme Ultimate* as an example of a statistical chart in *Historical Statistics* (Wei 1934) p. 157

tabular layout of historical data—numerical and narrative—as found in China's dynastic histories is a unique Chinese invention.[62] Liang's proposal to analyse the *Twenty-four Histories* empirically, using statistical methods, expands into a wave of historical studies but also into some apparently abstruse historiographies that interpret the *River Chart* (*hetu* 河圖), the *Diagram of the Supreme Ultimate* (*taiji tu* 太極圖) or more generally, Zheng Qiao's Song dynasty theory of combined diagrams and lists (*tupu xue* 圖譜學) as a kind of statistical chart (see Fig. 8.6).

---

[62]See Liang (1922). On Liang's ideas about applying statistics in historical research, see Song (2006).

The search for comparable aspects in the development of statistical diagrams with ancient Chinese illustrations emphasizes the importance of the specifically Chinese cultural heritage in the process of adaptation of a foreign knowledge system, even if the utility of traditional thought was never practically tested.

8.3 Surviving the 1949 (Statistical) Revolution

With the Communist Revolution the intellectual climate in China changed drastically and narrowed the vast range of research in statistics undertaken during the Republican period. This was mainly related to a change of reference from Western approaches to Russian ideology in the early 1950s. Socialist statistics were to be decidedly applied, not formalized, and they had to bear not only new but specifically Chinese characteristics. This ideological turn implied a rejection of formal approaches to statistics, because probabilistic theories stemming from the Anglo-Saxon tradition were labeled as bourgeois. For survey methodology a change was promoted by the CCP from the 1950s on, replacing the full enumeration censuses that had been the essential constituent of the state statistical agencies since the 1930s with the "typical survey" (*dianxing diaocha* 典型調查). This specific type of numerically laden report investigated information concerning a certain problem with a single, exemplary, "typical" sample and filtered all information through the local cadres in charge: "The typical survey of the Great Leap era came with a forceful rejection of the notion that social surveys should be constrained by questions of representativeness, for the task of the red expert was not to describe society as it was but to affirm a constructed vision of what society should be."[63]

This radical change of statistical theory and practice is what I would call China's "Statistical Revolution." It is best illustrated by comparing the prefaces and tables of content of two collections of statistical papers from 1934 and 1950.[64] The former came out of the creation of the first Statistical Society of China (*Zhongguo tongji xueshe* 中國統計學社) and contains articles related to mathematical considerations, such as the fitting of curves or the analysis of frequency distributions, with multiple references to the journal *Biometrika*, as well as articles on statistical investigations. The 1950 compilation is oriented towards Russia and Leninist theories, denigrates all earlier, bourgeois statistics as mere estimations (*guji* 估計), and argues against their formalist and objectivist elements.

As part of the anti-bourgeois campaign, many important theoretical statisticians from the Republican era were forced into self-criticism in the 1950s and relegated

---

[63] Garnaut (2013) p. 238. I discuss in more detail the conflicts between mathematical and political approaches to statistical concepts in twentieth-century China in an article currently in preparation for publication in the journal *Historia Mathematica*.

[64] See the tables of content in Figs. 8.7 and 8.8 from Chen (1934) and (Zhonghua quanguo zong gonghui tongji chu 1950) respectively.

目 錄

(以著作人姓之筆畫多寡為次序)

次數分配之分析——基本的問題············王仲武 [1]
中國歷代名人年壽之統計研究···············朱君毅 [33]
我國統計制度之研究 ·························吳大鈞 [43]
清末民政部戶口調查之新研究···············陳長蘅 [67]
我國統計學不發達之原因與鄭樵之圖譜學說···盛 俊 [95]
近六十年來中國農村人口增減之趨勢·········喬啓明 [101]
論分割數·····································楊西孟 [107]
波浪式曲線之配合問題···············諸一飛/鄭仲陶 [119]
中國之統計事業·····························劉大鈞 [145]
江西農田及每年米麥產量之估計···········劉治乾 [179]
上海工人生活程度的一個簡要分析···········蔡正雅 [193]
皮爾孫氏次數曲線之研究···············蔣招基 [205]
簡略生命表之編製法·······················羅志如 [231]

- Analysis of the distribution of index numbers– some basic problems
- Statistical research on the duration of life of some famous persons in China's past dynasties.
- Research about the statistical system of our country
- New research on the population census at the end of the Qing dynasty by the Ministry of Civil Administration
- Reasons why China has not developed statistical science and Zheng Qiao's 鄭樵 theory of illustrated registers
- Trends of increase and decline of the rural population in China since sixty years
- On partition values
- The problem of distribution of the wave equation
- Statistical entreprises in China
- Estimation of agricultural fields and rice and wheat harvests in the province of Jiangxi
- A concise analysis of the standards of living of workers in Shanghai
- Research on Pearson's frequency curves
- A simplified method for compiling life tables

Fig. 8.7 Table of contents of Chen (1934)

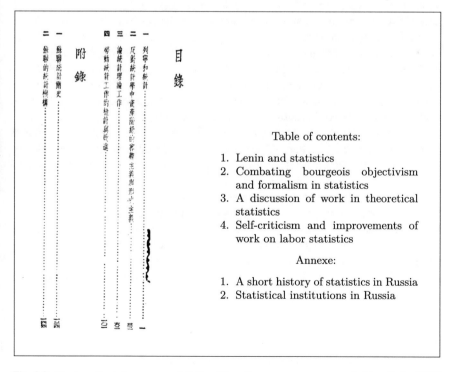

Table of contents:

1. Lenin and statistics
2. Combating bourgeois objectivism and formalism in statistics
3. A discussion of work in theoretical statistics
4. Self-criticism and improvements of work on labor statistics

Annexe:

1. A short history of statistics in Russia
2. Statistical institutions in Russia

Fig. 8.8 Newly collected papers on statistics (Zhonghua quanguo zong gonghui tongji chu 1950)

to minor responsibilities or entirely dismissed from office. Others adapted to the New China policies, such as Chen Qilu 陳其鹿 (1895–1981), who had studied at Harvard University and had based his 1925 book, *Elements of Statistics for Higher Commercial Schools* (*Tongji xue. Gaoji shangye xuexiao jiaokeshu* 統計學 高級 商業學校教科書) mainly on English language authors.[65] In 1957 Chen turned to translating a Russian book on *Diagrammatic Methods in Planning, Statistics and Accounting*,[66] while others, such as Zhu Junyi 朱君毅 (1892–1963), turned to history, strove for a consensus with the state ideology, or entirely disappeared from the scene. Representing different approaches to statistics—administrative, methodological, and mathematical—the fate of those statisticians who focused on the latter was particularly determined by the Communist Revolution, after which mathematics, as a pure science, was considered subversive to Party ideology. Jin Guobao 金國寶 (1893–1963), one of the targets of the post-1949 political changes, is a good example of a scientist (re-)acting scientifically under political constraints.[67]

Before 1949, Jin had a glorious career: he studied in Japan, edited journals, published the first Chinese translation of Lenin's writings, obtained a master's degree in statistics from Columbia University, became a professor of statistics upon his return in 1923 and headed the statistical department of the Ministry of Finance. In 1925 he was the first to introduce Irving Fisher's (1867–1947) theories of index numbers to the Chinese audience. In 1928, Cai Yuanpei 蔡元培 (1868–1940) sent him to Europe and the United States to collect statistical material and investigate different national statistical systems, which further enhanced his influential role in diverse government agencies and banks. In 1934 he published a textbook on mathematical statistics, *An Outline of Statistics* (*Tongjixue dagang* 統計學大綱), which appeared in a revised second edition in 1950, and was followed in 1951 by the mathematically more advanced *Higher Statistics* (*Gaoji tongjixue* 高級統計學) when Jin was professor of Statistics at Fudan University.

When formalist Anglo-American approaches to statistics came under harsh criticism from the Party in the 1950s, Jin was forced into self-criticism—specifically of his book *An Outline of Statistics*—in the February 1957 issue of the journal *Statistical Work* (*Tongji gongzuo* 統計工作). The phrase "Obsolete words and hackneyed expressions must be eliminated" (*chenyan wuqu* 陳言務去), provides the opener for an apologetic discourse about his mistake in copying from Pearson, Fisher and the other foreign theoretical statisticians upon whom his work had hitherto relied.[68] Item by item, he analyses the content of his book in light of the new ideology. The object of statistical sciences and the theoretical basis, mean values, indices and economic forecasting, were all themes that were subject to scrutiny

---

[65] Chen (1925).

[66] Beizuofu (1957).

[67] Detailed biographies of many important Chinese statisticians can be found in Gong (2000). For Jin Guobao, see also Ghosh (2014) p. 103–110.

[68] See Fig. 8.9 for his complete bibliography.

附錄戊 本書重要參考書

Fisher (I.)	The Making and Use of Index Numbers
Secrist (H.)	An Introduction to Statistical Methods
	Reading and Problems in Statistical Methods
Bowley (A.L.)	Elements of Statistics
Chaddock (R. E.)	Statistical Method
	Exercises in Statistical Methods
Yule (G.U.)	An Introduction to the Theory of Statistics
Jerome (H.)	Statistical Method
King (W. I.)	Elements of Statistical Method
Crum (W. L.)	An Introduction to the Methods of Economic Statistics
Moore (H. L.)	Forecasting the Yield and the Price of Cotton
Mills (F. C.)	Statistical Methods Applied to Economics and Business
Riggleman (J. R.)	Business Statistics
Babson (R. W.)	Business Barometers used in the Management of Business and Investment of Money
Davies (G. R.)	Introduction to Economic Statistics
Persons (W. M.)	Correlation of Time Series

Mitchell (W. C.) Index Numbers of Wholesale Prices in the United States and Foreign Countries (U. S. Department of Labour)
Journal of the Royal Statistical Society
Journal of the American Statistical Association
Darmois (G) Statistique Mathématique
Julin (G) Principes de Statistique théorique et appliquée
Aftalion (A) Cours de Statistique
Zizek (F) Grundri Grundrisz der Statistik
Meerwarth (F) Nationalökonomie und Statistik
Charlier (C. V. L.) Vorlesungen über die Grundzüge der mathematischen Statistik
Flüskamper (P) Statistik Theorie der Indexzahlen

社會月刊 (上海市社會局)
海關中外貿易統計年刊 (海關)
上海市工人生活費指數 (上海市社會局)
最近中國對外貿易統計圖解 (中國銀行)
經濟統計季刊 (南開大學經濟學院)
上海生活費指數 (固定稅則委員會)
棉花統計 (棉業統制委員會)
上海特別市工資和工作時間 (上海市社會局)
交通統計簡報 (交通部)
中國棉業及其貿易 (方顯廷)
經濟學季刊 (中國經濟學社)
中日貿易統計 (中國經濟學社中日貿易研究所)
統計月報 (立法院統計處及主計處統計局)
貨價季刊 (固定稅則委員會)

Fig. 8.9 Main sources of Jin (1934)

and were deemed suspicious for carrying formalist elements. Jin did not rebel, but proclaimed that he "would dedicate all of his remaining years and his entire energy to the Party and the People."[69]

---

[69] 我要尽我的余年, 把我的一切力量, 贡献给党和人民. See Jin Guobao's speech when attending the Second Plenary Conference of the Second National Committee of the Political Consultative Conference 政治协商会议第二届全国委员会第二次全体会议, published in the journal *People's Daily* (*Renmin ribao* 人民日報) 1956.02.10.

But just one year later, in 1958, Jin gave a talk at a meeting "Concerning the Problem of the Status and Use of Mathematical Statistics in Statistical Science in a Socialist Economy" (*Guanyu shuli tongji zai shehui jingji tongji kexue zhong de diwei yu zuoyong wenti de taolun* 关于数理统计在社会经济统计科学中的地位与作用问题的讨论), where he courageously insisted on the importance of mathematical theory.[70] He illustrated his argument by pointing out the existence of effective mathematical techniques to cope with small samples, such as t-tests for test-statistics that follow a Student's t-distribution, or the F-distribution and the ξ^2-Test, which allows for testing of the equality of variances of two independent samples and judging their significance in representing the entire population. Jin thus found a personal strategy through which to continue his theoretical work. This involved focusing on the kind of statistic that would conform to the official lines of thought, which were highly unsympathetic to pure mathematical research. He resists his critics by saying:

> Recently, rightist elements have made use of mathematical statistics to attack socialist economic statistics. Therefore, mathematical statistics have been sullied. But mathematical statistics is mathematical statistics and rightist elements are rightist elements, they are two separate affairs. We therefore certainly should not fear to talk [in the language of] mathematical statistics![71]
>
> 最近右派分子借数理統計来攻击社会径济統計，因此，数理統計有些骯髒肮脏旦数理統計是数理統計，右派分子 是右派分子，这是雨回事。我們絶不要因此而怕談数理扰舒。

Jin Guobao's case reflects well the duality between pure and applied mathematics which became central to official Communist Party discourse on science in the early 1950s: Jin is, on one side, a human being (re-)acting politically, on the other, a scientist, where thinking mathematically about numbers and acting politically exclude each other. Yet such ideal of scientific practice does not correspond well with the reality of mathematical knowledge construction described in this chapter. Statistical numbers, in particular, strongly reflect the tension but also the entanglements between politically charged ciphers and a numerical entity abstracted from its context of production. Be it the top down factors: the role of ideology, the impact of political and geopolitical constellations or such bottom up factors as the substantive disaffection of scientists with political dogmas and the amount of resistance that politically motivated methods met, the actual construction of applied and pure mathematical knowledge in early modern China bore political implications which make a good case for a larger project on historico-political epistemology in the first half of twentieth-century China.

---

[70]The summaries of the interventions were published 1958 in *Caijing yanjiu* 财经研究 (3) p. 57–62 & 65.

[71]Ibid., p. 65.

References

Anonymous (high official investigating governments 考察政治大臣) (around 1907). *Riben tongji shili* 日本統計釋例 (Explanation of Statistics in Japan). Beijing: Government Gazette Bureau (*Guanbao ju* 官報局).

Beizuofu 貝佐夫, P·A· (Byzov, P. A.) (1957). *Jihua, tongji yu hesuan de tushifa* 計划、統計与核算的圖示法 (Iconography in Planning, Statistics and Accounting). Translated by Zhang, Lei 章雷 and Chen Qilu 陳其鹿. Shanghai: Xin zhishi chubanshe 新知識出版社. Original in Russian, 1952 (2nd ed.).

Bennett, Adrian Arthur (1967). *John Fryer: The Introduction of Western Science and Technology into Nineteenth-Century China*, Volume 24 of *Harvard East Asian Monographs*. Cambridge, Mass.: Harvard University Press.

Borokh, Olga (2013). Chinese Tradition Meets Western Economics. In Ma Ying and Hans-Michael Trautwein (Eds.), *Thoughts on Economic Development in China*, 136–157. Routledge.

Bowley, Arthur Lyon (1920). *Elements of Statistics* (4th ed.). London: King.

Bréard, Andrea (2001). On Mathematical Terminology—Culture Crossing in 19th Century China. In Michael Lackner, Iwo Amelung, and Joachim Kurtz (Eds.), *New Terms for New Ideas: Western Knowledge & Lexical Change in Late Imperial China*, Volume 52 of *Sinica Leidensia*, 305–326. Leiden, Boston, Köln: Brill.

Bréard, Andrea (2002). Problems of Pursuit: Recreational Mathematics or Astronomy? In Yvonne Dold-Samplonius, Joseph W. Dauben, Menso Folkerts, and Benno van Dalen (Eds.), *From China to Paris: 2000 Years Transmision of Mathematical Ideas*, Volume 46 of *Boethius*, 57–86. Stuttgart: Steiner.

Bréard, Andrea (2006). Where Shall the History of Statistics in China Begin? In Gudrun Wolfschmidt (Ed.), *"Es gibt für Könige keinen besonderen Weg zur Geometrie" — Festschrift für Karin Reich*, Volume 59 of *Algorismus*, 93–100. Augsburg: Dr. Erwin Rauner Verlag.

Bréard, Andrea (2008). *Reform, Bureaucratic Expansion and Production of Numbers. Statistics in Early 20th Century China*. Habilitation (History of science), Technische Universität Berlin.

Bréard, Andrea and Annick Horiuchi (2014). Mathematics Education in East Asia in the Premodern Period. In Alexander Karp and Gert Schubring (Eds.), *Handbook on the History of Mathematics Education*, 153–174. New York, [etc.]: Springer.

Chen, Changheng 陳長蘅 (Ed.) (1934). *Tongji luncong* 統計論叢 (Collected Papers on Statistics), Volume 1 of *Zhongguo tongjixue congshu* 中國統計學叢書. Shanghai: Liming shuju 黎明書局.

Chen, Qilu 陳其鹿 (1925). *Tongji xue* 統計學 *Gaoji shangye xuexiao jiaokeshu* 高級商業學校教科書 (=Elements of Statistics for Higher Commercial Schools) (1st ed.). Xin xue zhi 新學制 (= New System Series). Shanghai: Shangwu yinshuguan 商務印書館. 4th edition 1928.

Chen, Yifu 陳毅夫 (1936). *Shehui diaocha yu tongjixue* 社會調查與統計學 (Social Case Study and Statistics). Wuxi: Minsheng yinshuguan 民生印書館.

Ding, Weizhi 丁衛志 and Chen Song 陳崧 (1995). *Zhong-Xi tiyong zhi jian—Wan Qing Zhong-Xi wenhuaguan shulun* 中西體用之間晚清中西文化觀述論 (Between China and the West, Essence and Use. A Commentary on Attitudes towards Chinese and Western Culture in the Late Qing Period). Beijing: Zhongguo shehui kexue chubanshe 中國社會出版社.

Elderton, W. Palin and Ethel M. Elderton (1909). *Primer of Statistics*. London: A&C Black Ltd.

Elman, Benjamin A. (2005). *On Their Own Terms. Science in China 1550–1900*. Cambridge, Mass. and London, England: Harvard University Press.

Fairbank, John K. (Ed.) (1978). *Late Ch'ing, 1800–1911*, Volume 10, Part 1 of *The Cambridge History of China*. Cambridge, London , New York, Melbourne: Cambridge University Press.

Fan, Huiguo 藩會國 (1937). Shuxue yu yingyong 數學與應用 (Mathematics and Applications). *Shuxue zazhi* 數學雜誌 *1*(3), 4–12.

Fan, Huiguo 范會國 (1944). *Lilun lixue* 理論力學 (Traité de mécanique rationelle). Longmen chuban gongsi 龍門出版公司.

Fu, Lanya 傅蘭雅 (Fryer, John) and Hua Hengfang 華蘅芳 (1896). *Jueyi shuxue* 決疑數學 (A Treatise on Probability) (周氏刊本 Zhou ed.). China: [n.p.]. Original: (Galloway 1839).

Galloway, Thomas (1839). *A Treatise on Probability: Forming the Article under that Head in the 7th Edition of the Encyclopaedia Britannica* (Reprint ed.). Edinburgh: Adam and Charles Black.

Galloway, Thomas (1843). On the Proper Motion of the Solar System (Abstract). *Abstracts of the Papers Communicated to the Royal Society of London 5*, 669–672.

Garnaut, Anthony (2013). Hard Facts and Half-truths: The New Archival History of China's Great Famine. *China Information 27*(2), 223–246.

Garrett, Henry Edward (1926). *Statistics in Psychology and Education*. New York, London: Longmans, Green and Co.

Ghosh, Arunabh (2014). *Making it Count: Statistics and State-Society Relations in the Early People's Republic of China, 1949–1959*. Doctor of philosophy, Graduate School of Arts and Sciences, Columbia University.

Gong, Jianyao 龔鉴尧 (2000). *Shijie tongji mingren chuanji* 世界統計名人傳記 (Biographies of World Famous Persons in Statistics). Beijing: Zhongguo tongji chubanshe 中國統計出版社.

Gu, Cheng 顧澄 (Trans.) (1910). *Zui xiao er cheng fa* 最小二乘法 (The Method of Least Squares). Peking: Board of Education. Original: (Merriman 1884).

Gu, Cheng 顧澄 (Trans.) (1913). *Tongjixue zhi lilun* 統計學之理論 (The Theory of Statistics). Shanghai: Wenming shuju 文明書局. Original: (Yule 1911).

Guo, Jinhai 郭金海 (2003). Jingshi Tongwenguan shuxue jiaoxue tanxi 京师同文館数学教学探析 (An Inquiry of Mathematics Teaching in the Imperial Tungwen College). *Ziran kexue yanjiu* 自然科学史研究 (Studies in the History of Natural Sciences) *2 Suppl.*, 47–60.

Guo, Shirong 郭世榮 (1989). Xifang zhuanru wo guo de di yi bu gailülun zhuanzhu—Jueyi shuxue 西方傳入我國的第一部概率論專著—決疑數學 (=Jue-Yi Shu-Xue, the First Book about Probability Introduced in China from the West). *Zhongguo keji shiliao* 中國科技史料 *10*(2), 90–96.

Index (1910). *Da Qing Guangxu Xuantong xin faling fenlei zongmu* 大清光緒宣統新法令分類總目 (Classified Index to New Administrative Methods and Regulations of the Guangxu and Xuantong Reign, Qing Dynasty). Shanghai: *Shangwu yinshuguan* 商務印書館.

Jin, Guobao 金國寶 (1934). *Tongjixue dagang* 統計學大綱 (An Outline of Statistics). Shanghai: Shangwu yinshuguan 商務印書館 (國立上海商學院叢書).

Kameichi, Nakaba 夏秋亀一 (1899). *Saishin tōkeigaku* 最新統計學 (The Most Recent Statistical Theories), Volume 40 of *Teikoku Hyakka Zensho* 帝國百科全書. Tokyo: Hakubunkan 博文館.

Keller, Agathe and Alexei Volkov (2014). Mathematics Education in Oriental Antiquity and Middle Ages. In Alexander Karp and Gert Schubring (Eds.), *Handbook on the History of Mathematics Education*, 55–84. New York, [etc.]: Springer.

King, Willford Isbell (1912). *The Elements of Statistical Method*. New York, London: Macmillan.

Levenson, Joseph Richmond (1968). *Confucian China and Its Modern Fate: A Trilogy*. Berkeley: University of California Press.

Li, Huang Xiaozhen 李黃孝貞 and Lu Zongwei 陸宗蔚 (Trans.) (1941). *Tongji fangfa* 統計方法 (= Statistical Methods, Applied to Economics and Business). Shanghai: Zhonghua shuju 中華書局. Original: (Mills 1924).

Li, Shuangbi 李双璧 (1995). Cong 'gezhi' dao 'kexue': Zhongguo jindai keji guan de yanbian guiji 从"格致"到"科学": 中国近代科技观的演变轨迹 (From 'Investigating Things and Extending Knowledge' to Science: Tracking the Conceptual Transformation in Chinese Modern Science and Technology). *Guizhou shehui kexue* 贵州社会科学 *137*(5), 102–107.

Li, Wenlin and Jean-Claude Martzloff (1998). Aperçu sur les échanges mathématiques entre la chine et la france (1880–1949). *Archive for History of Exact Sciences 53*, 181–200.

Li, Zhaohua 李兆华 (2012). Wan Qing suanxue keyi kaocha 晚清算學課藝考察 (A Study of Mathematical Examination Problems during the Late Qing). *Ziran kexue shi yanjiu* 自然科学史研究 (Studies in the History of Natural Sciences) *25*(4), 322–342.

Li, Zhiquan 李植泉 (Trans.) (1937). *Tongji xue yuanli* 統計學原理 (= Elements of Statistics). 4 vols. Shanghai: Shangwu yinshu guan 商務印書館. Original: (Bowley 1920).

Liang, Qichao 梁启超 (1922, November 28). Jiangyan. Lishi tongjixue (shiyi yue shi ri wei dongnan daxue shidi xuehui jiangyan) (fu biao) 講演.歷史統計學(十一月十日為東南大學史地學會講演) (附表) (Lectures: Historical Statistics (Lecture given on November 10 at the Society for Historical Geography of Dongnan University), with tables appended). *Chenbao fukan* 晨报副刊, 1–2, continued 11月29日, 1 and 11月30日, 1–2.

Liu, Hongwan 劉鴻萬 (1942). *Tongjixue gangyao* 統計學綱要 (The Elements of Statistics). Shanghai: Zhonghua Shuju 中華書局.

Lü, Yiran 呂一燃 (1995). Xue Fucheng yu Zhong Ying Dian Mian jiewu jiaoshe 薛福成與中英滇緬界務交涉 (Xue Fucheng and Chinese–British Diplomacy on the Question of the Boarder between Yunnan Province and Burma). *Zhongguo bianqiang shidi yanjiu* 中國邊彊史地研究 (China's Borderland History and Geography Studies) *2*, 57–72.

Martzloff, Jean-Claude (1993). Eléments de réflexion sur les réactions chinoises à la géométrie euclidienne à la fin du XVIIe siècle. *Historia Mathematica 20*, 160–179.

Meng, Sen 孟森 (1907). Jianyu tongji xu (fu biao) 監獄統計序(附表) (A Preface to Prison Statistics, with Tables Appended). *Fazhengxue jiaotong she zazhi* 法政學交通社雜誌 *2*, 103–107.

Meng, Sen 孟森 (1909). *Gaiding zengbu tongji tonglun* 改訂增補統計通論 (A General Discussion of Statistics) (3rd ed.). Shanghai: Shangwu yinshuguan 商務印書館. Original: (Yokoyama 1906).

Merriman, Mansfield (1884). *A Text-Book on the Method of Least Squares* (1st ed.). New York: J. Wiley.

Mills, Frederick Cecil (1924). *Statistical Methods, Applied to Economics and Business*. New York: Holt.

Ning, Encheng 甯恩承 (1935). *Tongji fangfa* 統計方法 (= The Elements of Statistical Method). Jingji congshu 經濟叢書. Shanghai: Dadong shuju 大東書局. Translation of (King 1912).

Niu, Yongjian 鈕永建 and Zhuonan Lin, 林桌男 (1903). *Tongji jiangyi lu* 統計講義錄 (Lecture Notes on Statistics). [N.p.]: Shizhong shuju 時中書局. Translation of (Yokoyama 1900).

Obituary (1851–1852). Report of the Council—Obituary of Thomas Galloway. *Monthly Notices Royal Astronomical Society XII*, 87–89.

Qin, Guwen 秦古温 (1934). *Jingji tongjixue conglun* 經濟統計學總論 (An Introduction to Economic Statistics) (3rd ed.). Jixue congshu 計學叢書. Qin Qingjun kuaiji shiwusuo 秦慶鈞會計事務所.

Qin, Guojing 秦國經, Tang Yinian 唐益年, and Ye Xiuyun 葉秀雲 (Eds.) (1996). *Zhongguo di yi lishi dang'an guan zang Qingdai guanyuan lüli dang'an quanbian* 中國第一歷史檔案館藏清代官員履歷檔案全編 (Complete Compendium of Records of Conduct for Qing Officials from the No1 Historical Archives), 30 vols. Shanghai: Huadong shifan daxue chubanshe 華東師範大學出版社. Reproduced from handwritten originals.

Richthofen, Freiherr von, Ferdinand (1877). *China. Ergebnisse eigener Reisen und darauf gegründeter Studien*, Volume 1 (Erster Band): Einleitender Theil. Berlin: Verlag von Dietrich Reimer.

Ruan, Yuan 阮元 and Luo Shilin 羅士琳 (Eds.) (光緒壬午春 1882). *Chouren zhuan sishiliu juan fu Xu Chouren zhuan liu juan* 疇人傳四十六卷續傳六卷 (Biographies of Astronomers and Mathematicians in 46 Scrolls; Continuation of Biographies of Astronomers and Mathematicians in 6 Scrolls) (Zhang 張氏重校刊 ed.). [N.p.]: Haiyan zhangjing changxing zhai 海鹽張敬常惺齋. Includes *Chouren jie* 疇人解 (Explanation of Astronomers and Mathematicians) by Tan Tai (Ed.) (清)談泰[撰]. Preface to *Chouren zhuan* 疇人傳 dated November 1799 (嘉慶四年十月), preface to *Xu Chouren zhuan* 續疇人傳 dated May 1840 (道光二十年夏四月).

Shen, Linyi 沈林一 (1901). *Ziqiang chuyi* 自強芻議 (A Rustic Opinion on the Self-Strengthening Movement), Volume 6 of *Lian qing xuan leigao* 練青軒類稿. [China?]: [n.p.].

Shen, Linyi 沈林一 and Li Kunfu 儷昆甫 (between 1875 and 1908). *Zhongxi qianbi duliangheng hekao* 中西錢幣度量衡合攷 (A Combined Investigation into the Currencies, Measures and Weights of China and the West), Volume 4, 5 of *Lian xuan lei gao* 練青軒類稿. [China?]: [n.p.].

Shen, Xuan 沈璿 (1933). Yingyong shuxue yu duncui shuxue zhi jingshen 應用數學與純粹數學之精神 (The Spirit of Pure and Applied Mathematics). *Xueyi* 學藝 *12*(3), 53–62.

Simon, Renée (Ed.) (1970). *Le P. Antoine Gaubil, S.J.: Correspondance de Pékin, 1722–1759.* Geneva: Librairie Droz.

Song, Xueqin 宋学勤 (2006). Liang Qichao dui lishi tongjixue de changdao yu shijian 梁启超对历史统计学的倡导与实践 (How Liang Qichao Advocated and Put into Practice Historical Statistics). Shixue lilun yanjiu 史学理论研究 *(Historiography)* 3, 31–41.

Tamura, Ichirō 田村市郎 (Trans.) (1927). *Tōkeigaku yōron* 統計学要論 (= The Elements of Statistical Method). Osaka: Ginkō mondai kenkyūkai 銀行問題研究会. Original: (King 1912).

Tang, Qingzeng 唐慶增 (1931, May). Tongjixue zhi jinxi guan 統計學之今昔觀 (Contemporary and Traditional Views on Statistics). *Zhengzhi jingji yu falü* 政治經濟與法律 *1*(1), 135–144.

XTXF (1911). *Da Qing Xuantong xin faling* 大清宣統新法令 (New Administrative Methods and Regulations of the Xuantong reign, Qing Dynasty). Shanghai: Shangwu yinshuguan 商務印書館. 35 vols. Indexed in (Index 1910).

Thurstone, Louis Leon (1925). *The Fundamentals of Statistics.* New York: The Macmillan Company.

Wang, Youjun 王幼軍 (2007). *Lapulasi gailü lilun de lishi yanjiu* 拉普拉斯概率理論的曆史研究 (Research on the History of Laplace's Probability Theory). Shanghai: Jiaotong daxue chubanshe 交通大學出版社.

Wei, Juxian 衛聚賢 (1934). *Lishi tongjixue* 歷史統計學 (Historical Statistics). Shanghai: Shangwu yinshuguan 商務印書館.

Whewell, William 胡威立 (1859). *Zhongxue* 重學 (Mechanics. Translated by Li Shanlan 李善蘭 and John Edkins 艾約瑟). Shanghai: Mohai shuguan 墨海書館 (Inkstone Press).

Xi, Gan 席淦 and Gui Rong 貴榮 (Eds.) (1880). *Suanxue keyi* 算學課藝 Mathematical Examination Problems, 4 scrolls 四卷. Beijing: Tongwen guan 同文館.

Xu, Yibao (2002). Chinese–U.S. Mathematical Relations, 1859–1949. In Karen Hunger Parshall and Adrian C. Rice (Eds.), *Mathematics Unbound: The Evolution of an International Mathematical Research Community, 1800–1945*, 287–310. AMS.

Yan, Dunjie 嚴敦傑 (1990). Ba jueyi shuxue shi juan 跋《決疑數學》十卷 (A Postscript to «The Mathematics of Probability in Ten Scrolls»). In Mei Rongzhao 梅榮照 (Ed.), *Ming Qing shuexueshi lunwenji* 明清數學史論文集, 421–444. Nanjing: Jiangsu Education Press 江蘇教育出版社.

Yang, Mo 楊模 (1910). *Xijin si zhe shishi huicun* 錫金四哲事實彙存 (Collected Facts about Four Wise Men of Wuxi and Jinkui Counties). China: [n.p.].

Yokoyama, Masao 橫山雅男 (1900). *Tōkeigaku kōgi* 統計学講義 (Lectures on Statistical Science) (增補 ed.). Tokyo: Kobayashi Matashichi 小林又七.

Yokoyama, Masao 橫山雅男 (1906). *Tōkei tsuron* 統計通論 (A General Discussion of Statistics) (8th ed.). Tokyo: Senshūgakkō 專修學校. Other editions: 1904[1], 1921[41].

Yule, George Udny (1911). *An Introduction to the Theory of Statistics.* London: Charles Griffin and Co.

Zhao, Erxun 趙爾巽 et al. (Eds.) (1981). *Qingshi gao* 清史稿 (Draft History of the Qing Dynasty). Taipei: Guofang yanjiuyuan 國防研究院.

Zhao, Wenrui 趙文鋭 (Trans.) (1924). *Tongjixue yuanli* 統計學原理 (= Primer of Statistics). Jingji congshu 經濟叢書. Shanghai: Shangwu yinshuguan 商務印書館. Original: (Elderton and Elderton 1909).

Zhonghua quanguo zong gonghui tongji chu 中华全国总工会统计处 (Ed.) (1950). *Xin tongji luncong* 新统计论丛 (Newly Collected Papers on Statistics). Beijing: Gongren chubanshe 工人出版社.

Zhu, Junyi 朱君毅 (Trans.) (1927). *Jiaoyu tongji xue gangyao* 教育統計學綱要 (The Fundamentals of Statistics). Shanghai: Shangwu yinshuguan 商務印書館. Original: (Thurstone 1925).

Zhu, Junyi 朱君毅 (Trans.) (1934). *Xinli yu jiaoyu zhi tongji fa* 心理與教育之統計法 (= Statistics in Psychology and Education). Daxue congshu 大學叢書. Shanghai: Shangwu yinshuguan 商務印書館. Original: (Garrett 1926).

Chapter 9
Visions of Modernity

Contents

Universal communicability of mathematical meaning is nowadays perfectly achieved. Even to such an extent that my colleagues in the Mathematics Department of the Université des Sciences et Techniques Lille 1 in France can understand Chinese math journal articles without being able to read the words between the lines filled with formula. But behind this seemingly monolithic universalism lies hidden a historical epistemology, which since pre-modern times in particular can only be understood through a global historical perspective. Although ideological, emotional discourses, as shown in the first chapter of this book, are part of this historical epistemology, negotiations and mediations on a very local level, as described in Chap. 2 for the case of the ellipse and in Chap. 4 for the case of translation, are equally important. The aim of these mediations was to productively connect two mathematical cultures for generating and representing new scientific content, or, as in the case of statistics depicted in Chap. 8, to find a consensus between science and the Chinese state. The outcome, a global-historical continuum between the rhetorical and algorithmic approach to mathematics and a formulaic one is certainly much more complex and differentiated than I have presented in this book. Not only because the processes described here have stretched across other geographic zones and across wider linguistic boundaries, in particular to Japan, but also because the very rich mathematical literature from the late Qing onwards has been very little studied.

© Springer International Publishing AG, part of Springer Nature 2019
A. Bréard, *Nine Chapters on Mathematical Modernity*, Transcultural
Research – Heidelberg Studies on Asia and Europe in a Global Context,
https://doi.org/10.1007/978-3-319-93695-6_9

Although mathematics today are universally communicable, and traditional Chinese mathematics seem to have been entirely superseded by "modern" mathematical discourse, one can still find traces of an alternative approach to mathematical objects that dwell upon China's past. I will discuss in the few remaining pages of this book what persists of China's efforts to revive its scientific past today (Sect. 9.1) and—finally—problematize the notion of "modernity" and its relevance to the kind of story narrated here (Sect. 9.3). I shall introduce this final discussion by asking what it means today to be a global modern mathematician, relating a recent event that has raised many questions about the possible cultural specificity of a Chinese-born American mathematician's pathbreaking approach to number theory (Sect. 9.2). All this remains a sketch and is intended as a stimulus for further inquiries into the entangled history of mathematics in China over the last 50 years.

9.1   The Comeback of "National Studies"

It has been reported that in recent university entrance examinations taken by the high school students of Hubei Province, there were questions reflecting the recent curriculum reforms emphasizing National Studies. In mathematics the following problems came from the *Nine Chapters*: "A bamboo with nine sections," a problem on proportional distribution in 2011; and in 2012, the "Procedure to extract [the cube root of the volume of] a circle" as a bonus question. Because they are closely related to the curriculum, these problems, according to the journalist who reported on the issue, "reveal conspicuously a mathematical culture and enhance patriotic feelings."[1]

Such news reflects a revival of the National Studies from the 1930s, resulting in a veritable "National Studies fever" (*Guoxue re* 國學熱) which erupted in the late 1980s. Sometimes characterized as artificial and void of substance, a Confucian fundamentalism[2] defining itself as identical to the Confucian *Six Arts*[3] or to sinology,[4] the phenomenon is far from homogenous in its ideological perspectives. In China today, one can observe a broad enthusiasm for the subject among the general public and academics, and the variety of definitions and debates surrounding the subject still need to be understood in their global historical context. "National Studies" are now a common section in bookstores, and they are widely debated on the Internet. Institutionalized since 1996, a national association, the China

---

[1] See Zhang (2012), online summary at: http://hbylzx.vicp.net/ReadNews.asp?NewsID=8865 consulted 27 January, 2017.

[2] Chen (2011).

[3] Cai (2013) p. 93.

[4] See Zhao (2015) p. 7: "Sinology education is an important part of ideological education in colleges and universities" (*Guoxue jiaoyu shi dangdai daxuesheng sixiang jiaoyu de zhongyao zucheng bufen* 国学教育是当代大学生思想教育的重要组成部分).

Guoxue Seminar (*Zhongguo guoxue yanjiuhui* 中国国学研究会), and many local organizations engage with activities related to National Studies. Even all the major universities now have a National Studies department[5] and hold regular conferences on concepts such as "Natural National Studies" (*Ziran guoxue* 自然國學) where the inclusion—and exclusion—of science is debated.[6] Indeed, whenever mathematics comes into play in debates about the content of teaching National Studies, the *Nine Chapters* is in the spotlight. They remain the flagship of Chinese mathematical culture, and their style is distinctly set off against the Western tradition in terms of deductive patterns of argumentation, its practical outlook, its synthetic character and monolithic status.[7]

That the 2015 Nobel Prize for medicine went to Tu Youyou 屠呦呦 (b. 1930)—who discovered artemisinin by initially skimming ancient Chinese medical texts systematically during the first stage of a project against malaria launched in 1967[8]—encouraged those proponents of National Learning who defend it as an important means of advancing and innovating scientific knowledge. For mathematics, there are not many examples where ancient Chinese sources have directly served in contemporary research—except for Wu Wenjun 吳文俊[9] and his student Zhou Xianqing 周咸青:

> The Chinese scientist who has received the most awards, the academician Wu Wenjun, has precisely profited from the constructivist nature of [Chinese] mathematics. He has discovered a mechanical method of proof. Early in the 1970s he has discovered that with a computer one can very quickly prove complicated theorems in geometry. Zhou Xianqing developed Wu's technique and compiled a valid general algorithm which proved 512 non-trivial theorems.[10]

Others are less confident about the usefulness of Natural National Learning for contemporary mathematics. "In spite of the *Nine Chapters on Mathematical Procedures'* immense scientific and cultural heritage," as the renowned intellectual historian and philosophy professor at Beijing University Chen Lai 陳來 recently said,[11] "we cannot make direct use of it. Under normal circumstances, we cannot apply again these canons, but, for example, what is contained in ancient science is a cosmology, a world view and a scientific outlook. As for today, this all plays

---

[5]For an overview of the curricula, see Zhao (2015) p. 9.

[6]The first of these conferences was held in Qingdao in 2013. For a summary, see Yu (2013).

[7]See Kang Kuanying's 亢宽盈 contribution on "The system of Natural National Studies considered from the characteristics of China's traditional mathematics" (*Cong Zhongguo chuantong shuxue de tezheng kan ziran guoxue tixi* 从中国传统数学的特征看自然国学体系) in Zhang et al. (2002) p. 8–9.

[8]See Tu (2011).

[9]See page 15.

[10]Translated from Zhang and Tong (2015) p. 93.

[11]Chen was named director of the newly founded National Studies Department of Beijing University in 1995.

a rectifying and developmental role."[12] Had Chen in 2013 already heard about a mathematician named Zhang Yitang, he would probably have cited his case as a good illustration of his ideas about the possibility of indirect impact on scientists' work through historical sources.

9.2   The Case of Zhang Yitang

The submission of an article entitled "Bounded gaps between primes"[13] to Princeton University and the Institute for Advanced Study's Journal *Annals of Mathematics* in 2013 shook the international mathematical community for two reasons. If the paper's content was correct, it would represent a major advance in analytical number theory about the distribution of large prime numbers. The other shock came from the very author himself, Zhang Yitang 張益唐, a close to sixty-year old part-time calculus teacher from New Hampshire University whom nobody had ever heard of and who had never published an article in his life except for a faulty online preprint. After only three weeks, the reviewers of Zhang's fifty-four-page long article confirmed that he had indeed proved the following result, establishing an upper bound for the gap between two consecutive prime numbers where p_n is the n-th prime:

$$\liminf_{n\to\infty}(p_{n+1} - p_n) < 7 \cdot 10^7$$

The technical correctness of Zhang's proof was thus established, but other questions remained open and puzzled scientists and journalists alike because the case of Zhang reveals a specific tension between an isolated educational background, individual creation and the dynamics of an international mathematical community. Here was this man, born in Shanghai in 1955, raised by his grandmother, working secluded from the world towards the solution of an unsolved and universally known problem, the twin prime conjecture.[14] Zhang had hardly been to school in his youth during the decade of the Cultural Revolution (ca. 1966–1977) before he entered Peking University at the age of twenty-three. There he learnt basic techniques in analytic number theory from his teacher Pan Chengdong 潘承洞 (1934–1997) who worked on the hitherto unsolved Goldbach conjecture.[15] Zhang eventually came to the United States for a PhD in algebraic geometry. After obtaining his degree he disappeared from the scene, surviving with minor jobs, including a stint in a sandwich bar, before reemerging in a spectacular fashion with a major

---

[12]Yu (2013) p. 91.

[13]Zhang (2014).

[14]Stating that there are an infinite number of pairs of twin primes, i.e. two consecutive primes where $p_{n+1} = p_n + 2$.

[15]Every even integer greater than 2 can be expressed as the sum of two primes.

breakthrough in pure mathematics. One wonders how Zhang became so interested in an unsolved number theoretical conjecture in the first place? What role did his educational background in China under Mao Zedong play? What kind of knowledge was available to Zhang in institutions in China where mathematics were taught on a higher level? What made Zhang a world-class mathematician of the twenty-first century? Or more specifically: How does number theory constitute itself by local specificities while imbedding results of universal character in larger theories over a long period of time?

More to the point, the case of Zhang Yitang displays extremes that are seemingly incompatible with each other: on the one hand, we have the domain of pure mathematics, which is as abstract and ahistorical as one can get in terms of academic disciplines; this is even more true for number theory after Peano's and Frege's paradigmatic input in the early twentieth century. Pure mathematics is a highly abstract game that is played with formal symbols. On the other hand, we have history, which is far from being pure, packed as it is with facts, opinions, contradictions, revisions, turns, archival dust and concrete evidence. At first glance, history does not seem to have any relevance in and to the highly selective game of pure mathematics. If you want to become a player in the game of pure mathematics, you need a high degree of talent combined with training and, of course, a lot of creative energy. It is implied that anyone can become a player, no matter where she or he comes from. And while I am not denying that Zhang Yitang is obviously highly talented, he was also exposed to the right kind of grooming and is certainly endowed with a substantial amount of creative energy. Zhang's access to the game of pure mathematics is, in fact, the result of a complex transcultural process. Without this specific process—or if you prefer, historical evolution—he would not have been able to achieve what he achieved with his recent breakthrough. In other words, numbers—pure mathematics—need a transcultural history because the game is played by human agents who can access the game table if and only if the global constellations allow them to do so.

A look at Zhang Yitang's biography shows the importance of the transcultural processes behind his intellectual trajectory and how these can illuminate our understanding of why he came to work and how he worked on a universal math problem. Zhang entered Beijing University at a time when most mathematics professors there were doctoral graduates from the United States and Europe. Zhang's teacher Pan Chengdong had obtained a postgraduate degree in 1961 advised by Szu-Hoa Min 閔嗣鶴 (Pinyin: Min Sihe, 1913–1973), who in turn had been a student of Edward Charles Titchmarsh (1899–1963) in Oxford, himself a student of the British number theorist Godfrey Harold Hardy (1877–1947). Such complex transcultural constellations in a young mathematician's genealogy emerged in China in the 1930s, when an important epistemic break in mathematical knowledge changed the institutional landscape of Republican China. In the context of a quest to save China through science,[16] pre-modern Chinese mathematics were (nearly) entirely

---

[16]See Chap. 1 p. 7.

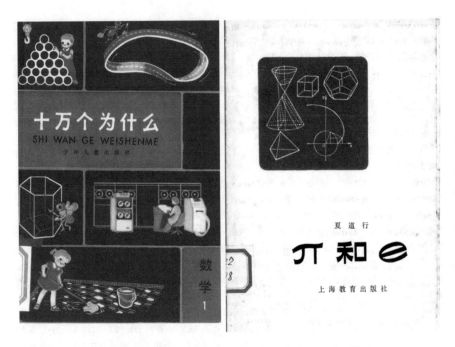

Fig. 9.1 *One Hundred Thousand Whys* (1962) and Xia Daoxing's *π and e* (1964)

superseded by an heterogenous imported knowledge system that also structured the newly established Mathematics Departments in China's universities.[17] Yet in spite of this epistemic *caesura*, pre-modern Chinese mathematics were not completely forgotten or disdained, and remained alive in various narrative frameworks.

In several interviews,[18] Zhang pointed out that in his youth he had studied in detail two mathematical books that had particularly aroused his passion for numbers. Both of these books are written in discursive style, and the technical parts are motivated by specific historical elements from China and complemented by outlooks to other geographic regions (Fig. 9.1):

1. At the age of 10 (in 1965), Zhang read the math volume in a successful popular science series *One Hundred Thousand Whys* (*Shiwan ge weishenme* 十

---

[17]Norbert Wiener (1894–1964), professor of mathematics at MIT who visited the Departments of Electrical Engineering and Mathematics at Qinghua University in Beijing (Fall, 1935–Summer, 1936) "provided a vivid description of how many of the Chinese faculty carried the distinct styles of the country in which they were trained." Wang (2002) p. 306n51. See Wiener (1956) p. 186 and Wei (1996).

[18]See for example http://news.xinhuanet.com/local/2015-08/25/c_1116370665.htm, consulted on January 25, 2017. See also Wilkinson (2015) and the film documentary *Counting from infinity* by George Csicsery, Zala Films 2015 http://www.zalafilms.com/films/countingindex.html.

万个为什么).[19] He learned about the Goldbach conjecture and the Riemann hypothesis[20] from this book, which was written by many renowned Chinese mathematicians, most of whom were products of transcultural processes of knowledge circulation.[21]

2. During the Cultural Revolution, which Zhang spent with his mother in the countryside in Hubei province, he studied Xia Daoxing's 夏道行 (b. 1930) π and e (π 和 e), published in 1964.[22]

In both readings, Zhang Yitang was exposed, to a certain extent, to the leitmotifs of twentieth-century writings on the history of mathematics in China. The latter book contains an extensive chapter on research into π by ancient Chinese mathematicians, ranging from early antiquity to Zu Chongzhi's value, which is precise to the sixth decimal:[23]

> Only more than 1000 years after Zu Chongzhi had discovered the precise ratio, the European A. Anthonisz (from the sixteenth to seventeenth cent.) rediscovered this value. This clearly shows that in the past the achievements of our country's mathematicians were outstanding![24]

In the beginning of the first book, in a chapter that answers the question where the symbols for arithmetic operations came from, one learns about a certain "mathematician Li Shanlan, well-known throughout the world by the 'Li Shanlan Identity'."[25] Unless you have read Chap. 5 of this book, my guess would be that you have not heard of this combinatorial identity before. Be that as it may, it is

---

[19]First published by the Shanghai-based Juvenile and Children's Publishing House (*Shaonian ertong chubanshe* 少年兒童出版社) in 1962.

[20]See Riemann (1859). The hypothesis concerns the location of zeros to the Riemann zeta function, defined as the absolutely convergent infinite series:

$$\zeta(s) = 1 + 2^{-s} + 3^{-s} + 4^{-s} + \cdots$$

or, in more compact notation:

$$\zeta(s) = \sum_{n=1}^{\infty} \frac{1}{n^s}$$

where s is a fixed complex number with a real part greater than 1.

[21]The entry on the Goldbach conjecture, for example, was written by the number theorist Wang Yuan 王元 (b. 1930). Former president of the Chinese Mathematical Society, he works as a member of the Chinese Academy of Sciences at the Institute of Mathematics. In the attempt to solve the Goldbach conjecture, Wang's research focuses on the sieve method and its applications, Diophantine approximations and equations and applications of number theory.

[22]After the Cultural Revolution, Xia obtained his professorship at Fudan University in 1978 and went to the United States in 1984. He mainly focused on operator theory and algebraic topology.

[23]Chapter 2, "Research on π by Chinese mathematicians in the past" (*Wo guo gudai shuxuejia guanyu π de yanjiu* 我国古代数学家关于π的研究) in Xia (1964) p. 10–17.

[24]*Idem* p. 17.

[25]Page 2 of Shaonian ertong chubanshe 少年儿童出版社, 1962 ed.

Zhang Yitang who is now an internationally renowned mathematician; not unlike Li Shanlan he developed an exceptional talent on his own through a youth spent reading books, solving problems in retreat and unconsciously absorbing a patriotic spirit. Although he reached out to a universal of modern mathematics, Zhang's outlook remained essentially Chinese, and his scientific creativity and force of intuition, I suspect, is not unrelated to the kind of self-training he followed with the two above-mentioned books. In search for institutional support, he first found it in Beijing and later in the West. Indeed, one cannot help but see in Zhang a hybrid modern academic with a hint of the Qing literati, unacknowledged by the examination system, working in seclusion and without recognition on a major unsolved math problem. These are just a few superficial parallels with Li Shanlan's trajectory, but they show that the universality of modern mathematics is only an illusion. Even modern mathematics needs a history that takes into account the articulation of a mathematician's local, scientific and personal culture with his or her mathematical formulas.

9.3   On "Mathematical Modernity"

The aim of this book was to portray strategies of scientific modernization in their global historical setting. Yet, as you might have observed, the very concept of "modernity" has not yet been mentioned in any of the quotations provided in the previous eight chapters. I have neither mentioned it as an analytical category, nor attempted to date, with respect to the existing historiography, the beginning of modern mathematics in China. It is time, therefore, to discuss "mathematical modernity" in China.

Eurocentric narratives of the history of science equate modernity with modern science which began with the scientific revolution, thus framing it as a uniquely European event.[26] In the case of China, historians have ascribed the beginning of scientific modernity to various periods of Chinese history, either by taking Western influence as a criterion—the arrival of the Jesuits around 1600 or the arrival of the Protestant missionaries after the Opium Wars in the mid-nineteenth century—or by defining as a turning point in science one of the early twentieth-century revolutions that broke with the past: the 1911 Republican Revolution that brought about major institutional changes or the May Fourth New Culture Movement (1915–1925) that did away with the classical written language.

As for mathematics more specifically, "mathematical modernity" inside of China was certainly unrelated to the concept as it was in vogue during the 1960s outside of China, where it was understood as a kind of "New Math" based on set theory and structures which resulted from a conflict between different visions of mathe-

---

[26] As in Wootton (2015).

matics.[27] No such metamathematical debates existed in China where at best we can look at alternative landmarks for modernity, such as the new social conditions of the mathematician or methodological ruptures and conceptual changes. One might, for example, draw a superficial line between the Chinese rhetorically based algorithmic texts and those written in the formulaic or translated semi-formulaic language of Western mathematics. However, there is a danger in identifying this kind of watershed between the traditional and the modern, particularly when speaking to contemporary mathematicians: anything before axiomatization in their eyes is not a solid theory, cannot be considered mathematics and thus does not belong to the history of science—full stop. Shiing-Shen Chern 陳省身 (Pinyin: Chen Shengshen, 1911–2004), China's most famous differential geometer who enjoyed an international career, had an equally critical definition of modern mathematics. It excluded China's past from the history of mathematics entirely, since their mathematics did not linearly and cumulatively evolve into a modern theory-building discipline, although late Qing actors displayed a high degree of originality in hybridizing old and new elements:

> The advancement of mathematics is of continuous nature. Modern mathematics is the extension of traditional mathematics. What makes it different is that it has new concepts, new methods, and new problems, that is all.
> The history of the development of mathematics has a definite pattern, and that is the influence of the rupture between experience and logic. For example, the laws of arithmetic operations with numbers follow experience, whereas a rigorous number theory is logic.[28]

If alternatively, we define "mathematical modernity" in China as a perceived rupture with the past, one can only conclude that mathematics in China has never been truly modern. Chinese statesmen, such as the reformer Deng Xiaoping 鄧小平 (1904–1997), shared that view when they declared science part of the major reform targets or the "Four Modernizations." The idea was introduced as early as 1963 by the first premier of the People's Republic, Zhou Enlai 周恩來 (1898–1976). Zhou consulted a large number of Chinese scientists, among them the differential geometer Su Buqing 苏步青(1902–2003), before delivering his programmatic speech at a Shanghai Conference on Science and Technology:

> The main requirements for realizing the modernization of science and technology are: search the truth through the facts, make steady progress, advance [intellectual and material modernization] side by side, try hard to catch up!
> 实现科学技术现代化的主要要求是：实事求是，循序前进，齐头并进，迎头赶上。[29]

It was only after the Cultural Revolution and Mao's death that Deng Xiaoping officially launched the "Four Modernizations" in December 1978. For mathematical research, the reforms implied a strong focus upon computer science, which, para-

---

[27] See Mehrtens (1990) p. 7.

[28] Translated from Chern (1948) p. 11.

[29] Cao (1963). The "Four Modernizations" concerned agriculture, industry, science and technology, and national defence.

doxically, allowed Wu Wen-Tsun and a few other mathematicians to return—albeit in an abstract way—to the algorithmically based Chinese tradition.

All the chapters in this book have provided examples that illustrate the numerous ways in which the specificities of a Chinese mathematical tradition played a role in late nineteenth- and early twentieth-century mathematical research. The ways in which they continued and still continue to do so up to the present day, need to be examined in depth. Impressions of the persistence of a more calculatory approach to proving geometrical theorems,[30] for example, would need to be confirmed by studies of contemporary mathematical research practices. As for mathematical education, the stereotypical idea of an authoritarian style of Chinese education and high-achieving kids, the "paradox of the Chinese learner," has in recent years revived interest in the Chinese problem-oriented approach among other didactic methods of mathematics.[31] This has contributed to a nationally defined epistemological territory that could easily be integrated into the larger movement of "National Studies." It is not unusual to find quick abacus calculation back on the curriculum of private Confucian schools in China today.[32]

This goes some way towards explaining why I doubt that we need to pin down a beginning and an end of "mathematical modernity" for China, where modernity, an often self-evident notion for historians, was not even part of the discourses about mathematics before the mid-twentieth century. Modernization, on the other hand, I would suggest, was and still is an ongoing process in China, manifesting itself in an oscillating but on average steady parting from the authoritative model of the *Nine Chapters*. According to the transmitted mathematical literature, this trend showed its first shoots in the Song dynasty attempts to reorganize mathematical knowledge from the *Nine Chapters* in a creative way. In 1867 when Li Shanlan declared that he had erected a new flag next to the *Nine Chapters*, he took a giant step forward and gained some of the mathematical freedom that Georg Cantor had formulated in 1925 as the motto of a new generation of mathematicians. A history of mathematical "modernity" in China, as I have therefore assumed in this book, should be a history that analyses the actors' strategies of liberation from a canonical model, thereby allowing the renewal of mathematics through a diversification of its objects, through conceptual changes, a transformation of the language in which it is written, and a modification of the arguments in which its correctness is demonstrated. Contact with foreign mathematical (and astronomical) cultures was a driving force in this kind of "modernization," but the integration of China into a global community did not mean that imported bodies of scientific knowledge were simply accepted. On the

---

[30]Private communication with Prof. Juan Alvarez-Paiva.

[31]On the "Chinese way" of learning mathematics today, see Fan et al. (2004) and the critical review (Bréard 2012) of the collection of papers written by a group of educators and researchers interested in the culture-specific kind of mathematics education in China today.

[32]Such valorization of ethnoepistemology can equally be observed in India, where "Vedic mathematics," a set of versified calculation tricks presumably as old as the Vedas, has been introduced in the school curriculum. See (Michaels 2015).

contrary, it spurred efforts to construct a mathematical alternative, another kind of "counter-modernity," based on the essential characteristics of Chinese mathematics.

References

Bréard, Andrea (2012). Review of: Fan Lianghuo, Wong Ngai-Ying, Cai Jinfa and Li Shiqi (eds.), *How Chinese Learn Mathematics: Perspectives from Insiders*, River Edge, NJ [etc.]: World Scientific (Series on Mathematics Education; 1), 2004 (reprinted 2006), 592 pp. *East Asian Science, Technology and Medicine 35*, 143–146.

Cai, Tiequan 蔡铁权 (2013). Zhongguo chuantong wenhua yu chuantong shuxue, shuxue jiaoyu de yanjin 中国传统文化与传统数学、数学教育的演进 (Research on the Evolution of Traditional Mathematics and Mathematics Education in the Context of Chinese Traditional Culture). *Quanqiu jiaoyu zhanwang* 全球教育展望 (Global Education) *42*(8), 91–100.

Cao, Xinghua 曹兴华 (1963, January 31). Zai Shanghai shi juxing de kexue jishu gongzuo huiyi shang Zhou zongli chanshu kexue jishu xiandaihua de zhongda yiyi 在上海市举行的科学技术工作会议上周总理阐述科学技术现代化的重大意义 (At the Conference on Scientific and Technological Work, Held in Shanghai, Premier Zhou Set Forth the Major Importance of Modernization of Science and Technology). *Renmin ribao* 人民日报 (People's Daily).

Chen, Jiaming (2011). La fièvre des études nationales. Le phénomène, le débat et quelques réflexions. *Perspectives Chinoises 1*, 22–32.

Chern, Shiing-Shen 陳省身 (1948). Xiandai shuxue 現代數學 (Modern Mathematics). *Sixiang yu shidai* 思想與時代 *51*, 11–15.

Fan, Lianghuo, Ngai-Ying Wong, Jinfa Cai, and Shiqi Li (Eds.) (2004). *How Chinese Learn Mathematics: Perspectives from Insiders* (2nd ed.), Volume 1 of *Series on Mathematics Education*. River Edge, NJ: World Scientific.

Mehrtens, Herbert (1990). *Moderne Sprache Mathematik. Eine Geschichte des Streits um die Grundlagen der Disziplin und des Subjekts formaler Systeme*. Frankfurt am Main: Suhrkamp.

Michaels, Axel (2015). Mathematics and Vedic Mathematics. Paper presented at the International Conference Scientification and Scientism in the Humanities, held at the Center for the Study of Social Systems, Jawaharlal Nehru University, New Delhi 25–26 November 2015.

Riemann, Bernhard (1859, November). Über die Anzahl der Primzahlen unter einer gegebenen Größe. *Monatsberichte der Königlichen Preußischen Akademie der Wissenschaften zu Berlin*, 671–680.

Tu, Youyou (2011). The Discovery of Artemisinin (qinghaosu) and Gifts from Chinese Medicine. *Nature Medicine 17*(10), 1217–1220.

Wang, Zuoyue (2002). Saving China through Science: The Science Society of China, Scientific Nationalism, and Civil Society in Republican China. *Osiris 17*, 291–322.

Wei, Hongsen (1996). Norbert Wiener at Qinghua University. In Fan Dainian and Robert S. Cohen (Eds.), *Chinese Studies in the History and Philosophy of Science and Technology*, 447–451. Dordrecht: Springer.

Wiener, Norbert (1956). *I Am a Mathematician*. New York: Doubleday.

Wilkinson, Alec (2015, February 2). The Pursuit of Beauty. Yitang Zhang Solves a Pure-Math Mystery. *The New Yorker*. https://www.newyorker.com/magazine/2015/02/02/pursuit-beauty; accessed 9-September-2017.

Wootton, David (2015). *The Invention of Science: A New History of the Scientific Revolution*. New York: HarperCollins.

Xia, Daoxing 夏道行 (1964). *Pi he e* π 和 e (π *and e*). Shanghai: Shanghai jiaoyu chubanshe 上海教育出版社.

Yu, Li 于丽 (2013). Zhenxing ziran guoxue chuancheng zhonghua guibao 振兴自然国学 传承中华瑰宝 (=The Revival of Traditional Culture). *Qingdao huabao* 青岛画报 *(Qingdao Pictorial) 7*, 90–91.

Zhang, Jing 章兢 and Tong Tiaosheng 童调生 (2015). Guoxue ying baokuo ziran kexue 国学应包括自然科学 (Studies of Chinese Ancient Civilization Should Include Natural Science). *Xiangtan daxue xuebao (Zhexue shehui kexue ban)* 湘潭大学学报 (哲学社会科学版) (Philosophy and Social Sciences) *39*(3), 91–95.

Zhang, Lei 张蕾 (2012, June 9). "Xin kaogao" zhong "Guoxue" Chuci, Honglou meng, Jiu zhang suan shu ru ti "新高考"重"国学"楚辞、红楼梦、九章算术入题 (The *New Entrance Examinations* emphasizing *National Learning* put in problems from the *Songs of Chu*, the *Dream of the Red Chamber* and the *Nine Chapters on Mathematical Procedures*). *Sanxia shangbao* 三峡商报.

Zhang, Yitang (2014). Bounded Gaps between Primes. *Annals of Mathematics (2) 179*(3), 1121–1174.

Zhang, Yicheng 张以诚, Liu Zhanglin 刘长林, Shang Hongkuan 商宏宽, Li Sizhen 李似珍, Song Zhenghai 宋正海, Ma Xiaotong 马晓彤, Sun Guanlong 孙关龙, Kang Kuanying 亢宽盈, Yan Chunyou 严春友, Zhang Jinfeng 张进峰, Li Zhichao 李志超, Mei Zutong 姜祖桐, Chen Guangzhu 陈光柱, Li Shihui 李世辉, and Zhou Yongqin 周永琴 (2002). Ziran guoxue shi fou neng xingcheng tixi (bitan) 自然国学是否能形成体系(笔谈) (Can National Natural Science Form a System, Informal Notes). *Taiyuan shifan xueyuan xuebao (Renwen kexue ban)* 太原师范学院学报 (人文科学版) *1*, 1–15.

Zhao, Xingyue 赵星月 (2015). Gaoxiao guoxue jiaoyu xianzhuang fenxi yu duice yanjiu 高校国学教育现状分析与对策研究 (Present Situation and Countermeasures of Sinology Education in Colleges and Universities). *Qiqihaer daxue xuebao (Zhexue shehui kexue ban* 齐齐哈尔大学学报 (哲学社会科学版) (Journal of Qiqihar University, Philosophy & Social Science Edition) *4*, 7–11.

Appendix A
A Timeline of Mathematics from the Late Ming to the People's Republic of China

© Springer International Publishing AG, part of Springer Nature 2019
A. Bréard, *Nine Chapters on Mathematical Modernity*, Transcultural
Research – Heidelberg Studies on Asia and Europe in a Global Context,
https://doi.org/10.1007/978-3-319-93695-6

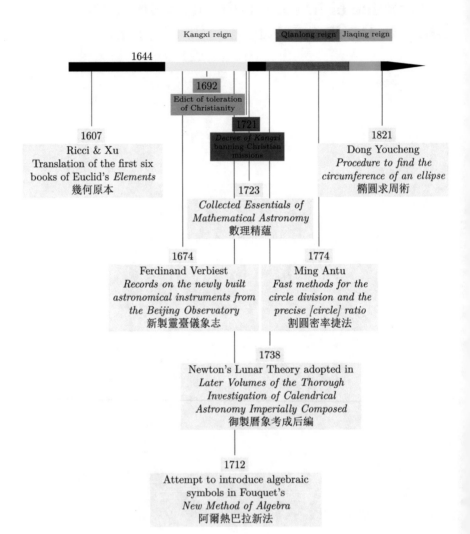

Ming Dynasty   *Qing Dynasty*

Kangxi reign   Qianlong reign   Jiaqing reign

1644

1692
Edict of toleration
of Christianity

1721
*Decree of Kangxi
banning Christian
missions*

1607
Ricci & Xu
Translation of the first six
books of Euclid's *Elements*
幾何原本

1821
Dong Youcheng
*Procedure to find the
circumference of an ellipse*
橢圓求周術

1723
*Collected Essentials of
Mathematical Astronomy*
數理精蘊

1674
Ferdinand Verbiest
*Records on the newly built
astronomical instruments from
the Beijing Observatory*
新製靈臺儀象志

1774
Ming Antu
*Fast methods for the
circle division and the
precise [circle] ratio*
割圓密率捷法

1738
Newton's Lunar Theory adopted in
*Later Volumes of the Thorough
Investigation of Calendrical
Astronomy Imperially Composed*
御製曆象考成后編

1712
Attempt to introduce algebraic
symbols in Fouquet's
New Method of Algebra
阿爾熱巴拉新法

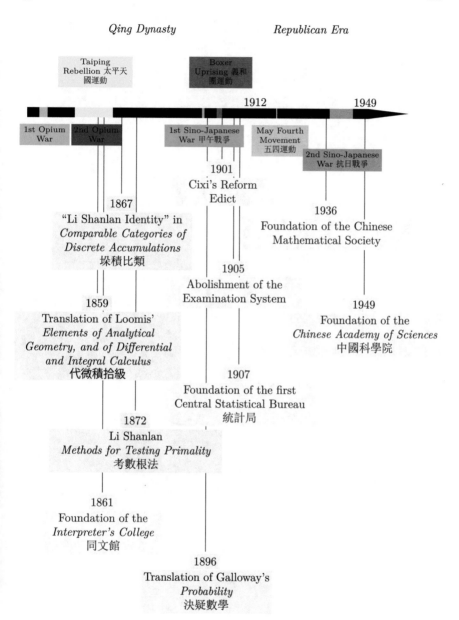

Appendix B
Translation of Li Shanlan's *Methods for Testing Primality* (*Kao shugen fa* 考數根法), 1872

Scores of original Chinese sources have been used in this book. Since none of them are yet available in a Western language, it is important to provide them in translation. I have selected one exemplary text, which illustrates the kind of mathematical discourse prevalent in nineteenth-century China and is central to Chap. 3. In order to make the mathematical content more accessible to the readers, symbolic translations have been added in square brackets, as well as semantic additions for ease of comprehension.

> Methods for Testing Primality, part 14 of the *Mathematics from the Studio Devoted to the Imitation of the Ancient Chinese Tradition* by Li Shanlan, from Haining.
>
> > In general, a number, that when measured by another number cannot be exhausted, and can only exhaustively be measured by the unit, is called a prime number (*shugen* 數根). See the *Elements of Geometry*. But, if we take any number and if we want to decide whether it is a prime number or not then there is no method in antiquity. I have thought about this carefully and for a long time, and I have found four methods for testing this in order to complement what in the [*Elements of*] *Geometry* was not yet complete.
>
> The number that is to be tested makes the original number (*benshu* 本數) [N]. 2 and 3 make the auxiliary numbers (*yongshu* 用數) [a].
>
> > We do not need further numbers. When using only the 2 and the 3, it is convenient for multiplication and division in finding the smallest [factors].
>
> All the powers of 1 and the auxiliary numbers make the positive numbers.
>
> > The first power of 2 is 2, the second power is 4, the third power is 8, the fourth power is 16.
>
> > The first power of 3 is 3, the second power is 9, the third power is 27, the fourth power is 81, the fifth power and beyond can be deduced analogically (*ke leitui* 可類推).[1]

---

[1]For the significance of this expression in Li Shanlan's modes of argumentation, see Chap. 5, p. 132.

© Springer International Publishing AG, part of Springer Nature 2019
A. Bréard, *Nine Chapters on Mathematical Modernity*, Transcultural
Research – Heidelberg Studies on Asia and Europe in a Global Context,
https://doi.org/10.1007/978-3-319-93695-6

From all the positive numbers subtract the original number. The remainders make the negative numbers.

> As for example, the original number is 87. 86 is the negative number for 1.[2] If you take 2 as the auxiliary number, then 85 is the negative number of the first power,[3] 83 the negative number of the second power,[4] 79 the negative number of the third power, 71 the negative number of the fourth power. If you take 3 as the auxiliary number, then 84 is the negative number of the first power, 78 the negative number of the second power, 60 the negative number of the third power, 6 the negative number of the fourth power.[5] The others can be deduced analogically.

What remains when the original number measures all the positive numbers makes all the remainders.

> As for example, the original number is 211. Then 45 makes the remainder of the eighth power of 2,[6] 90 makes the remainder of the ninth power of 2,[7] 32 makes the remainder of the fifth power of 3,[8] 96[9] makes the remainder of the sixth power of 3.[10] The others can be deduced analogically.

All the numbers above are used in all the [four] methods alike.

The first one is called "Method for testing primality by repeated multiplication and reduction to One."
The method: One takes a power with the auxiliary number [as its base] $[a^n]$ such that it is either bigger than the original number $[a^n > N]$, or bigger than half the original number $[a^n > N/2]$. From this subtract the original number $[a^n - N]$, and make what remains the multiplication norm. Multiply the multiplication norm by itself, or multiply it twice with itself and measure it with the original number $[(a^n - N)^2 \equiv r_1 \pmod{N}]$. If it is not exhausted [if $r_1 \neq 0$], then again multiply this [what is not exhausted] with the multiplication norm and measure it with the original number $[(a^n - N)r_1 \equiv r_2 \pmod{N}]$. If it is not exhausted [if $r_2 \neq 0$], then again multiply this [what is not exhausted] with the multiplication norm and measure it with the original number $[(a^n - N)r_2 \equiv r_3 \pmod{N}]$. In this manner proceed subsequently until what remains is one of the positive numbers or the negative numbers, then stop $[(a^n - N)r_{m-2} \equiv r_{m-1} \pmod{N}$ where $r_{m-1} = \pm a^c]$. Then calculate how many times in total you have used the multiplication norm [m times]. With this number of times [m] multiply the exponent of the auxiliary number [n] and make this the generalized number of times [mn]. If what remains [when measuring $(a^n - N)^m$ with N] is either 1 or -1 [if $(a^n - N)^m \equiv \pm 1 \pmod{N}$], then the generalized number of times is the definite number of times [$mn =: d$]. If [what remains] is a power [with the auxiliary number as its base] or a negative power [if $(a^n - N)^m \equiv \pm a^c \pmod{N}$], then subtract the exponent from the generalized number of times and make it the definite number of times [$mn - c =: d$]. With the definite number of times measure the original number [$N \equiv b \pmod{d}$]. If what remains is not 1 [if $b \neq 1$], then the original number

---

[2] That is, $1 \equiv -86 \pmod{87}$.

[3] That is, $2^1 \equiv -85 \pmod{87}$.

[4] That is, $2^2 \equiv -83 \pmod{87}$, etc.

[5] That is, $3^4 \equiv -6 \pmod{87}$.

[6] $2^8 \equiv 45 \pmod{211}$.

[7] $2^9 \equiv 90 \pmod{211}$.

[8] $3^5 \equiv 32 \pmod{211}$.

[9] The text erroneously has "93." See Li (1872a) p. 13B.

[10] $3^6 \equiv 96 \pmod{211}$.

is not a prime number. If what remains is 1 [if $b = 1$, i.e. if $N \equiv 1 \pmod{d}$], then we have to consider which two numbers [say p and q] are the factors whose product makes the definite number of times [$d = pq$]. If the product of the factors is even, then it [pq] will make the subsequently added number (*dijia shu* 遞加數) [$s = pq$]. If it is odd, double it [$2pq$] to make it the subsequently added number [$s = 2pq$]. Then, put down 1, add once the subsequently added number, again add once the subsequently added number. In this manner, subsequently add to subsequently eliminate from the original number. When it is precisely exhausted, then stop [when $1 + s + s + \cdots = 1 + rs = N$]. If, by taking all the numbers smaller than the divisor [$< r$], it is still not exhausted, than the original number is a prime number [if N is not divisible by $1 + ts$ for $t < r$].

As for example, the original number is 31. The fourth power of the auxiliary number 2 is 16. It is bigger than half the original number. Subtracting it from the original number, the remainder 15 makes the multiplication norm. Multiplying it by itself gives 225. Measuring it with the original number, the remainder is 8 [$(31 - 2^4)^2 \equiv 8 \pmod{31}$]. It is the third power of the auxiliary number [$2^3 = 8$]. Take the number of times, 2, that the multiplication norm was multiplied, multiply this with 4, the exponent [in $2^4 = 16$], to obtain 8 as the generalized number of times. Subtract from it 3, the exponent of the remainder [in $2^3 = 8$] to obtain 5. Make it the definite number of times. When measuring the original number with it, what remains is 1 [$31 \equiv 1 \pmod{5}$]. The definite number of times, 5, is the product of 1 and 5. 5 is an odd number. Double it to make it the subsequently added number [10]. Add 1 to it to obtain 11. Eliminate it from the original number. One obtains 2 and what is not exhausted makes 9 [$31 = 2 \cdot 11 + 9$]. The obtained number is already smaller than the divisor [$9 < 11$ or $11 \nmid 31$], therefore one can determine that the original number is a prime number.

As for example, the original number is 1093. The sixth power of the auxiliary number 3 is 729. It is bigger than half the original number. Subtracting it from the original number, the remainder 364 makes the multiplication norm [$1093 - 3^6 = 364$]. Multiplying it by itself gives 132496. Measuring it with the original number, the remainder is 243 [$(1093 - 3^6)^2 \equiv 243 \pmod{1093}$]. It is the fifth power of the auxiliary number [$3^5 = 243$]. Take the number of times, 2, that the multiplication norm was used, multiply this with 6, the exponent [in $3^6 = 729$], to obtain 12. Subtract from it 5, the exponent of the remainder [in $3^5 = 243$] to obtain 7. Make it the definite number of times. When measuring the original number with it, what remains is 1 [$1093 \equiv 1 \pmod{7}$]. The definite number of times, 7, it is the product [of 1 and 7]. Double it to make it the subsequently added number [14]. Add 1 to it to obtain 15, again add to it to obtain 29. Subsequently eliminate from the original number. None is exhaustive. Again add to it to obtain 43. Eliminate it from the original number. One obtains 25 and it still is not exhausted [$1093 = 25 \cdot 43 + 18$] . The obtained number is already smaller than the divisor [$18 < 43$ or $43 \nmid 1093$], therefore one can determine that the original number is a prime number.

The two numbers above are prime numbers. When measuring them, what remains is 1, eliminating from them is not exhaustive.

As for example, the original number is 341. The ninth power of the auxiliary number 2 is 512. It is bigger than the original number. Subtracting the original number from it, the remainder 171 makes the multiplication norm. Multiplying it by itself gives 29241. Measuring it with the original number, the remainder is 256 [$(2^9 - 341)^2 \equiv 256 \pmod{341}$]. It is the eighth power of the auxiliary number [$2^8 = 256$]. Take the number of times, 2, that the multiplication norm was used, multiply this with 9, the exponent [in $2^9 = 512$], to obtain 18. Subtract from it 8, the exponent of the remainder [in $2^8 = 256$] to obtain 10. Make it the definite number of times. When measuring the original number with it, what remains is 1 [$341 \equiv 1 \pmod{10}$]. The definite number of times, 10, it is the product of 2 and 5. Double 5 to obtain 10 and make it the subsequently added number. Add 1 to it to obtain 11. Eliminate it from the original number. One obtains 31 and it is entirely exhausted [$341 = 31 \cdot 11 + 0$]. Therefore 341 is not a prime number.[11]

As for example, the original number is 91. The fourth power of the auxiliary number 3 is 81. It is bigger than half the original number. Subtracting it from the original number, the remainder is 10. Multiplying it by itself gives 100. Measuring it with the original number, the remainder is 9 [$(91 - 3^8)^2 \equiv 9 \pmod{91}$]. It is the second power of the auxiliary number [$3^2 = 9$]. Take the number of times, 2, that the multiplication norm was used, multiply this with 4, the exponent [in $3^4 = 81$], to obtain 8. Subtract from it 2, the exponent of the remainder [in $3^2 = 9$] to obtain 6. Make it the definite number of times. When measuring the original number with it, what remains is 1 [$91 \equiv 1 \pmod 6$]. The definite number of times, 6, it is the product of 2 and 3. Double 3 to obtain 6 and make it the subsequently added number. Add 1 to it to obtain 7. Eliminate it from the original number. One obtains 13 and it is entirely exhausted [$91 = 13 \cdot 7 + 0$]. Therefore 91 is not a prime number.

The two numbers above are not prime numbers. When measuring them, one obtains 1, eliminating from them is entirely exhaustive.

As for example, the original number is 323. The eighth power of the auxiliary number 2 is 256. It is bigger than half the original number. Subtracting it from the original number, the remainder 67 makes the multiplication norm. Multiplying it by itself and measuring it with the original number, the remainder is 290 [$(323 - 2^8)^2 \equiv 290 \pmod{323}$]. Multiplying it with the multiplication norm and measuring it with the original number, the remainder is 50 [$(323 - 2^8) \cdot 290 \equiv 50 \pmod{323}$]. Multiplying it with the multiplication norm and measuring it with the original number, the remainder is 120 [$(323 - 2^8) \cdot 50 \equiv 120 \pmod{323}$]. Multiplying it with the multiplication norm and measuring it with the original number, the remainder is 288 [$(323 - 2^8) \cdot 120 \equiv 288 \pmod{323}$]. Multiplying it with the multiplication norm and measuring it with the original number, the remainder is 239 [$(323 - 2^8) \cdot 288 \equiv 239 \pmod{323}$]. Multiplying it with the multiplication norm and measuring it with the original number, the remainder is 186 [$(323 - 2^8) \cdot 239 \equiv 186 \pmod{323}$]. Multiplying it with the multiplication norm and measuring it with the original number, the remainder is 188 [$(323 - 2^8) \cdot 186 \equiv 188 \pmod{323}$]. Multiplying it with the multiplication norm and measuring it with the original number, the remainder is 322 [$(323 - 2^8) \cdot 188 \equiv 322 \pmod{323}$]. It is the negative number of 1 [$1 \equiv -322 \pmod{323}$]. Take the number of times, 9, that the multiplication norm was used, multiply this with 8, the exponent [in $2^8 = 256$], to obtain 72. Make it the definite number of times. When measuring the original number with it, what remains is 35 [$323 \equiv 35 \pmod{72}$]. Therefore, 323 is not a prime number.

---

[11] 341 is a counter-example, the smallest actually (see Jeans 1898 p. 174), to show that the converse of Fermat's Little Theorem does not hold, i.e. the converse of:

Let p be a prime number. Then: $a^p \equiv a \pmod p$

is not valid. Here, 341 | ($2^{340} - 1$), or $2^{341} \equiv 2 \pmod{341}$, but 341 is not prime. Such numbers N that are not prime but for which $2^N - 2$ is divisible by N are the now so-called pseudoprimes. Li Shanlan himself has several other examples of pseudoprimes in his text. The other three found among his examples are $N = 4681$, $N = 63973$ and $N = 1398101$ (see pages 246, 247, and 251), but for none of them does Li explicitly point out their being a counterexample for the inverse of Fermat's Little Theorem. That he has not published his "Theorem" as reported by Wylie was interpreted by some historians as a sign that Li became aware of the problem that his statement was not correct. See for example Han and Siu (2008) p. 947.

The definite number of times, 72, it is the product of 8 and 9. Double 9 to obtain 18. Add 1 to it to obtain 19. Eliminate it from the original number. One obtains 17 and it is entirely exhausted [$323 = 17 \cdot 19 + 0$].

As for example, the original number is 14209. The ninth power of the auxiliary number 3 is 19683. It is bigger than the original number. Subtracting from it the original number, the remainder 5474 makes the multiplication norm. Multiplying it by itself and measuring it with the original number, the remainder is 12104 [$(14209 - 3^8)^2 \equiv 12104$ (mod 14209)]. Multiplying it with the multiplication norm and measuring it with the original number, the remainder is 729 [$(14209 - 3^8) \cdot 12104 \equiv 729$ (mod 14209)]. It is the sixth power of the auxiliary number [$3^6 = 729$]. Take the number of times, 3, that the multiplication norm was used, multiply this with 9, the exponent [in $3^9 = 19683$], to obtain 27. Subtract from it 6, the exponent of the remainder [in $3^6 = 729$] to obtain 21. Make it the definite number of times. When measuring the original number with it, what remains is 13 [$14209 \equiv 13$ (mod 21)]. Therefore, 14209 is not a prime number. The definite number of times is the product of 3 and 7. Double 3 to obtain 6 and make it the subsequently added number. Add 1 to it to obtain 7, again add to it to obtain 13. Eliminate it from the original number. One obtains 1093 and it is entirely exhausted [$14209 = 1093 \cdot 13 + 0$].

The two numbers above are not prime numbers. When measuring them, one does not obtain 1.[12]

The other one is called "Method for testing primality by the Celestial Element [method][13] and reduction to One."
The method: One takes a power with the auxiliary number [as its base] [a^n] such that it is either bigger than the original number [$a^n > N$], or bigger than half the original number [$a^n > N/2$]. With the original number [N], search to obtain unity [i.e. solve $a^n \cdot x \equiv 1$ (mod N)] and make the celestial element [x] the multiplication norm. As in the previous [method] repeatedly multiply and repeatedly measure [$a^{mn} \cdot x^m \equiv 1$ (mod N)] until one obtains one of the positive numbers or the negative numbers [$x^m = a^c$], then stop [$a^{mn+c} \equiv 1$ (mod N)]. Add the exponent of what remains to the generalized number of times and make it the definite number of times [$mn + c =: d$]. The rest is as in the previous method.

As for example, the original number is 103. The seventh power of the auxiliary number 2 is 128. Reduce to one, do so by using the *Dayan* procedure (*Dayan shu* 大衍術).[14]

128	103
1元	
25	3
1元	4元
1	
33元	

---

[12]Translated entirely from Li (1872a).

[13]In Song and Yuan dynasty algebra the term "Celestial Element" (*tian yuan* 天元) was used to designate the unknown in a polynomial equation.

[14]Literally the "procedure of the great extension," the name of this procedure allowing to solve indeterminate equations, usually remains untranslated in Western language literature. For a detailed account of the history of this procedure in China and its historiography in Europe, see Libbrecht (1973) part V, p. 213–413, Martzloff (2006) p. 316–319 and Shen (1986).

Take the two numbers, align them right and left. Establish the celestial element below to the left. First, subtract right from left, one obtains the second level on the left [128 − 103 = 25]. Subtract its fourfold value from the right [103 − 4 · 25 = 3]. One obtains the second level on the right. Subtract its eightfold value from the left, one obtains the third level on the left [25 − 8 · 3 = 1]. Above makes 1, below makes 33 celestial elements. Then, 33 makes the multiplication norm [128 · 33 ≡ 1 (mod 103)]. The multiplication norm multiplied with itself [$33^2 = 1089$], when measured with the original number, it is not exhausted by 59 [$33^2 \equiv 59$ (mod 103)]. Multiply this with the multiplication norm, when measured with the original number, it is not exhausted by 93 [59 · 33 ≡ 93 (mod 103)]. Multiply this with the multiplication norm, when measured with the original number, it is not exhausted by 82 [93 · 33 ≡ 82 (mod 103)]. Multiply this with the multiplication norm, when measured with the original number, it is not exhausted by 28 [82 · 33 ≡ 28 (mod 103)]. Multiply this with the multiplication norm, when measured with the original number, it is not exhausted by 100 [28 · 33 ≡ 100 (mod 103)]. Multiply this with the multiplication norm, when measured with the original number, it is not exhausted by 4 [100·33 ≡ 4 (mod 103)]. It is the second power of the auxiliary number [$2^2 = 4$]. Take the number of times, 7, that the multiplication norm was used, multiply this with 7, the exponent [in $2^7 = 128$], to obtain 49. Add it to 2, the exponent of the remainder [in $2^2 = 4$] to obtain 51. Make it the definite number of times. When measuring the original number with it, what remains is 1 [103 ≡ 1 (mod 51)].

The definite number of times is the product of 3 and 17. Double 3 to obtain 6, furthermore add 1 to obtain 7. Also 17, double it to obtain 34, add 1 to obtain 35. When eliminating [7 and 35 respectively] from the original number, it is never exhausted. Therefore one can determine that 103 is a prime number.

The number above is a prime number. When measuring it, what remains is 1, eliminating from it is not exhaustive.

As for example, the original number is 63973. The tenth power of the auxiliary number 3 is 59049. Reduce to one, do so by using the *Dayan* procedure.

59049	63973
1元	
4885	4924
12元	-1元
10	39
1637元	-13元
1	9
6561元	-4924元

Take the two numbers, align them right and left. Establish the celestial element below to the left. Subtract left from right, one obtains the second level on the right [63973 − 59049 = 4924]. Subtract its elevenfold value from the left. One obtains the second level on the left [59049 − 11 · 4924 = 4885]. This, in turn, subtract from the second level on the right, one obtains the third level on the right [4924 − 4885 = 39]. Subtract its 125-fold value from the second level on the left. One obtains the third level on the left [4885 − 125 · 39 = 10]. Subtract its threefold value from the third level on the left, one obtains the fourth level on the right [39 − 3 · 10 = 9]. This, in turn, subtract from the third level on the left, one obtains the fourth level on the left [10 − 9 = 1]. Above makes 1, below makes 6561 celestial elements.[15] Then, 6561 makes the multiplication norm. 6561 is the eighth power of the auxiliary number [$3^8 = 6561$]. It is not necessary to multiply it again. Adding the exponent 8 to the exponent 10, one obtains 18 as the definite number of times. When measuring the original number with it, what remains is 1 [$63973 \equiv 1 \pmod{18}$]. The definite number of times is the product of 3 and 6, it is also the product of 2 and 9. Adding 1 to 6, one obtains 7. Eliminating it from the original number, one obtains 9139 [$63973 = 9139 \cdot 7 + 0$]. Doubling 6 and adding 1, one obtains 13. Eliminating it from the original number, one obtains 4921 [$63973 = 4921 \cdot 13 + 0$]. Doubling 9 and adding 1, one obtains 19. Eliminating it from the original number, one obtains 3367 [$63973 = 3367 \cdot 19 + 0$]. Doubling 18 and adding 1, one obtains 37. Eliminating it from the original number, one obtains 1729 [$63973 = 1729 \cdot 37 + 0$]. All entirely exhaust [the original number]. Therefore, 63973 is not a prime number.

The number above is not a prime number. When measuring it, although one obtains 1, eliminating from it is entirely exhaustive.

As for example, the original number is 4681. The ninth power of the auxiliary number 3 is 19683. Reduce to one, do so by using the *Dayan* procedure.

19683	4681
1元	
959	845
1元	-4元
114	47
5元	-39元
20	7
83元	-205元
6	1
493元	-698元

Take the two numbers, align them right and left. Establish the celestial element below to the left. Quadruple the number on the right and subtract it from the left. One obtains the second level on the left [$19683 - 4 \cdot 4681 = 959$]. Subtract its fourfold value from the right, one obtains the second level on the right [$4681 - 4 \cdot 959 = 845$].[16] This, in turn, subtract from the second level on the left, one obtains the third level on the left [$959 - 845 = 114$]. Subtract its sevenfold value from the second level on the right, one obtains the third level on the right [$845 - 7 \cdot 114 = 47$]. Subtract its twofold value from the third level on the left, one obtains the fourth level on the left [$114 - 2 \cdot 47 = 20$]. Subtract its twofold value from the third level on the right, one obtains the fourth level on the right [$47 - 2 \cdot 20 = 7$]. Subtract its twofold value from the fourth level on the left, one obtains the fifth level on the left [$20 - 2 \cdot 7 = 6$]. This, in turn, subtract from the fourth level on the right, one obtains the fifth level on the right [$7 - 6 = 1$]. Above makes 1, below makes 698 celestial elements. Then, 698 makes the multiplication norm.

---

[15]This number is obtained by a recurrent series of operations where $a_i = q_i a_{i-1} + a_{i-2}$ with $a_0 = 1$ and the q_i denoting the factor in each subsequent substraction. Here: $q_1 = 1$, $q_2 = 11$, $q_3 = 1$, $q_4 = 125$, $q_5 = 3$, and $q_6 = 1$, thus giving $a_1 = q_1 a_0 = 1$, $a_2 = q_2 a_1 + a_0 = 12$, $a_3 = q_3 a_2 + a_1 = 13$, $\cdots$, and $a_6 = q_6 a_5 + a_4 = 6561$.

[16]The original table erroneously gives the number 835.

The multiplication norm multiplied with itself [698^2], when measured with the original number, it is not exhausted by 380 [$698^2 \equiv 380 \pmod{4681}$]. Multiply this with the multiplication norm, when measured with the original number, it is not exhausted by 3104 [$380 \cdot 698 \equiv 3104 \pmod{4681}$]. Multiply this with the multiplication norm, when measured with the original number, it is not exhausted by 3970 [$3104 \cdot 698 \equiv 3970 \pmod{4681}$]. Multiply this with the multiplication norm, when measured with the original number, it is not exhausted by 4589 [$3970 \cdot 698 \equiv 4589 \pmod{4681}$]. Multiply this with the multiplication norm, when measured with the original number, it is not exhausted by 1318 [$4589 \cdot 698 \equiv 1318 \pmod{4681}$]. Multiply this with the multiplication norm, when measured with the original number, it is not exhausted by 2488 [$1318 \cdot 698 \equiv 2488 \pmod{4681}$]. Multiply this with the multiplication norm, when measured with the original number, it is not exhausted by 4654 [$2488 \cdot 698 \equiv 4654 \pmod{4681}$]. It is the third power of the negative number of the auxiliary number [$4681 - 3^3 = 4654$]. Take the number of times, 8, that the multiplication norm was used, multiply this with 9, the exponent [in $3^9 = 19683$], to obtain 72. Add it to the exponent of the remainder [3 in $4681 - 3^3 = 4654$] to obtain 75. Make it the definite number of times. When measuring the original number with it, what remains is 31 [$4681 \equiv 31 \pmod{75}$]. Therefore, 4681 is not a prime number. The definite number of times is the product of 5 and 15. Doubling 15 and adding 1, one obtains 31. Eliminating it from the original number, one obtains 151 and it is entirely exhausted [$4681 = 151 \cdot 31 + 0$].

The number above is not a prime number. When measuring it, one does not obtain 1.

The other one is called "Method for testing primality with recurring decimals" (*xiao shu huihuan* 小數迴環).
[The method:] In general, when the original number makes the divisor and we divide 1 with it, it generates repeating and never ending decimal numbers. Among the recurring numbers are those that alternate between positive and negative and there are those that have positive and no negative [numbers] (其迴環數有正負相間者。有有正無負著。).[17] One observes after how many positions they are recurring.[18] With the number of positions one replaces the "definite number of times" in the previous methods. The rest is all as in the previous methods.

As for example, the original number is 271. Make it the divisor and divide 1. One obtains 36900.36900, recurring and never ending. The number of [recurrent] positions is 5. With 5 measure the original number, the remainder is 1 [$271 \equiv 1 \pmod 5$]. The number of [recurrent] positions is the product of 1 and 5. When doubling 1, one obtains

---

[17]It is unclear what Li Shanlan had in mind with this phrase. Even Hua Hengfang 華蘅芳, in his *Study on Recurring Decimals* (*Xunhuan xiaoshu kao* 循環小數考) (Hua 1893) p. 35A, did not find an interpretation of this passage:

> Above I quoted from Li Shanlan's procedures. What he calls the recurring numbers (*huihuan shu* 迴環數), are [what I call] the repeating numbers (*xunhuan shu* 循環數). The original number is the denominator. The definite number of times are the repeating number of positions. According to Li's procedures, it seems that one can from the denominator obtain the number of positions. Only when he says that "among the recurring numbers are those that alternate between positive and negative" I have not yet been able to obtain verification.

[18]For N prime other than 2 or 5, the length of the repetend is either $N - 1$ (if 10 is a primitive root modulo N) or a factor of $N - 1$. Yet there is no hint that Li Shanlan had any knowledge of such number theoretical considerations, and he indeed has "observed" the period of the repeating decimal by counting. The examples that follow seem to imply this approach.

2, when doubling 5, one obtains 10. They both make the subsequently added number (*dijia shu* 遞加數). Put down 1, and with each of the two numbers subsequently add to subsequently eliminate from the original number. When [the result] is smaller than the divisor, then stop. Because none can exhaust [the original number], 271 is a prime number.

The number above is a prime number. When measuring it, one obtains 1, eliminating from it is not exhaustive.

As for example, the original number is 481. Make it the divisor and divide 1. One obtains 207900.207900, recurring and never ending. The number of [recurrent] positions is 6. With it measure the original number, the remainder is 1 [$481 \equiv 1 \pmod 6$]. The number of [recurrent] positions is the product of 2 and 3. When doubling 3, one obtains 6. It makes the subsequently added number. [Taking it] twice and adding 1, one obtains 13. Again, [taking it] six times and adding 1, one obtains 37. With these eliminate from the original number. They are all entirely exhaustive [$481 \equiv 0 \pmod{13}$ and $481 \equiv 0 \pmod{37}$]. Therefore, 481 is not a prime number.

As for example, the original number is 11111. Make it the divisor and divide 1. One obtains 90000.90000, recurring and never ending. The number of [recurrent] positions is 5. With 5 measure the original number, the remainder is 1 [$11111 \equiv 1 \pmod 5$]. The number of [recurrent] positions is the product of 1 and 5. When doubling 5, one obtains 10. It makes the subsequently added number. [Taking it] four times and adding 1, one obtains 41. With this eliminate from the original number, one obtains 271, it is entirely exhausted [$1111 = 271 \cdot 41 + 0$]. Therefore, 1111 is not a prime number.

The numbers above are not prime numbers. When measuring them, although one obtains 1, eliminating from them is entirely exhaustive.

As for example, the original number is 13837. Make it the divisor and divide 1. One obtains 72270000.72270000, recurring and never ending. The number of [recurrent] positions is 8. With it measure the original number, the remainder is 5 [$13837 \equiv 5 \pmod 8$]. The number of [recurrent] positions is the product of 2 and 4, it is also the product of 1 and 8. When making 4 the subsequently added number, one subsequently adds 1 until after 25 times one obtains 101 [$25 \cdot 4 + 1 = 101$]. With this eliminate from the original number, one obtains 137, it is entirely exhausted [$13837 = 137 \cdot 101 + 0$]. Or, making 8 the subsequently added number and subsequently adding 1 until after 17 times one obtains 137 [$17 \cdot 8 + 1 = 137$]. With this eliminate from the original number, one obtains 101, it is entirely exhausted [$13837 = 101 \cdot 137 + 0$].

The number above is not a prime number. When measuring it, one does not obtain 1.[19]

The other one is called "Method for testing primality by normalizing primes and factorization steps" (*zhungen fenji* 準根分級).
This method is convenient for numbers with many positions. The method: From the original number subtract 1. Half this, it makes the total of all factors. Find out which smallest prime numbers multiplied with each other make the product of the total of all factors. Take all these primes to make them the order of the exponents in the normalizations [$(N − 1)/2 = p_1 p_2 p_3 \cdots p_n$ where all p_i are prime and $p_1 > p_2 > p_3 > \cdots > p_n$]. Thus, with the auxiliary number normalize the biggest of the prime [factors] by using the method of super-multiplication[20] and substitutive multiplication (*chaocheng bucheng fa* 超乘補乘法).[21]

---

[19]Translated from Li (1872b) p. 15A–18B.

[20]That is, raising a power to a power by multiplying the exponents.

[21]Li does not say explicitly what the method is about, but it becomes clear from the first numeric example below. The first one shows that in order to avoid the calculation of 2^{23} to determine r_1

Multiply [the auxiliary number] as many times to make the first step [a^{P_1}]. Measure it with the original number [$a^{P_1} \equiv r_1$ (mod N)]. In case the remaining number makes 1 or -1, then one does not have to multiply any more. If one does not obtain 1, then, with the remaining number normalize the next larger prime [factor] by using the method of super-multiplication to substitute multiplication. Multiply [the auxiliary number] as many times to make the second step [$r_1^{P_2}$]. Measure it with the original number [$r_1^{P_2} \equiv r_2$ (mod N)]. In case the remaining number makes 1 or -1, then one does not have to multiply any more. If one does not obtain 1, then, with the remaining number normalize the third [largest] prime [factor], by multiplying again [$r_2^{P_3} \equiv r_3$ (mod N)]. This way multiply up to the total factorization, then stop [$r_{n-1}^{P_n} \equiv r_n$ (mod N)]. If one still does not obtain 1, then the original number is not a prime number. If in the last of all the steps one has obtained 1 or -1, then one again uses the subsequently added number to eliminate from the original number in order to determine if the original number is a prime number or not. If one obtains one of the powers of the auxiliary number or one of its negative numbers, then the original number is not a prime number. When in the sequence of multiplications one happens to obtain 1 before having completed all the steps,[22] then one should look at the definite number of times. If it is not the same as the number of steps, then the original number is not a prime number. If it is the same, one again [as in the previous methods] uses the subsequently added number to determine [if] it [(the original number) is a prime number or not].

As for example, the original number is 1289. Subtract 1 and half it. One obtains 644 as the total number. It is the product of the four prime numbers 2, 2, 7, 23 multiplied with each other. Take 2 as the auxiliary number. Using the method of super-multiplication: 2 multiplied by itself, one obtains 4, again multiplied by itself one obtains 16, again multiplied by itself one obtains 256, again multiplied by itself, take away what is filled by the original number, one obtains 1086 [$2^{16} \equiv 1086$ (mod 1289)]. Then use what remains after subtraction of the multiplication norm, 16 times. 16, subtract 23, lacking 7. Then, with the seventh power of the auxiliary number [2^7], 128, substitute for the multiplication [$2^{16} \cdot 2^7 \equiv 1086 \cdot 128$ (mod 1289)]. Measuring it with the original number, one obtains 1085 [$2^{23} \equiv 1085$ (mod 1289)]. It makes the remainder of the first step [$r_1 = 1085$]. Then, to make the multiplication norm, multiply it by itself, take away what is filled by the original number, one obtains 368 [$1085^2 \equiv 368$ (mod 1289)]. Multiplied [by itself] and again multiplied [by it] take away what is filled by the original number, one obtains 714 [$(1085^2)^3 \equiv 368^3$ (mod 1289) $\equiv 714$ (mod 1289)]. Then use what remains after subtraction of the multiplication norm, 6 times. 6, subtract 7, lacking 1. Then, with the multiplication norm substitute for the multiplication [$1085^6 \cdot 1085 \equiv 714 \cdot 1086$ (mod 1289)]. Measuring it with the original number, one obtains 1 [$1085^7 \equiv 1$ (mod 1289)]. One does not have to multiply any more. The first step, 23, doubling it, one obtains 46. The second step, 7, doubling it, one obtains 14. They all make the subsequently added numbers. Subsequently [add] adding one to make the divisors and eliminate from the original number.[23] None of the obtained

---

in $2^{23} \equiv r_1$ (mod 1289), one first finds $2^{16} \equiv 1086$ (mod 1289), and with $2^{23} = 2^{16} \cdot 2^7$ more easily simplifies $2^{16} \cdot 2^7 \equiv 1086 \cdot 2^7$ (mod 1289).

[22] I have omitted from the text in Li (1872c) p. 15B the line "then the original number is not a prime number. If one obtains one of the powers of the auxiliary number or one of its negative numbers" (則本數非數根。若得用數之諸方積或負數者), which seems to be erroneously copied again into the text from the previous sentence. See also Yan (1954) p. 13 for a similar amendment.

[23] Find t_1 such that $1 + t_1 \cdot 46 = 1289$ and find a t_2 such that $1 + t_2 \cdot 14 = 1289$; here $t_1 = 28$ and $t_2 = 92$.

numbers smaller than the divisors are exhaustive.[24] Therefore, one can determine that 1289 is a prime number.

As for example, the original number is 131071. Subtract 1 and half it. One obtains 65535 as the total number. It is the product of the four prime numbers 3, 5, 17, 257 multiplied with each other. Take 2 as the auxiliary number. Using the method of super-multiplication: 2 multiplied by itself, one obtains 4, again multiplied by itself one obtains 16, again multiplied by itself one obtains 256, again multiplied by itself one obtains 65536, again multiplied by itself, take away what is filled by the original number, one obtains 32768 [$2^{32} \equiv 32768 \pmod{131071}$]. This is the fifteenth power of the auxiliary number [$2^{15} = 32768$]. Altogether one has super-multiplied 5 times, therefore one uses what remains after subtraction of the multiplication norm, 32 times. Subtract the exponent of what has not been exhausted, 15, remains 17. Make it the definite number of times. It is equal to the second step. By doubling the definite number of times, one obtains 34. Make it the subsequently added number. Subsequently [add] adding one to make the divisor and eliminate from the original number.[25] None of the obtained numbers smaller than the divisor is exhaustive.[26] Therefore, one can determine that 131071 is a prime number.

The numbers above are prime numbers. The definite numbers of times equals a step, eliminating from it [i.e. from the original number] is not exhaustive.

As for example, the original number is 1398101. Subtract 1 and half it. One obtains 699050 as the total number. It is the product of the six prime numbers 2, 5, 5, 11, 31, 41 multiplied with each other. Take 2 as the auxiliary number. Using the method of super-multiplication: 2 multiplied by itself, one obtains 4, again multiplied by itself one obtains 16, again multiplied by itself one obtains 256, again multiplied by itself one obtains 65536, again multiplied by itself, take away what is filled by the original number, one obtains 1024 [$2^{32} \equiv 1024 \pmod{1398101}$]. This is the tenth power of the auxiliary number [$2^{10} = 1024$]. Altogether one has super-multiplied 5 times, therefore one uses what remains after subtraction of the multiplication norm, 32 times. Subtract the exponent of what has not been exhausted, 10, remains 22. Make it the definite number of times. It is equal to the second step, 11, and the final step, 2. Then, with the definite number of times make the subsequently added number. Adding one, one obtains 23. Eliminate from the original number, one obtains 60787 [$(1+22) \cdot 60787 = 1398101$].[27] It is precisely exhausted. Also, four times the definite number of times and added 1, one obtains 89. When eliminating it from the original number, one obtains 15709 [$(1+4 \cdot 22) \cdot 15709 = 1398101$].[28] It is precisely exhausted. Therefore, 1398101 is not a prime number.

The number above is not a prime number. Although the definite number of times equals a step, eliminating from it [i.e. from the original number] is precisely exhaustive.

As for example, the original number is 13447. Subtract 1 and half it. One obtains 6723 as the total number. It is the product of the five prime numbers 3, 3, 3, 3,

---

[24]None of the $1 + t \cdot 46$ divides 1289 for $t < t_1$, and none of the $1 + t \cdot 14$ divides 1289 for $t < t_2$.

[25]Find t_1 such that $1 + t_1 \cdot 34 = 131071$; here $t_1 = 3855$.

[26]None of the $1 + t \cdot 34$ divides 131071 for $t < t_1$.

[27]Here, Li Shanlan, according to what he describes in his first method, should correctly have searched for t such that $1 + t_1 \cdot 22 = 1398101$. He would have found $t_1 = 63550$.

[28]In spite of the error in the previous step, Li has found here a $t < t_1$ such that $1 + t \cdot 22$ divides the original number and thus comes to the correct conclusion that N is not prime.

83, multiplied with each other. Take 2 as the auxiliary number. Using the method of super-multiplication: 2 multiplied by itself, one obtains 4, again multiplied by itself one obtains 16, again multiplied by itself one obtains 256, again multiplied by itself, take away what is filled by the original number, one obtains 11748 [$2^{16} \equiv 11748$ (mod 13447)]. Again, multiplied by itself, take away what is filled by the original number, one obtains 8943 [$2^{32} \equiv 11748^2$ (mod 13447) $\equiv 8943$ (mod 13447)]. Again, multiplied by itself, take away what is filled by the original number, one obtains 7940 [$2^{64} \equiv 8943^2$ (mod 13447) $\equiv 7940$ (mod 13447)]. Put it on top [of the calculation surface]. Altogether one has super-multiplied 6 times, therefore one uses the multiplication norm, 64 times. Subtract 83, lacking 19. Then, put down the 19th power of the auxiliary number, 524288. Taking away what is filled by the original number, one obtains 13302 [$2^{19} \equiv 13302$ (mod 13447)]. With this substitute the multiplication of what is on top [of the calculation surface]. Measuring this with the original number, one obtains 5142 [$2^{64} \cdot 2^{19} \equiv 7940 \cdot 13302$ (mod 13447) $\equiv 5142$ (mod 13447)]. This makes the first step, one does not obtain 1. Then, to make the multiplication norm with 5142, multiply it by itself and again multiplied, when measuring the original number with this, one obtains 8009 [$5142^3 \equiv 8009$ (mod 13447)]. This makes the second step, one still does not obtain 1. Then, to make the multiplication norm with 8009, multiply it by itself and again multiplied, when measuring the original number with this, one obtains 3697 [$8009^3 \equiv 3697$ (mod 13447)]. This makes the third step, one still does not obtain 1. Then, to make the multiplication norm with 3697, multiply it by itself and again multiplied, when measuring the original number with this, one obtains 3844 [$3697^3 \equiv 3844$ (mod 13447)]. This makes the fourth step, one still does not obtain 1. Then, to make the multiplication norm with 3844, multiply it by itself and again multiplied, when measuring the original number with this, one obtains 8 [$3844^3 \equiv 8$ (mod 13447)]. This makes the fifth step. One has already filled the total number but one still does not obtain 1. Therefore, 13447 is not a prime number. 8 makes the third power of the auxiliary number. Therefore, subtract 3 from the total number, one obtains 6720. This number contains the prime [factor] 7. By doubling 7 one obtains 14. Make it the subsequently added number. 8 times and adding 1, one obtains 113 [$8 \cdot 14 + 1 = 113$]. When eliminating it from the original number, one obtains 119 [$13447 = 119 \cdot 113 + 0$]. It is precisely exhausted. [The number 6720] also contains the prime [factor] 3. By doubling 3 one obtains 6. Make it the subsequently added number. Adding 1, one obtains 7. When eliminating it from the original number, one obtains 1921 [$13447 = 1921 \cdot 7 + 0$]. It is precisely exhausted.

As for example, the original number is 16637. Subtract 1 and half it. One obtains 8318 as the total number. It is the product of the two prime numbers 2, 4159 multiplied with each other. Take 2 as the auxiliary number. Using the method of super-multiplication: 2 multiplied by itself, one obtains 4, again multiplied by itself one obtains 16, again multiplied by itself one obtains 256, again multiplied by itself, take away what is filled by the original number, one obtains 15625 [$2^{16} \equiv 15625$ (mod 16637)]. Again, multiplied by itself, take away what is filled by the original number, one obtains 9287 [$2^{32} \equiv 9287$ (mod 16637)]. Again, multiplied by itself, take away what is filled by the original number, one obtains 2161 [$2^{64} \equiv 2161$ (mod 16637)]. Again, multiplied by itself, take away what is filled by the original number, one obtains 11561 [$2^{128} \equiv 11561$ (mod 16637)]. Again, multiplied by itself, take away what is filled by the original number, one obtains 11700 [$2^{256} \equiv 11700$ (mod 16637)]. Again, multiplied by itself, take away what is filled by the original number, one obtains 764 [$2^{512} \equiv 764$ (mod 16637)]. Again, multiplied by itself, take away what is filled by the original number, one obtains 1401 [$2^{1024} \equiv 1401$ (mod 16637)]. Again, multiplied by itself, take away what is filled by the original number, one obtains 16272 [$2^{2048} \equiv 16272$ (mod 16637)]. Again, multiplied by itself, take away what is filled by the original number, one obtains 129 [$2^{4096} \equiv 129$ (mod 16637)]. Put it on top [of the calculation surface]. Altogether one has super-multiplied 12 times, therefore one uses the multiplication norm, 4096 times. Subtract 4159, lacking 63. Then, put down the remainder of the sixth super-multiplication, 2161 [$2^{64} \equiv 2161$ (mod 16637)]. Add it to the original number, half it, one obtains 9399 [$(2161 + 16637)/2 = 9399$]. This is the remainder of the 63rd power of the auxiliary number after having taken away what is filled by the original number [$2^{64} \cdot 2^{-1} = 2^{63} \equiv 9399$ (mod 16637)]. With this substitute the multiplication of what is on top [of the calculation surface]. Measuring this with the original number, one obtains 14607 [$2^{4096} \cdot 2^{63} = 2^{4159} \equiv 129 \cdot 9399$ (mod 13447) $\equiv$ 14607 (mod 16637)]. This makes the first step. Then, the remaining number, multiply it by itself [14607^2]. When measuring the original number with this, one obtains 11561 [$14607^2 \equiv 11561$ (mod 16637)]. This makes the second step. One has already filled the total number but not obtained 1. Therefore, 16637 is not a prime number. The final remainder, 11561, is not a power [of the auxiliary number], neither is it a negative number. When one wants to know the prime [factors] of the original number, one should apply the *Dayan* procedure. Take the fifteenth power of the auxiliary number, 32768, with the original number reduce to one. The layout of the formulas is as below.

32768	16637
1元	
16131	506
1元	-1元
445	61
32元	-33元
18	7
263元	-822元
4	3
1907元	-2729元
1	
4636元	

The two numbers, align them right and left. Establish the celestial element below to the left. Subtract right from the left, one obtains the second level on the left [32768 − 16637 = 16131]. This, in turn, subtract it from the right [on top]. One obtains the second level on the right [16637 − 16131 = 506]. Subtract its 31-fold value from the second level on the left, one obtains the third level on the left [16131 − 31 · 506 = 445]. This, in turn, subtract it from the second level on the right, one obtains the third level on the right [506 − 445 = 61]. Subtract its 7-fold value from the third level on the left, one obtains the fourth level on the left [445 − 7 · 61 = 18]. Subtract its 3-fold value from the third level on the right, one obtains the fourth level on the right [61 − 3 · 18 = 7]. Subtract its twofold value from the fourth level on the left, one obtains the fifth level on the left [18 − 2 · 7 = 4]. This, in turn, subtract it from the fourth level on the right, one obtains the fifth level on the right [7 − 4 = 3]. This, in turn, subtract it from the fifth level on the left, one obtains the sixth level on the left [4 − 3 = 1]. Above makes 1, below makes 4636 celestial elements. Then, 4636 makes the multiplication norm. Multiply the final remainder with it [11561 · 4636], when measured with the original number, there is a remainder of 9019 [11561 · 4636 $\equiv$ 9019 (mod 16637)]. Multiply this with the multiplication norm, when measured with the original number, there is a remainder of 3303 [9019 · 4636 $\equiv$ 3303 (mod 16637)]. Multiply this with the multiplication norm, when measured with the original number, there is a remainder of

6668 [3303 · 4636 ≡ 6668 (mod 16637)]. Multiply this with the multiplication norm, when measured with the original number, there is a remainder of 1302 [6668 · 4636 ≡ 1302 (mod 16637)]. Multiply this with the multiplication norm, when measured with the original number, there is a remainder of 13478 [1302·4636 ≡ 13478 (mod 16637)]. Multiply this with the multiplication norm, when measured with the original number, there is a remainder of 12073 [13478 · 4636 ≡ 12073 (mod 16637)]. Multiply this with the multiplication norm, when measured with the original number, there is a remainder of 3560 [12073 · 4636 ≡ 3560 (mod 16637)]. Multiply this with the multiplication norm, when measured with the original number, there is a remainder of 256 [3560 · 4636 ≡ 256 (mod 16637)]. It is the eighth power of the auxiliary number [$2^8 = 256$]. Take the number of times, 8, that the multiplication norm was used, multiply this with 15, the exponent [in $2^{15} = 32768$], also, add the remaining exponent to it. One obtains 128 [$8 · 15 + 8 = 128$]. When subtracting it from the total number, one obtains 8190 [$8318 - 128 = 8190$]. This number contains the prime [factor] 13. It also contains the prime [factor] 7. By doubling 13 one obtains 26. Make it the subsequently added number. 5 times and adding 1, one obtains 131 [$5 · 26 + 1 = 131$]. When eliminating it from the original number, one obtains 127 [$16637 = 127 · 131 + 0$]. It is precisely exhausted. Or, by doubling 7 one obtains 14. Make it the subsequently added number. 9 times and adding 1, one obtains 127 [$9 · 14 + 1 = 127$]. When eliminating it from the original number, one obtains 131 [$16637 = 131 · 127 + 0$]. It is precisely exhausted.

The numbers above are not prime numbers. When multiplying in sequence the total number is filled, but one does not obtain 1.

In the first method, when the multiplication norm, or, in the second and fourth method, when the exponent used when searching to obtain unity with the celestial element become bigger, then the number of times one successively multiplies and successively measures becomes lesser. This is why powers of the auxiliary number in the tens or hundreds are likely to be in use. But when the positions of its product are too many in number, one cannot use the positive numbers[29] but must use the remaining numbers (shengshu 勝數). The remaining numbers are the rest of what does not fill the original number. Also, in general, when it [the remaining number] is smaller than the used power product[30] and bigger than all the power products of the original number, one has to exhaustively search for the remaining numbers. Proceeding one by one, successively multiplying and successively measuring, from the remainders one obtains the remaining numbers for each exponent. It is no different from the principles for obtaining the positive numbers.[31]

The method for finding the remaining number. Starting from the first power of the auxiliary number, if its product is smaller than the original number [$a^1 < N$] one always uses the positive numbers. If from the beginning it is bigger than the original number, then subtract it from the original number and make it the first remaining number. Multiply it with the auxiliary number, again multiply it. If it fills the original

---

[29]That is, the powers of the auxiliary number a^i.

[30]Lit. fangji 方積. In Chinese, exponentiation where the exponent n is a positive integer is expressed as a repeated multiplication of the base b, thus the result b^n is the product of multiplying n bases b with each other.

[31]What Li discusses in this paragraph corresponds to a simplification of the calculations in order to avoid large powers p. With $p = p_1 p_2$ for example, by first calculating $a^{p_1} \equiv r_1$ (mod N), one can replace the calculation of r_2 in the congruence $a^{p_1 p_2} \equiv r_2$ (mod N) by solving $r_1^{p_2} \equiv r_2$ (mod N) instead.

number then eliminate it to make the second and the third remaining number. In this manner successively multiply it with the auxiliary number. If it fills the original number, eliminate it until one arrives at the power of the remaining number that one is using, then stop. One obtains all the remaining numbers. With all the positive numbers one aligns them in order. When successively multiplying and successively measuring, one has to investigate the correctness of each remainder (*yu shu* 餘數) obtained.[32]

---

[32]Translated from Li (1872c) p. 15A–20B.

Appendix C
On Conics (Some Technicalities)

Contents

C.1 Binomial Expansions

Xia Luanxiang's 夏鸞翔 (1823–1864) expansion of $f(x) = \sqrt{a+x}$ in his work *A Myriad Images—A Single Origin* (*Wan xiang yi yuan* 萬象一原) corresponds to the series[1]:

$$a^{\frac{1}{2}} + \frac{1}{2}\frac{a^{\frac{1}{2}}}{(a+x)}x + \frac{1\cdot3}{2\cdot4}\frac{a^{\frac{1}{2}}}{(a+x)^2}x^2 + \frac{1\cdot3\cdot5}{2\cdot4\cdot6}\frac{a^{\frac{1}{2}}}{(a+x)^3}x^3 + \cdots$$

It is probably derived by expansion of the equivalent function:

$$f(x) = \sqrt{a} \cdot \frac{1}{\sqrt{1 - \frac{x}{a+x}}}$$

---

[1]Xia (1898) p. 3B, shown in Fig. 2.3.

© Springer International Publishing AG, part of Springer Nature 2019
A. Bréard, *Nine Chapters on Mathematical Modernity*, Transcultural
Research – Heidelberg Studies on Asia and Europe in a Global Context,
https://doi.org/10.1007/978-3-319-93695-6

which only converges for $|x| < |a + x|$, a fact which Xia does not mention explicitly.[2] Only at the beginning of the general, and rhetorically formulated, procedure for finding the nth root (*Zhu chengfang qiu gen* 諸乘方求根), Xia speaks about "taking a slightly smaller number than the surface [itself] as the 'borrowed surface'."[3] If I am not mistaken, the "slightly smaller number" here corresponds to a where $a < a + x$ and x is a relatively small quantity. With this interpretation Xia's recursively defined terms are as follows:

$$s_1 = \sqrt[n]{a}$$

$$s_2 = s_1 \cdot \frac{x}{a \cdot n}$$

$$s_3 = s_2 \cdot \frac{x \cdot (n - 1)}{a \cdot 2n}$$

$$s_4 = s_3 \cdot \frac{x \cdot (2n - 1)}{a \cdot 3n}$$

and thus lead to the expansion of $f(x) = \sqrt[n]{a + x} = a^{\frac{1}{n}} \left(1 + \frac{x}{a}\right)^{1/n}$ as:

$$\sqrt[n]{a + x} = s_1 + s_2 - s_3 + s_4 - \cdots = a^{\frac{1}{n}} + a^{\frac{1}{n}} \frac{1}{1 \cdot n \cdot a} x - a^{\frac{1}{n}} \frac{(n - 1)}{2 \cdot n^2 \cdot a^2} x^2$$

$$+ a^{\frac{1}{n}} \frac{(n - 1)(2n - 1)}{2 \cdot 3 \cdot n^3 \cdot a^3} x^3 - \cdots$$

which converges for $|x| < a$. Elias Loomis (1811–1889) gives the same expansion of $f(x)$ for $n = 2$:[4]

$$\sqrt{a + x} = a^{\frac{1}{2}} + \frac{1}{2} a^{-\frac{1}{2}} x - \frac{1}{2 \cdot 4} a^{-\frac{3}{2}} x^2 + \frac{1 \cdot 3}{2 \cdot 4 \cdot 6} a^{-\frac{5}{2}} x^3 -, etc.$$

Lu Jing 盧靖 (1856–1928) transcribes most rhetoric procedures of Xia Luanxiang's text into formulaic language, and gives two expansions each of $\sqrt[n]{a + x}$ and $\sqrt[n]{a - x}$, without mentioning under which conditions his formulas are valid.[5]

---

[2] According to the title, Xu (1901) is concerned with discussing errors in Xia's text. Unfortunately, I have not been able to consult a copy of this book in order to understand whether convergence criteria were part of Xu's discussion.

[3] Xia (1898) p. 3A.

[4] In the original text (Loomis 1851) p. 141 and in the Chinese translation (Li and Wylie 1859) scroll 11, p. 5A.

[5] See Lu (1902) formulas 7–10, p. 8B–9A.

Formulas 7 and 10 correspond to the generalized binomial series for $|x| < a$:

$$\sqrt[n]{a+x} = \sqrt[n]{a}\sqrt[n]{1+x/a} = \sqrt[n]{a}\sum_{k=0}^{+\infty}\binom{\frac{1}{n}}{k}(x/a)^k = \sum_{k=0}^{+\infty}\binom{\frac{1}{n}}{k}a^{\frac{1}{n}-k}x^k$$

$$\sqrt[n]{a-x} = \sqrt[n]{a}\sqrt[n]{1-x/a} = \sqrt[n]{a}\sum_{k=0}^{+\infty}\binom{\frac{1}{n}}{k}(-1)^k(x/a)^k = \sum_{k=0}^{+\infty}\binom{\frac{1}{n}}{k}(-1)^k a^{\frac{1}{n}-k}x^k$$

whereas formulas 8 and 9 (see Fig. 2.4) are the expansions for the case $|x| < |a+x|$:

$$\sqrt[n]{a+x} = \sqrt[n]{a}\frac{1}{\sqrt[n]{1-\frac{x}{a+x}}} = \sqrt[n]{a}\sum_{k=0}^{+\infty}\binom{-\frac{1}{n}}{k}(-1)^k\left(\frac{x}{a+x}\right)^k$$

$$\sqrt[n]{a-x} = \sqrt[n]{a}\frac{1}{\sqrt[n]{1+\frac{x}{a-x}}} = \sqrt[n]{a}\sum_{k=0}^{+\infty}\binom{-\frac{1}{n}}{k}\left(\frac{x}{a-x}\right)^k$$

C.2 The Circle

In his *Procedures for Curves* (*Zhiqu shu* 致曲術), Xia Luanxiang gives two series to "find the arc with the sine" (*zhengxian qiu hubei* 正弦求弧背). They allow to calculate the arc z of a circle of radius r when the sine $s = \sin z$ is known. Thereby, one procedure, as he claims, stems from the Jesuit missionary Jartoux. It is indeed equivalent to the series (C.1) indicated by Ming Antu 明安圖 (ca. 1692–ca. 1763) as his eighth of altogether nine procedures[6]:

$$z = s + \frac{s^3}{r^2 \cdot 3!} + \frac{s^5 \cdot 3^2}{r^4 \cdot 5!} + \frac{s^7 \cdot 3^2 \cdot 5^2}{r^6 \cdot 7!} + \cdots \tag{C.1}$$

Yet Xia's other, "newly determined" procedure[7] is different and gives the square of the arc z:

$$z^2 = s^2 + \frac{s^4 \cdot 4}{r^2 \cdot 4 \cdot 3} + \frac{s^6 \cdot 4 \cdot 16}{r^4 \cdot 6 \cdot 5 \cdot 4 \cdot 3} + \frac{s^8 \cdot 4 \cdot 16 \cdot 32}{r^6 \cdot 8 \cdot 7 \cdot 6 \cdot 5 \cdot 4 \cdot 3} + \cdots \tag{C.2}$$

---

[6]See Ming (1774) vol. 1, p. 6B–7A. Later, Dong Youcheng 董祐誠 (1791–1823) gives the same procedure in similar form, see Dong (1819) scroll 1, p. 4B.

[7]See Xia (1908) p. 1A–2A.

Squaring Jartoux's series (C.1) for the arc based on the sine as given by Ming Antu, we obtain indeed the power series as in Eq. (C.2):

$$z^2 = (s + \frac{s^3}{r^2 \cdot 3!} + \frac{s^5 \cdot 3^2}{r^4 \cdot 5!} + \frac{s^7 \cdot 3^2 \cdot 5^2}{r^6 \cdot 7!} + \cdots)^2 =$$

$$= s^2 + \frac{2s^4}{r^2 \cdot 3!} + \frac{s^6 \cdot 4 \cdot 32}{r^4 \cdot 6!} + \frac{s^8 \cdot 2 \cdot 4 \cdot 16 \cdot 32}{r^6 \cdot 8!} + \cdots$$

This is probably what Xia did to produce a procedure for z^2.[8] Also, investigating into an eventual relation to Loomis's work, no such formula can be found neither in the original text (Loomis 1851) nor in the Chinese translation (Li and Wylie 1859). The sole example related to the rectification of the arc of a circle is based on its tangent t and yields z, not its square. By integration of the differential of the arc z[9]:

$$dz = \frac{dt}{1 + t^2}$$

and with:

$$\frac{1}{1 + t^2} = 1 - t^2 + t^4 - t^6 +, etc.$$

one obtains the following series for the arc z:

$$\int dz = z = t - \frac{t^3}{3} + \frac{t^5}{5} - \frac{t^7}{7} + \frac{t^9}{9} -, etc.$$

Had Xia followed Loomis's method and calculated the arc with the sine as indicated with the tangent, he would have arrived at exactly the same conclusion as his teacher Ming Antu 明安圖 (ca. 1692–ca. 1763), and indirectly as Jartoux, but not at his own original contribution (C.2) yielding the square of the arc z^2. This is shown by the following calculations.

For any arc z and $s = \sin z$ Loomis obtained for the differential of the arc of a circle of radius r[10]:

$$dz = \frac{r \, ds}{\sqrt{r^2 - s^2}}$$

---

[8] See his comment translated on page 24.
[9] See Loomis (1851) p. 248 and Li and Wylie (1859) scroll 18, p. 2A–3A.
[10] See Loomis (1868) p. 172.

Integration then gives the length of the arc z:

$$z = \int \frac{r}{\sqrt{r^2 - s^2}} \, ds$$

Loomis could easily do this term by term by developing the function in s into a MacLaurin series:

$$z = \int \frac{1}{\sqrt{1 - \frac{s^2}{r^2}}} \, ds = \int \left\{ 1 + \frac{1}{2} \cdot \frac{s^2}{r^2} + \frac{3}{8} \cdot \frac{s^4}{r^4} + \frac{5}{16} \cdot \frac{s^6}{r^6} + \cdots \right\} ds$$

$$= s + \frac{1}{2 \cdot 3} \cdot \frac{s^3}{r^2} + \frac{3}{5 \cdot 8} \cdot \frac{s^5}{r^4} + \frac{5}{7 \cdot 16} \cdot \frac{s^7}{r^6} + \cdots = s + \frac{s^3}{r^2 \cdot 3!} + \frac{s^5 \cdot 3^2}{r^4 \cdot 5!} + \frac{s^7 \cdot 3^2 \cdot 5^2}{r^6 \cdot 7!} + \cdots$$

The resulting series is indeed equivalent to the expression in (C.1).

In an inverse fashion, Xia gives two procedures to calculate any arc z of a circle of radius r with the versed sine $x = \text{versin } z$. Here, Xia first gives what he claims to be newly established for finding z, before crediting the "Westerner Jartoux" for the second procedure yielding z^2. As for Jartoux's formula, it is given as the sum $z^2 = \sum_{i=1}^{\infty} s_i$, whereby the s_i are defined recursively:

$$s_1 = 2rx$$

$$s_2 = \frac{s_1 \cdot x}{2r} \cdot \frac{4}{3 \cdot 4}$$

$$s_3 = \frac{s_2 \cdot x}{2r} \cdot \frac{16}{5 \cdot 6}$$

$$s_4 = \frac{s_3 \cdot x}{2r} \cdot \frac{36}{7 \cdot 8}$$

$$s_5 = \cdots$$

This corresponds, when written in explicit terms for the sake of comparability, to the following series:

$$z^2 = 2rx + \frac{2rx^2 \cdot 4}{r \cdot 4!} + \frac{2rx^3 \cdot 4 \cdot 16}{2r^2 \cdot 6!} + \frac{2rx^4 \cdot 4 \cdot 16 \cdot 36}{4r^3 \cdot 8!} + \cdots \qquad (C.3)$$

The first procedure, "newly determined," is again defined recursively, but the sum here yields z and not the square of the arc z^2 as in (C.3):

$$s_1 = \sqrt{2rx}$$

$$s_2 = \frac{s_1 \cdot x}{2r} \cdot \frac{1}{2 \cdot 3}$$

$$s_3 = \frac{s_2 \cdot x}{2r} \cdot \frac{9}{4 \cdot 5}$$

$$s_4 = \frac{s_3 \cdot x}{2r} \cdot \frac{25}{6 \cdot 7}$$

$$s_5 = \cdots$$

$$z = \sqrt{2rx} + \frac{x \cdot \sqrt{2rx} \cdot 1}{2r \cdot 3!} + \frac{x^2 \cdot \sqrt{2rx} \cdot 1 \cdot 9}{4r^2 \cdot 5!} + \frac{x^3 \cdot \sqrt{2rx} \cdot 1 \cdot 9 \cdot 25}{8r^3 \cdot 7!} + \cdots \quad (C.4)$$

As above, when we square Xia's procedure for the arc based on the versed sine in Eq. (C.4) we obtain:

$$z^2 = (\sqrt{2rx} + \frac{x \cdot \sqrt{2rx} \cdot 1}{2r \cdot 3!} + \frac{x^2 \cdot \sqrt{2rx} \cdot 1 \cdot 9}{4r^2 \cdot 5!} + \frac{x^3 \cdot \sqrt{2rx} \cdot 1 \cdot 9 \cdot 25}{8r^3 \cdot 7!} + \cdots)^2$$

$$= 2rx + \frac{2rx^2}{r \cdot 3!} + \frac{2rx^3 \cdot 4 \cdot 16}{2r^2 \cdot 6!} + \frac{2rx^4 \cdot 8 \cdot 9 \cdot 32}{4r^3 \cdot 8!} + \cdots$$

which, in that case, does correspond to the correct procedure (C.3) ascribed to Pierre Jartoux 杜德美 (1668–1720).

Since Xia gives no details how he exactly proceeded in finding his own, "newly determined" procedures, we have no certain means to reconstruct his methodology. What is nevertheless interesting for us are his claims of originality and his reference to Loomis's *Differential and Integral Calculus*. Assuming that his reference to Loomis's work indicates his veritable methodological source, and taking into account Xia's commentary quoted on page 24 we can argue that Xia indeed applied a similar method as the one indicated by Loomis for finding the length of any arc of a circle when the tangent is known, in order to give a procedure that yields the arc and not its square.[11] We can thus try to reconstruct how Xia might have obtained his series in (C.4).

With z representing any arc of a circle of radius r and $x = \text{versin } z$ we have its differential[12] as:

$$dx = \frac{\sin z}{r} dz.$$

Since $\sin z = \sqrt{(2r - x)x}$ we obtain:

$$dz = \frac{r}{\sqrt{(2r - x)x}} dx.$$

---

[11] See Loomis (1851) p. 248.

[12] Deduced in Loomis (1868) p. 172.

By integration, the length of the arc z can be found through expansion into a MacLaurin series:

$$z = \int \frac{r}{\sqrt{(2r-x)x}} \, dx = \int \frac{r}{\sqrt{2rx}} \frac{1}{\sqrt{1-\frac{x}{2r}}} \, dx$$

$$= \int \frac{r}{\sqrt{2rx}} \cdot \left[1 + \frac{1}{2} \cdot \frac{x}{2r} + \frac{3}{8} \cdot \left(\frac{x}{2r}\right)^2 + \frac{5}{16} \cdot \left(\frac{x}{2r}\right)^3 + \frac{35}{128} \cdot \left(\frac{x}{2r}\right)^4 + \cdots \right] dx$$

$$= \int \frac{r}{\sqrt{2rx}} + \frac{\sqrt{2r} \cdot x^{1/2}}{2 \cdot 4 \cdot r} + \frac{3 \cdot \sqrt{2r} \cdot x^{3/2}}{2 \cdot 4 \cdot 8 \cdot r^2} + \frac{5 \cdot \sqrt{2r} \cdot x^{5/2}}{2 \cdot 8 \cdot 16 \cdot r^3} + \cdots \, dx$$

$$= \sqrt{2rx} + \frac{x \cdot \sqrt{2rx}}{2r \cdot 3!} + \frac{x^2 \cdot \sqrt{2rx} \cdot 3^2}{4r^2 \cdot 5!} + \frac{x^3 \cdot \sqrt{2rx} \cdot 3^2 \cdot 5^2}{8r^2 \cdot 7!} + \cdots$$

Here, applying Loomis's model has indeed brought about Xia's "newly determined" series in (C.4) for the arc of a circle based on its versed sine.

C.3   The Ellipse

After quoting the procedures for the circumference of the entire ellipse by Xiang Mingda 項名達 (1789–1850) and Dai Xu 戴煦 (1805–1860), Xia gives a "newly determined" procedure for calculating the arc z of an ellipse from the point $(0, b)$ to any point (x, y) on the ellipse defined by the equation[13]:

$$y^2 = (1 - e^2)(a^2 - x^2)$$

Here, Xia does not quote what he supposedly found in Loomis.[14] It is very likely though that Xia did base his procedure on Loomis, since the latter obtains[15]:

$$z = X_0 - \frac{e^2}{2a} X_2 - \frac{e^4}{2 \cdot 4a^3} X_2 - \frac{3e^6}{2 \cdot 4 \cdot 6a^5} X_2 -, \text{etc.}, \qquad (C.5)$$

---

[13] The eccentricity e of an ellipse is defined as $e = \sqrt{\frac{a^2-b^2}{a^2}}$, whereby a and b are the semi-major and the semi-minor axis of the ellipse.

[14] See the "Procedure for finding the arc of an ellipse with the sine" (*Tuo zhengxian qiu tuo hubei shu* 橢正弦求橢弧背術) in Xia (1908) p. 3A–5B.

[15] As for the procedure given as an application of the integral calculus, see Loomis (1868) p. 251–252 and its Chinese translation in Li and Wylie (1859) vol. 18, p. 5B–7A.

where X_0 represents the arc of a circle with radius a from the point $(0, a)$ to any point (x, y) on the circle, and where:

$$X_2 = \frac{a \cdot X_0}{2} - \frac{x}{2}\sqrt{a^2 - x^2},$$

$$X_4 = \frac{3a^2 \cdot X_2}{4} - \frac{x^3}{4}\sqrt{a^2 - x^2},$$

$$X_6 = \frac{3a^3 \cdot X_4}{6} - \frac{x^5}{6}\sqrt{a^2 - x^2}, \text{etc.}$$

Given the fact that Xia likes to present the previous procedures for the arc of the circle with the terms of the series defined in recursive manner, it is surprising that he does not reproduce here Loomis's scheme but develops each of the X_{2i} into a MacLaurin series, by expressing the terms of each again in recursive manner. In the end, he obtains a sum of infinite series $z = \sum_{j=0}^{\infty} T_j$ where $T_j = \sum_{i=0}^{\infty} s_{i,j}$[16]:

$$z = \left(x + \frac{1^2}{3!a^2}x^3 + \frac{1^2 \cdot 3^2}{5!a^4}x^5 + \frac{1^2 \cdot 3^2 \cdot 5^2}{7!a^6}x^7 + \cdots \right) -$$

$$- \left(\frac{c^2}{2 \cdot 3a^4}x^3 + \frac{c^2}{2 \cdot 2 \cdot 5a^6}x^5 + \frac{3c^2}{2 \cdot 2 \cdot 4 \cdot 7a^8}x^7 + \frac{3 \cdot 5c^2}{2 \cdot 2 \cdot 4 \cdot 6 \cdot 9a^{10}}x^9 + \cdots \right) -$$

$$- \left(\frac{c^4}{2 \cdot 4 \cdot 5a^8}x^5 + \frac{c^4}{2 \cdot 4 \cdot 2 \cdot 7a^{10}}x^7 + \frac{3c^4}{2 \cdot 4 \cdot 2 \cdot 4 \cdot 9a^{12}}x^9 \right.$$

$$\left. + \frac{3 \cdot 5c^4}{2 \cdot 4 \cdot 2 \cdot 4 \cdot 6 \cdot 11a^{14}}x^{11} + \cdots \right) -$$

$$- \left(\frac{c^6}{2 \cdot 4 \cdot 6 \cdot 7a^{12}}x^7 + \frac{c^6}{2 \cdot 4 \cdot 6 \cdot 2 \cdot 9a^{14}}x^9 + \frac{3c^6}{2 \cdot 4 \cdot 6 \cdot 2 \cdot 4 \cdot 11a^{16}}x^{11} + \cdots \right) - \cdots$$

Xia actually expresses the terms $s_{i,j}$ of each series T_j recursively. The recursive definition of the coefficients certainly allows more easily to identify the numeric patterns and the regularity of their composition. Xia gives the following terms $s_{i,j}$ for the first four series T_0 to T_3 (each in one line above):

$$s_{0,0} = x$$

$$s_{1,0} = s_{0,0} \cdot \frac{x^2}{a \cdot a} \cdot \frac{1 \cdot 1}{2 \cdot 3}$$

---

[16]By c we denote half the distance between the focal points of the ellipse (*ban xin cha* 半心差), thus $c^2 = a^2 - b^2$ or $c^2 = e^2 \cdot a^2$. For an affirmative reconstruction of Xia's use of Loomis's work, see Gao and Wang (2012) p. 255.

$$s_{2,0} = s_{1,0} \cdot \frac{x^2}{a \cdot a} \cdot \frac{3 \cdot 3}{4 \cdot 5}$$

$$s_{3,0} = s_{2,0} \cdot \frac{x^2}{a \cdot a} \cdot \frac{5 \cdot 5}{6 \cdot 7}$$

$$s_{0,1} = \frac{c^2 \cdot x^3}{a^4} \cdot \frac{1}{2 \cdot 3}$$

$$s_{1,1} = s_{0,1} \cdot \frac{x^2}{a \cdot a} \cdot \frac{1 \cdot 3}{2 \cdot 5}$$

$$s_{2,1} = s_{1,1} \cdot \frac{x^2}{a \cdot a} \cdot \frac{3 \cdot 5}{4 \cdot 7}$$

$$s_{3,1} = s_{2,1} \cdot \frac{x^2}{a \cdot a} \cdot \frac{5 \cdot 7}{6 \cdot 9}$$

$$s_{0,2} = \frac{c^4 \cdot x^5}{a^8} \cdot \frac{1 \cdot 1}{2 \cdot 4 \cdot 5}$$

$$s_{1,2} = s_{0,2} \cdot \frac{x^2}{a \cdot a} \cdot \frac{1 \cdot 5}{2 \cdot 7}$$

$$s_{2,2} = s_{1,2} \cdot \frac{x^2}{a \cdot a} \cdot \frac{3 \cdot 7}{4 \cdot 9}$$

$$s_{3,2} = s_{2,2} \cdot \frac{x^2}{a \cdot a} \cdot \frac{5 \cdot 9}{6 \cdot 11}$$

$$s_{0,3} = \frac{c^6 \cdot x^7}{a^{12}} \cdot \frac{1 \cdot 1 \cdot 1}{2 \cdot 4 \cdot 6 \cdot 7}$$

$$s_{1,3} = s_{0,3} \cdot \frac{x^2}{a \cdot a} \cdot \frac{1 \cdot 7}{2 \cdot 9}$$

$$s_{2,3} = s_{1,3} \cdot \frac{x^2}{a \cdot a} \cdot \frac{3 \cdot 9}{4 \cdot 11}$$

$$s_{3,3} = s_{2,3} \cdot \frac{x^2}{a \cdot a} \cdot \frac{5 \cdot 11}{6 \cdot 13}$$

before concluding for the sum of each of them respectively:

Following this way, all following [terms] are like this. Determine them until you arrive below the units [desired for precision], then stop. By adding them up, one obtains the first/second/third/fourth total number [T_0, T_1, T_3, T_4 respectively].
順是以下皆如是。求至單位下止。乃相并為總第一/二/三/四數。

And finally:

Determine them [i.e. the terms $s_{i,j}$ of each series] like this by superposing layer by layer. Determining [their sum], one obtains the total numbers [T_j]. Then, with all the total numbers, adding the positive and subtracting the negative makes the length of an elliptic arc.
如是疊次求之。求得總數。降至單位下止。乃以諸總數正負并減為橢圓弧背。[17]

C.4   Constructing the Ellipse

Due to a strong interest in Newton's lunar theory, the compilers of the *Later Volumes of the Thorough Investigation of Calendrical Astronomy Imperially Composed* (*Yuzhi lixiang kaocheng houbian* 御製曆象考成後編, 1738) not only introduced elliptic orbits for the sun, moon and planets, but "in addition to the lunisolar tables, J. Richer's observation of Mars at Cayenne, Kepler's equation, the theories of procession, parallax, refraction, and the obliquity of ecliptic were introduced."[18]

In this astronomical context, where planetary positions and trajectories were important objects of calculation, the ellipse was not introduced as a conic section, but formally defined by its two foci, which also served in the geometric construction of its circumference. One well-known construction with practical applications is the gardener's construction with a nail and a thread. In China it has already been explained and illustrated (see Fig. C.1) by Ferdinand Verbiest (chin. 南懷仁, 1623–1688) in his *Records on the Newly Built Astronomical Instruments from the Beijing Observatory* (*Xinzhi lingtai yixiang zhi* 新製靈臺儀象志, 1674). The same plate also depicts an instrument, the elliptical compass, or trammel, developed by Guidobaldo Dal Monte (1545–1607) in his *Planisphaeriorum universalium theorica* (1579).[19]

The authors of the *Later Volumes* decide to include another construction method for the ellipse, the now so-called "director circle" construction (Fig. C.2), because, they argue:

[17] Translated from Xia (1908) p. 5A–5B.

[18] Han (2001) p. 152.

[19] See Dal Monte (1579) p. 99–128. The book was probably part of Nicolas Trigault's collection in the Beitang library, see Verhaeren (1949) p. 415, item 1429. On Guidobaldo Del Monte's contributions to projective geometry, see Aterini (2016).

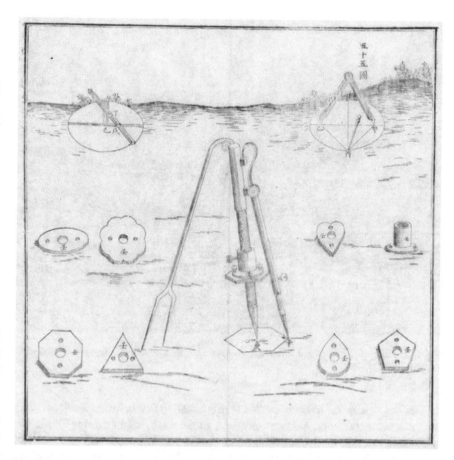

Fig. C.1 Elliptical compass (upper left) and gardener construction (upper right) in F. Verbiest's *Records on the newly built astronomical instruments from the Beijing Observatory* (1674)

with this [construction] it is most exquisite and pertinent to elucidate the principles of the ellipse![20]

The construction described in the *Later Volumes* and illustrated here in Fig. C.3 proceeds by a circle with a radius equal to the major axis of the desired ellipse, i.e. a circle of radius $2a$, and with a centre located in one of the two focal points, say the focal point F_2. For any point A of the circle, one then finds a point on the ellipse as the intersection of the segment AF_2 with the perpendicular bisector of the segment AF_1.

According to this method, successively, degree by degree, construct a point. By connecting them one accomplishes the circumference of the ellipse.[21]

---

[20]Qianlong 乾隆 (1773) scroll 1, p. 56B.

[21]*Idem.*

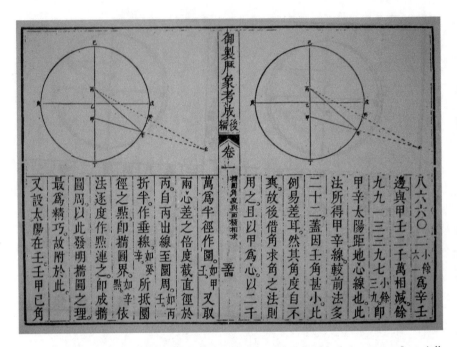

Fig. C.2 *Later Volumes of the Thorough Investigation of Calendrical Astronomy Imperially Composed* (1738)

Since the *Later Volumes* are the result of a collaboration between Jesuit and Chinese astronomers,[22] not the translation of a single work, but based upon a number of sources and informed scholars, it is not always possible to identify precisely the underlying writings. But unlike the well-known and common gardener construction, the director circle construction is rather rare in European sources. Probably the first occurence is in Claude Mydorge's (1585–1647) *Prodromi catoptricorum et dioptricorum*,[23] but it seems not to have been available to the Jesuits in Beijing, who could at best consult only a synopsis of the propositions in Mydorge's work on conics, compiled by the French mathematician Marin Mersenne (1588–1648).[24] It is thus more likely that the French Jesuit Claude François Milliet Dechales

---

[22] As Han (2001) p. 152 points out, this includes astronomers of the Observatory in Paris with whom Ignatius Kögler, and other Jesuits in Beijing in charge of the compilation of the *Later Volumes* kept close contact.

[23] See Mydorge (1660) Liber secundus Probl. XII, Proposition XXI, p. 96–97.

[24] See Verhaeren (1949) p. 652, item 2237. The 1644 edition of Mersenne's *Universæ Geometriæ* did contain a section on the four books of Mydorge's *Conicorum*, but it only gave a short synopsis of the relevant proposition. See Mersenne (1644) p. 345:

Datis, positione, ellipseos umbilicis, and alterutro vertice, ellipsim in codem plano per puncta describere.

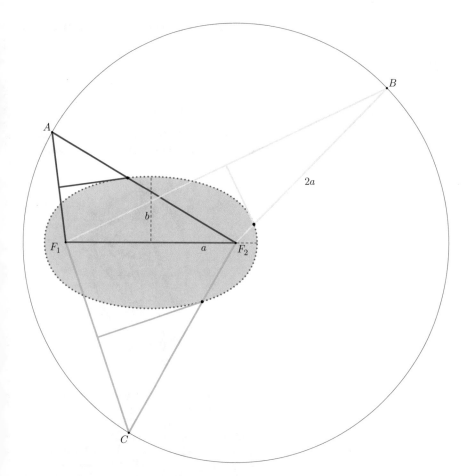

Fig. C.3 Construction of the ellipse as described in the *Later Volumes of the Thorough Investigation of Calendrical Astronomy Imperially Composed* (1738)

(1621–1678) has played a role. In his *Course in Mathematics*, in a treatise on conic sections, of which a copy was available in the Jesuit library in Beijing,[25] Dechales describes and proofs the correctness of the above described construction (see Fig. C.4).[26]

---

[25] See Verhaeren (1949) p. 365–366, item 1259.

[26] See Dechales (1674) p. 828. I am grateful to Eisso Atzema who pointed out these Latin sources to me when listening to a talk I gave at the AMS-MAA Joint Mathematics Meeting in New Orleans in 2007.

Fig. C.4 Claude François
Milliet Dechales, *De
sectionibus conicis* (Dechales
1674)

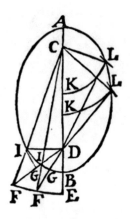

PROPOSITIO L I.

Problema,

Ellypfin defcribere.

Praxis per funiculum vix in Charta adhiberi
poteft , eò quòd nec clavos vmbilicis , nec funes

adhibere poffumus : proponatur linea AB,pro fu-
tura Ellypfis diametro , in qua determinentur
vmbilici C & D, fit BE æqualis BD, & ex puncto C, vt centrô, intervallô CE, circulus defcri-
batur, eritque CE æqualis axi maiori AB,ducan-
tur quæcumque lineæ C F , iunganturque D F,
quæ dividantur bifariàm in puncto G,educantur-
que perpendiculares GI, dico puncta I, pertinere
ad Ellypfin.

References

Aterini, Barbara (2016). Guidobaldo Del Monte (1545–1607). In Michela Cigola (Ed.), *Distin-
guished Figures in Descriptive Geometry and Its Applications for Mechanism Science: From
the Middle Ages to the 17th Century*, Volume 30 of *History of Mechanism and Machine Science*,
153–180. Cham et al.: Springer.

Dal Monte, Guidobaldo (1579). *[Guidiubaldi e Marchionibus Montis] Planisphaeriorum universalium theorica*. Pisauri [Pesaro]: apud Hieronymum Concordiam.

Dechales, Claude François Milliet (1674). De sectionibus conicis. In *Cursus seu Mundus mathematicus*, Volume 3 (Tomus Tertius), Tractatus XXXI, 793–863. Lugduni [Lyon]: Anisson.

Dong, Youcheng 董祐誠 (1819). Geyuan lianbili shu tujie 割圜連比例術圖解 (Diagrammatic Explanations of Continued Proportions in Circle Division). In *Dong Fangli yishu* 董方立遺書 (The Posthumous Works of Dong Fangli), vol. 1 (1823 ed.). Reprint in (Guo et al. 1993) 5:435–460.

Gao, Hongcheng 高红成 and Wang Rui 王瑞 (2012). Tuoyuan jifen zai Zhongguo de yi ge lishi zhuji 椭圆积分在中国的一个历史注记 (A Historical Note on Elliptic Integral in China). *Shuxue de shixian yu renshi* 数学的实践与认识 (Mathematics in Practice and Theory) *42*(17), 251–257.

Guo, Shuchun 郭书春 et al. (Eds.) (1993). *Zhongguo kexue jishu dianji tonghui: Shuxue juan* 中國科學技術典籍通彙: 數學卷 (Comprehensive Collection of Ancient Classics on Science and Technology in China: Mathematical Books), 5 vols. Zhengzhou: Henan jiaoyu chubanshe 河南教育出版社.

Han, Qi (2001). The Compilation of the *Lixiang kaocheng houbian*, its Origin, Sources and Social Context. In Luís Saraiva (Ed.), *History of Mathematical Sciences: Portugal and East Asia II*, 147–152. Lisboa: EMAF-UL.

Han, Qi and Man-Keung Siu (2008). On the Myth of an Ancient Chinese Theorem about Primality. *Taiwanese Journal of Mathematics 12*(4), 941–949.

Hua, Hengfang 華蘅芳 (1893). Xunhuan xiaoshu kao 循環小數考 (A Study on Recurring Decimals). In Hua Hengfang 華蘅芳 (Ed.), *Suancao congcun* 算草叢存 (*Collection of Mathematical Workings*) (Xingsu xuan suangao 行素軒算稿 1882 ed.), Volume 7. Wuchang.

Jeans, J. H. (1898). The Converse of Fermat's Theorem. *Messenger of Mathematics*, 174–174.

Li, Shanlan 李善蘭 (1872a). Kao shugen fa 考數根法 (Methods for Testing Primality). *Zhongxi wenjian lu* 中西聞見錄 (Peking Magazine) *2*, 13A–17A.

Li, Shanlan 李善蘭 (1872b). Kao shugen fa 考數根法 (Methods for Testing Primality). *Zhongxi wenjian lu* 中西聞見錄 (Peking Magazine) *3*, 15A–18B.

Li, Shanlan 李善蘭 (1872c). Kao shugen fa 考數根法 (Methods for Testing Primality). *Zhongxi wenjian lu* 中西聞見錄 (Peking Magazine) *4*, 15A–20B.

Li, Shanlan 李善蘭 and Alexander Wylie 偉烈亞力 (1859). *Dai weiji shiji* 代微積拾級 (Elements of Analytical Geometry and of the Differential and Integral Calculus) 18 scrolls. Shanghai: Mohai shuguan 墨海書館. Original by Elias Loomis 羅密士 (Loomis 1851).

Libbrecht, Ulrich (1973). *Chinese Mathematics in the Thirteenth Century. The Shu-shu chiu-chang of Ch'in Chiu-shao*. Cambridge (Mass.): MIT Press.

Loomis, Elias (1851). *Elements of Analytical Geometry and of the Differential and Integral Calculus*. New York: Harper & Brothers.

Loomis, Elias (1868). *Elements of Analytical Geometry and of the Differential and Integral Calculus* (19th ed.). New York: Harper & Brothers.

Lu, Jing 盧靖 (1902). *Wan xiang yi yuan yanshi* 萬象一原演式 (*Formulas for A Myriad Images—A Single Origin*), 1+9 卷 scrolls (Mianyang Lu shi 沔陽盧氏刊本 ed.). [Hubei?].

Martzloff, Jean-Claude (2006). *A History of Chinese Mathematics*. Berlin, Heidelberg: Springer. Corrected second printing of the first English edition of 1977.

Mersenne, Marin (1644). *Universæ geometriæ mixtæque mathematicæ synopsis, et bini refractionum demonstratarum tractatus*. Paris: Antonium Bertier.

Ming, Antu 明安圖 (1774). *Geyuan milü jiefa* 割圜密率捷法 (Fast Methods for the Circle Division and the Precise [circle] Ratio) (Luo Shilin 羅士琳, Guanwosheng shi huigao 觀我生室彙稿, 1839 ed.). Reprint in (Guo et al. 1993) 4:865–943.

Mydorge, Claude (1660). *Prodromi catoptricorum et dioptricorum sive Conicorum operis ad abdita radii reflexi et refracti mysteria praevii et facem praeferentis. Libri quatuor priores D. A. L. G.* Paris: P. Guillemot. First published 1631–1639.

Qianlong 乾隆 (Ed.) (1773). *Yuzhi lixiang kaocheng houbian* 御製曆象考成後編 (Later Volumes of the Thorough Investigation of Calendrical Astronomy Imperially Composed), Volume 6 (子部六) of *Chizao tang Sikuquanshu huiyao* 摛藻堂四庫全書薈要. Originally compiled in 1738.

Shen, Kangshen 沈康身 (1986). 《Shushu jiuzhang》 dayan lei suanti zhong de shulun mingti 《数书九章》 大衍类算题中的数论命题 (=Analysis on Shushu Jiuzhang to Its

Propositions about Theory of Numbers). *Hangzhou daxue xuebao (Ziran kexue ban)* 杭州大学
学报 (自然科学版) (Journal of Hangzhou University, Natural Science ed.) *4*, 421–434.

Verhaeren, Hubert (1949). *Catalogue de la Bibliothèque du Pé-T'ang*. Beijing: Imprimerie des
Lazaristes. Reprint Paris: Les Belles Lettres, 1969.

Xia, Luanxiang 夏鸞翔 (1898). Wan xiang yi yuan 萬象一原 (A Myriad Images—A Single
Origin). In Liu Duo 劉鐸 (Ed.), *Gujin suanxue congshu* 古今算學叢書 (Compendium of
Mathematics, Old and New), Volume 3 (象數第三). Shanghai suanxue shuju 上海算學書局.
Preface dated 1862.

Xia, Luanxiang 夏鸞翔 (Guangxu 1875–1908). *Zhiqu shu* 致曲術 (Procedures for Curves) (螢雲
雷齋 Zhiyun leizhai ed.).

Xu, Yi 徐异 (光緒辛丑年 1901). Wan xiang yi yuan jiaokan ji 萬象一原校勘記 (Notes
and Corrections to *A Myriad Images—A Single Origin*). In *Yan Yiting suangao* 沿沂亭算稿
(Mathematical Manuscripts by Yan Yiting), Volume 2.

Yan, Dunjie 嚴敦傑 (1954). Zhongsuanjia de sushulun 中算家的素數論 (The Chinese
Mathematicians' Theories of Prime Numbers). *Shuxue tongbao* 數學通報 *5*, 12–15.

Index

Aerrebala xinfa 阿爾熱巴拉新法, *see*
    Foucquet, Jean-Francois, *New*
    Method of Algebra
Ai Yuese 艾約瑟, *see* Edkins, John
Algorithms
    versified, 144, 145, 147, 149
Analogy
    ἀναλογία, 113
    comparable categories (*bilei* 比類), 112,
        113, 116, 118, 140, 147, 149
    induce by (*leitui* 類推), 108, 112, 119, 124,
        126, 132
Arabic numerals, 97–100, 102
Archimedes (ca. 287–212 BCE)
    in China, 35
    De Sphaera et Cylindro (*Yuanshu* 圓
        書), 33, 35
    quadrature of the ellipse, 35
Arithmetic triangle, 72, 151
    generalized, 127
    Jia Xian triangle, 68, 120, 151n
        combinatorial interpretation, 121
        related to hexagrams, 153
    Pascal triangle, 120

Barber, W. T. A. 巴修理, 98
Ba Xiuli 巴修理, *see* Barber, W. T. A.
Bilei 比類, *see* analogy, comparable
    categories
Billingsley, Henry (d. 1606), 53
Book of Changes, see Classic of Changes
    (*Yijing* 易經)

Cai Xiyong 蔡錫勇 (1850–1896), 201
Cai Yuanpei 蔡元培 (1868–1940), 218
Celestial element (*tianyuan* 天元)
    as an algebraic method, 10, 84, 88, 89, 91,
        96, 155
    as the unknown, 92, 93
Census, 178
Ceyuan haijing 測圓海鏡, *see* Li Ye 李冶
    (1192–1279), *Sea Mirror of Circle*
    Measurements
Cheng Dawei 程大位 (1533–1606)
    Unified Lineage of Mathematical Methods
        (*Suanfa tongzong* 算法統宗), 148
Chen Houyao 陳厚耀 (1648–1722)
    Meaning of the Methods of Combination
        and Alternance (*Cuozong fayi* 錯綜
        法義), 119
Chen Qilu 陳其鹿 (1895-1981), 218
Chen Song 陳崧
    critique of Li Shanlan, 133, 134
Chern Shiing-Shen 陳省身 (1911–2004),
    209n, 233
Chinese
    characters, 77, 99
    as a universal language, 97
Chinese origins of Western science (*Xixue*
    zhongyuan 西學中源), 2, 4, 205
Chinese Remainder Theorem, 64
Chinese studies for fundamental
    principles—Western studies
    for practical applications (*Zhong*
    ti xi yong 中體西用), 172, 196,
    198–200, 202
Church Missionary Society, 79

© Springer International Publishing AG, part of Springer Nature 2019
A. Bréard, *Nine Chapters on Mathematical Modernity*, Transcultural
Research – Heidelberg Studies on Asia and Europe in a Global Context,
https://doi.org/10.1007/978-3-319-93695-6

Printed in the United States
By Bookmasters